# Diagnosing Hemp and Cannabis Crop Diseases

# Diagnosing Hemp and Cannabis Crop Diseases

**Shouhua Wang, PhD**

*State Plant Pathologist*
*Nevada Department of Agriculture*
*Sparks, Nevada, USA*

**CABI**

CABI is a trading name of CAB International

CABI
Nosworthy Way
Wallingford
Oxfordshire OX10 8DE
UK

CABI
WeWork
One Lincoln St
24th Floor
Boston, MA 02111
USA

Tel: +44 (0)1491 832111
Fax: +44 (0)1491 833508
E-mail: info@cabi.org
Website: www.cabi.org

Tel: +1 (617)682-9015
E-mail: cabi-nao@cabi.org

A catalogue record for this book is available from the British Library, London, UK.

**Library of Congress Cataloging-in-Publication Data**

Names: Wang, Shouhua, (Plant pathologist), author.
Title: Diagnosing hemp and cannabis crop diseases / Shouhua Wang.
Description: Boston, MA : CAB International, [2021] | Includes bibliographical references and index. | Summary: "Extensively illustrated with over 300 colour photographs, this book describes how to diagnose hemp and cannabis crop diseases. It is aimed as a practical guide for farmers and growers, as well as researchers in plant pathology"-- Provided by publisher.
Identifiers: LCCN 2021036112 (print) | LCCN 2021036113 (ebook) | ISBN 9781789246070 (hardback) | ISBN 9781789246087 (ebook) | ISBN 9781789246094 (epub)
Subjects: LCSH: Cannabis--Diseases and pests--Identification. | Hemp--Diseases and pests--Identification. | Handbooks and manuals.
Classification: LCC SB295.C35 W36 2021  (print) | LCC SB295.C35  (ebook) | DDC 633.7/9--dc23
LC record available at https://lccn.loc.gov/2021036112
LC ebook record available at https://lccn.loc.gov/2021036113

References to Internet websites (URLs) were accurate at the time of writing.

ISBN-13: 9781789246070 (hardback)
        9781789246087 (ePDF)
        9781789246094 (ePub)

DOI: 10.1079/9781789246070.0000

Commissioning Editor: Rebecca Stubbs
Editorial Assistant: Lauren Davies
Production Editor: James Bishop

Typeset by SPi, Pondicherry, India

# Contents

# Preface

Hemp and cannabis, both belonging to *Cannabis sativa*, have emerged as some of the most valuable crops because of their multiple functionalities – industrial, medicinal and recreational uses. Like all other crops, diseases and pests attack these crops at the high expense of hemp growers and cannabis cultivators. In certain cases, an entire hemp field fails to produce desired products due to unexpected diseases. As a new and highly regulated crop, research on *Cannabis* crop diseases is scarce and growers often do not have science-based information to guide production and disease management practices. Since 2014, I have had opportunities to work on *Cannabis* crop diseases and witnessed many diseases and pests both in the field and in indoor production facilities. These diseases, which have been observed in fields and then researched in the lab, shed amazing insight on *Cannabis* pathology. These original findings along with other reports on hemp and cannabis diseases and pests are incredibly valuable to *Cannabis* growers and this is what motivated me to write this book.

When growers encounter a plant disease or pest problem, diagnosis is the first step. A disease can only be treated effectively when it is correctly diagnosed. Over the past two decades of my clinical plant pathology practice and research, I realized that diagnosis has its own science and art, has unique processes in defining causes and is a collaborative effort between growers and clinical plant pathologists. Also, it occurred to me that there should be a book that systemically illustrates how to diagnose plant diseases and has all the information needed for growers and lab diagnosticians to perform diagnoses for a specific crop. Furthermore, there is a wealth of disease information generated in diagnostic labs, which includes field observations, unique symptoms, microscopic details and even molecular data. These clinical data captured at every step of diagnosis are often neglected in literature but are valuable to growers and diagnosticians. Thus, I was inspired to use hemp/cannabis as a model crop to illustrate how, as a farmer, cultivator, crop consultant, or lab diagnostician, to diagnose a disease and then lead to effective disease management.

This book is written for anyone who is interested in learning about *Cannabis* crop diseases and serves as a field and laboratory guide to diagnosing diseases and pest problems. The content is arranged from general sections to specific disease sections. The general sections cover the disease concepts related to *Cannabis* plants, the art of plant diagnostics, setting up a diagnostic lab, and commonly used diagnostic protocols and procedures. The specific sections describe the diseases and pests that have been found from *Cannabis* crops and how to diagnose each of them. All sections are written in detail, accompanied by pictures and illustrations. Although this book is mainly on *Cannabis* crop diseases, readers can use the concepts, principles, strategies and methods described in this book to diagnose diseases of other crops.

Preparing this book took me much longer than the writing itself. From concept formation to image collection, all these took time and came from day-to-day diagnosis. As a clinical plant pathologist, I am honoured to be working with growers and other clientele who have trusted me with challenging and interesting disease cases, from which I have been able to develop the concepts and procedures presented in this book. I am also grateful to my former and current lab members who have assisted me in diagnosing tens of thousands of plant disease samples, from which some important diagnostic data have been collected and presented in this book. As the sole author of this book, it has not been an easy task to cover all the subjects of diseases, especially when all pictures and illustrations used in this book are original. For this attainment, I am deeply grateful to the late Prof. Wei-Fan Chiu who brought me into nematology and virology research during my master and doctorial programmes, to Drs Robert D. Riggs and Rose C. Gergerich, who offered me further research opportunities in their nematology and virology laboratories, and to Dr Wei Yan, who provided a unique research opportunity in his world-class biomedical research laboratory where I earned my second doctoral degree in Cellular and Molecular Pharmacology and Physiology. These researches have greatly benefited my diagnostic career and offered valuable materials for this book. In preparation of this book's typescript, I thank my former colleague Gary Cross who gave me encouragement throughout the writing, Dr Hong Chen who critically reviewed and edited Chapter 12, Dr Weimin Ye who helped identify nematode specimens, and my daughter, who drew some illustrations and edited my entire first draft. Finally, I especially thank Rebecca Stubbs, Ali Thompson, Lauren Davies, James Bishop and other CABI members for their contributions to making this book a reality.

**Shouhua Wang**
12 February 2021

# 1 The *Cannabis* Plant

*Cannabis*, a genus name referring to both hemp and marijuana (or hereafter cannabis), has emerged as a global cash crop with diverse applications such as medicinal and recreational use, human consumption of seed and oil and the industrial use of fibre and other products. *Cannabis* plants have been cultivated and used for several thousand years (Schluttenhofer and Yuan, 2017; Brand and Zhao, 2017), but only recently has hemp and cannabis cultivation dramatically increased in the USA, Canada and other countries. Because *Cannabis* is considered a multi-functional crop, it has captured the attention of a broad range of scientists, including agriculturists, chemists, biomedical researchers and clinical scientists. At the time of writing, there are 22,424 research articles or books related to *Cannabis sativa* in the PubMed database (https://pubmed.ncbi.nlm.nih.gov/), compared with 35,704 for rice, 27,346 for wheat, 35,789 for corn, 27,346 for soybean and 10,341 for potato. A draft haploid genome sequence of 534 Mb and a transcriptome of 30,000 genes for *C. sativa* is also available (van Bakel *et al.*, 2011), which provides a base for further research in *Cannabis* plant biology and pathology.

## Classification

The nomenclature of *Cannabis* can be traced back to 1753 when Carl Linnaeus first described *Cannabis sativa* as a single species in the genus *Cannabis* in his book *Species Plantarum* (Linnaeus, 1753; Pollio, 2016). This book is considered the starting point for giving every plant species a binomial name comprised of two Latin words. For example, hemp is named *Cannabis sativa*, where *Cannabis* is the genus name and *sativa* is the species name. The binomial nomenclature has been widely used in naming living organisms such as plants, animals and microorganisms. *Cannabis* is a genus of flowering plants in the family Cannabaceae and

it contains the most known species, *Cannabis sativa* (Table 1.1). In 1785, the French biologist Jean-Baptiste Lamarck proposed a new species *C. indica* based on plant samples he received from India (Erkelens and Hazekamp, 2014). When compared with the European *C. sativa*, the *Cannabis* plants from India appeared to have smaller and narrower leaves, as well as much firmer stem. In Lamarck's view, *C. indica* was a psychoactive non-fibre producing species of *Cannabis*, different from the European *C. sativa* in terms of morphological characteristics and physiological effects. Thus, the genus of *Cannabis* temporarily contained *C. sativa*, the species mostly cultivated in the western continents, and *C. indica*, a wild species mainly growing in India (Erkelens and Hazekamp, 2014). However, this taxonomic treatment by Lamarck only remained intact for about 50 years. In 1838, Lindley rejected the two-species classification and restored *C. sativa* as the only species in the *Cannabis* genus (Lindley, 1838) and since then *Cannabis* had been considered a monospecific genus. In 1924, a new species, *Cannabis ruderalis*, was identified in wild areas of south-eastern Russia (Janischevsky, 1924) and 50 years later Schultes *et al.* (1974) reinstated the species *C. indica*. Thus, *C. sativa*, *C. indica* and *C. ruderalis* are commonly seen in literature. However, these proposed species may not have solid taxonomic foundations (Pollio, 2016). There are still debates and disagreements on the classification of many types of *Cannabis* plants. One study compared 157 *Cannabis* accessions of diverse geographical origins for allozyme variations at 17 gene loci and the results support a polytypic concept for *Cannabis* genus, which recognizes species of *C. sativa*, *C. indica* and *C. ruderalis* as well as seven other putative taxa (Hilig, 2005). Others consider that the genus *Cannabis* comprises only *C. sativa* L. with highly polymorphic subspecies *sativa*, *indica* and *ruderalis*.

DOI: 10.1079/9781789246070.0001

**Table 1.1.** Classification from Kingdom Plantae down to species *Cannabis sativa* L. (adapted from USDA, 2020).

---

**Kingdom** Plantae – Plants
  **Subkingdom** Tracheobionta – Vascular plants
    **Superdivision** Spermatophyta – Seed plants
      **Division** Magnoliophyta – Flowering plants
        **Class** Magnoliopsida – Dicotyledons
          **Subclass** Hamamelididae
            **Order** Urticales
              **Family** Cannabaceae – Hemp family
                **Genus** *Cannabis* L. – hemp
                  **Species** *Cannabis sativa* L. –
                    hemp or marijuana (cannabis)

---

## Characteristics

### Hemp versus marijuana

Botanically, both hemp and marijuana belong to *C. sativa*, but they differ by use and chemical compositions (Table 1.2). Marijuana or cannabis generally refers to a group of distinct cultivars or varieties within the *C. sativa* species that are cultivated and used as psychotropic drugs, for either medicinal or recreational purposes, while hemp refers to another set of cultivars or varieties cultivated mainly for fibre, seed or oil (Fig. 1.1). Industrial hemp is bred to maximize fibre, seed and oil with very low levels of THC (delta-9 tetrahydrocannabinol) while marijuana varieties are bred for high levels of THC. Hemp plants and their products can be used for food and beverages, nutritional supplements, personal care products, fabrics and textiles, paper, construction materials and other industrial goods. Because of the significant difference in their uses, hemp and marijuana are cultivated differently. The majority of hemp crops are planted in regular farmlands ranging from a few acres to a thousand acres, while marijuana plants are mostly cultivated in secured indoor facilities. Hemp and marijuana also have separate statutory definitions in US laws (Congressional Research Service, 2019).

## Sex

*Cannabis* is a genus of annual, dioecious and flowering plants, some of which can attain heights up to 8 m (Edwards and Whittington, 1992). In most populations, there are more female plants than male plants. Plants undergo vegetative growth in early season, then turn to flower production when days are shortened. During the vegetative stage, it may be difficult to determine the sex of the plant. In general, the males flower earlier, but there is an overlap in flowering periods so female plants can be fertilized to produce seeds. Male plants generally die after anthesis. *C. sativa* plants have 20 chromosomes ($2n = 20$). Female plants have a homogametic pair of sex chromosomes (XX), while male plants have a heterogametic pair of chromosomes (XY). Using AFLP (amplified fragment length polymorphisms) method, Flachowsky *et al.* (2001) demonstrated that there are an abundant number of potential markers associated with male sex and segregated with male plants. This research supports the presence of a male sex chromosome in *Cannabis*. As the sex of most dioecious plants can be easily determined only during the flowering period, a quick method to determine the sex of plants would help the breeding process and aid in developing a population of only male or female plants based on desired products to be harvested. For example, marijuana cultivation encourages female-only plants to maximize female flower production, while male plants are good for fibre production. One of the best methods to select male or female plants at an early stage is to use genetic markers linked to sex. For example, using RAPD (random amplified polymorphic DNA) technique, two novel male-specific molecular markers called MADC5 and MADC6 were identified in hemp; these markers were suggested to be linked to the Y chromosome and can be used to quickly distinguish male and female (Törjék *et al.*, 2002).

## Inflorescence

The inflorescence is the main product of marijuana plants and cannabidiol (CBD)-based hemp plants (Fig. 1.2). Hundreds of specialized metabolites such as cannabinoids, terpenes and flavonoids are produced and accumulated in the glandular trichomes of female inflorescences (Andre *et al.*, 2016). It is important to understand the inflorescence architecture and florogenesis in female *Cannabis* plants as it is the most valuable organ produced by the plant. Spitzer-Rimon *et al.* (2019) used three medical cultivars of *C. sativa* L., 'NB130', 'NB140' and 'NB150', as a model system to study the morphophysiological and genetic mechanisms governing flower and inflorescence development. Under a long photoperiod,

**Table 1.2.** Differences between hemp and marijuana (adapted and modified from Congressional Research Service, 2019)

| | Hemp | Marijuana |
|---|---|---|
| **Scientific name** | *Cannabis sativa* | *Cannabis sativa* |
| **Statutory definition** | The term 'hemp' means 'the plant *Cannabis sativa* L. and any part of that plant, including the seeds thereof and all derivatives, extracts, cannabinoids, isomers, acids, salts, and salts of isomers, whether growing or not, with a delta-9 tetrahydrocannabinol [delta-9 THC] concentration of not more than 0.3 percent on a dry weight basis'. <br> – **Agricultural Marketing Act of 1946, Subtitle G, SEC. 297A** | The term 'marijuana' means 'all parts of the plant *Cannabis sativa* L., whether growing or not; the seeds thereof; the resin extracted from any part of such plant; and every compound, manufacture, salt, derivative, mixture, or preparation of such plant, its seeds or resin'. Such term does not include the mature stalks of such plant, fiber produced from such stalks, oil or cake made from the seeds of such plant, any other compound, manufacture, salt, derivative, mixture, or preparation of such mature stalks (except the resin extracted therefrom), fiber, oil, or cake, or the sterilized seed of such plant which is incapable of germination. <br> – **21 U.S.C. § 802 (16), Title 21 – Food and Drugs** |
| **US Laws governing *Cannabis sativa*** | **Agricultural Marketing Act of 1946** (AMA, 7 U.S.C. 1621 et seq.) <br> **Federal Food, Drug, and Cosmetic Act** (FFDCA; 21 U.S.C. §§ 301 et seq.) | **Controlled Substances Act** (CSA, 21 U.S.C. §§ 801 et seq.) <br> **Federal Food, Drug, and Cosmetic Act** (FFDCA; 21 U.S.C. §§ 301 et seq.) |
| **Primary Federal Regulatory Agencies** | US Department of Agriculture (USDA), Food and Drug Administration (FDA) | US Drug Enforcement Administration (DEA), Food and Drug Administration (FDA) |
| **THC threshold defined** | No more than 0.3% delta-9 THC on a dry-weight basis | Not specified |
| **Phytocannabinoids** | More than 90 different cannabinoids have been reported in the literature (Andre *et al.*, 2016) | More than 90 different cannabinoids have been reported in the literature (Andre *et al.*, 2016) |
| **Physiological effect** | Not psychoactive | Psychoactive |
| **Plant part harvested** | Fibre, seed and flower | Flower |
| **Major use of products** | Regular or functional foods, cosmetics, fabrics and industrial products | Recreational and medicinal products |
| **Common growing conditions** | Farmlands and greenhouses | Secured indoor facilities |
| **Size of mature plants** | 3.0–4.6 m for fibre crops, 1.8–2.7 m for seed crops and 1.2–2.4 m for flower or oil crops | 1.2–2.4 m |

the main shoot of the cannabis plants branched monopodially, producing alternate branching shoots. The monopodial plant contained numerous phytomers, each of which included an internode with one large leaf and an axillary shoot. There were two bracts located on each side of the leaf petiole base, each subtending a solitary flower. This observation confirms that, under long photoperiod growth conditions, the main and axillary meristems produce subtending bracts and flower primordia, which suggests that the plants were in a reproductive stage. In all three cultivars examined, solitary flowers were observed in the leaf axis during growth under the long photoperiod. In two cultivars ('NB130' and 'NB150'), these flowers reached anthesis. The results contradict the common belief that the long photoperiod is 'non-inductive' or 'vegetative'. The study observed the development of solitary flowers and bracts in shoot internodes and suggested that induction of solitary flowers is age-dependent and controlled by internal signals rather than by photoperiod. After transition to short photoperiod, plants started to branch and develop a compound raceme. Solitary flowers at the leaf axis began to develop and stigmata were visible. At the same time, plants continued to produce phytomers, each consisting of a reduced leaf, two bracts, two solitary flowers and an axillary shoot. Thus, the study suggests that short photoperiod causes a dramatic change in shoot apex architecture to form a compound racemose inflorescence structure rather than flower induction (Spitzer-Rimon *et al.*, 2019).

### Hemp root

The root system stabilizes soil structure and supports the plant. It explores the soil and uptakes water and nutrients. A healthy hemp plant develops a strong root system to support growth (Fig. 1.3). There are three root parameters commonly used to characterize plant root systems. Root Length Density (RLD), defined as cm of length of root per $cm^3$ of soil, is a parameter used to determine root morphology (Vamerali *et al.*, 2003) and the crop's potential for nutrient and water uptake. Root diameter (RD) is another important characteristic of the root system that also affects nutrient and water uptake. Root biomass (RB) is a measurement used to determine the costs associated with root construction and root maintenance (Bouma *et al.*, 2000). Amaducci *et al.* (2008) conducted a study on the root system of a fibre hemp crop and characterized hemp roots under different growing conditions. The study revealed that RLD in the depth of the first 10 cm of soil was highest, almost 5 cm per $cm^3$ of soil, but it decreased progressively with depth. Roots were found in depths of 130 cm or sometimes even 200 cm. Root diameter was about 190 µm in the upper soil layer, increased with soil depth until reaching 100 cm of depth and then remained at 300 µm thereafter. Similar to RLD, root biomass in the first soil layer was highest and

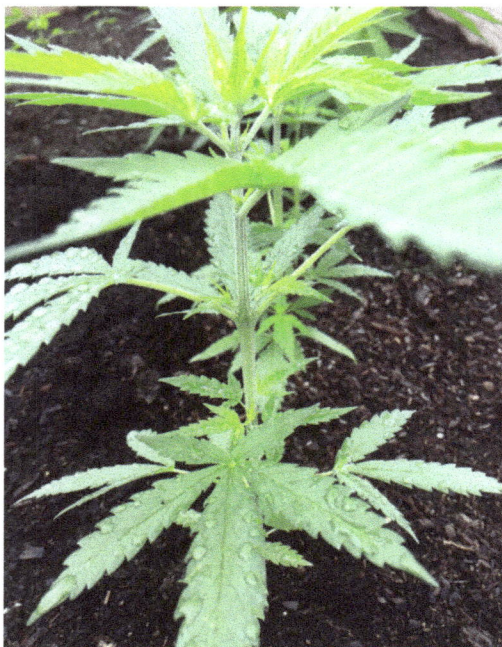

**Fig. 1.1.** A seedling plant of hemp (*Cannabis sativa*).

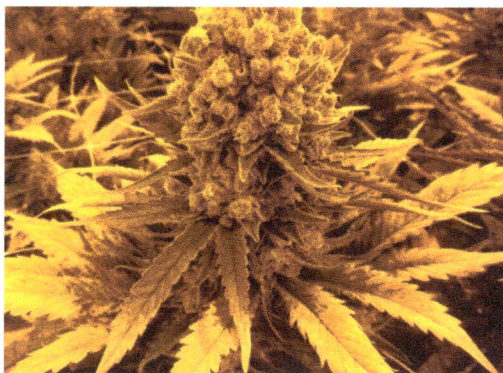

**Fig. 1.2.** Flower buds produced on a female plant of *Cannabis sativa*.

**Fig. 1.3.** A healthy root system of the hemp plant. Note the fresh white primary roots and extensive 1st, 2nd and 3rd orders of lateral roots.

about 50% of the root biomass was found in the first 20 cm or 50 cm of depth. The study also indicated that plant population did not affect these root parameters.

## Seed

Hemp seeds are dark-coloured, relatively defined in shape and may vary by weight (Fig. 1.4). *C. sativa* seeds can be categorized into three types: regular, feminized, or autoflowering (Congressional Research Service, 2019). Regular seeds produce both male and female plants at about a 50/50 ratio, while feminized seeds, also referred to as 'female seeds', produce only female plants. Feminized seeds are obtained by encouraging female plants to produce viable, genetically identical seeds without being fertilized by a male plant. Autoflowering seeds are cross-bred or selectively bred to produce female plants containing less or zero THC.

Sacilik *et al*. (2003) studied physical properties of hemp seeds at different moisture contents. In the moisture range from 8.62% to 20.88%, the physical properties of hemp seeds changed linearly when seed moisture content increased. The dimensions of the hemp seed increased by 8.44% (major axis),

6.51% (medium axis) and 14.17% (minor axis). The sphericity, surface area and thousand-seed mass increased from 0.795 to 0.808, 9.4 mm$^2$ to 10.3 mm$^2$, and 15.3 g to 16. 9 g, respectively. The bulk and true densities decreased from 557.5 kg/m$^3$ to 512.3 kg/m$^3$ and 1043.0 kg/m$^3$ to 894.8 kg/m$^3$, respectively. The angle of repose and terminal velocity increased linearly from 24.6° to 27.7° and 5.5 m/s to 6.4 m/s, respectively, whereas porosity decreased from 46.5% to 42.7%. Knowledge of these physical properties of hemp seeds may help the design of equipment for harvesting, processing and storing the seed. Suriyong *et al*. (2015) compared the impact of different storage conditions on hemp seed quality and concluded that storing seeds at 15°C was most efficient at maintaining seed viability, though temperatures of 4°C or –4°C were still promising conditions for maintaining high seed quality within a year.

## Pathology

Like any other crops, a *Cannabis* crop can be affected to various degrees by fungi, bacteria, viruses, nematodes, insects, mites, or abiotic factors. There are some reports on hemp and cannabis diseases dating back to the middle or late 20th century, but the literature is scattered. Unlike many other crops on which a compendium has recorded the most reported diseases in one place and for which specific IPM programmes are available for major diseases, hemp and cannabis production lacks such a resource. Since the 2014 farm bill (Agricultural Act of 2014, P.L. 113-79) and the 2018 farm bill (Agriculture Improvement Act of 2018, P.L. 115-334) relaxed the restrictions on US hemp production and marketing, many states began to introduce hemp as an alternative crop; some states also legalized medicinal and recreational marijuana use and allow cannabis cultivation under a licence. The surge of hemp and marijuana cultivation has encountered many disease and pest issues that are new to growers and diagnosticians. Pathogens or arthropods, when found to be associated with *Cannabis* crops, may not be new species or strains, but their roles in disease development, their interactions with *Cannabis* plants and their management are new topics.

Punja *et al*. (2019) isolated a number of plant pathogens infecting marijuana plants growing in Canada. These pathogens affected the roots, crown,

**Fig. 1.4.** Typical shapes of hemp seeds.

foliage and the inflorescences (buds) and caused substantial damage to plants. Using a PCR assay based on the internal transcribed spacer (ITS) region of ribosomal RNA gene, they identified a number of root-infecting pathogens, including *Fusarium oxysporum*, *Fusarium solani*, *Fusarium brachygibbosum*, *Pythium dissotocum*, *Pythium myriotylum* and *Pythium aphanidermatum*. These fungal and oomycete pathogens caused discoloration of the crown and pith tissues, root browning, stunting and yellowing of plants, and eventual death of some plants. On the foliage, they found powdery mildew predominantly caused by *Golovinomyces cichoracearum*. On inflorescences, several fungal species were found to cause bud rot, including *Penicillium olsonii*, *P. copticola*, *Botrytis cinerea*, *Fusarium solani* and *F. oxysporum*.

In the USA, many new reports of diseases are based on hemp crops. For example, a list of diseases caused by oomycetes, fungi, phytoplasmas and possible viruses were found from hemp and cannabis crops in Nevada (Schoener *et al.*, 2019). The most destructive diseases include: (i) marijuana vascular wilt caused by *Fusarium oxysporum* and *F. solani*; (ii) hemp stem canker caused by *F. oxysporum* and *F. solani*; (iii) hemp root rot associated with five *Fusarium* species: *F. oxysporum*, *F. solani*, *F. redolens*, *F. tricinctum* and *F. equiseti*; (iv) hemp crown rot caused by *Pythium aphanidermatum*; (v) marijuana powdery mildew caused by a species closely related to *Golovinomyces ambrosiae*; (vi) latent infection of hemp seeds by *Alternaria infectoria* and *A. tenuissima*; (vii) latent infection of hemp seeds by *Rhizopus oryzae*; (viii) hemp witches' broom caused by a species closely related to clover proliferation phytoplasma; and (ix) hemp leaf roll caused by *Beet curly top virus*. In certain cases, the majority of crop plants were destroyed by *Fusarium* root rot (Schoener *et al.*, 2017), *Pythium* crown rot (Schoener *et al.*, 2018) and phytoplasma witches' broom (Schoener and Wang, 2019). In one field, the hemp crop was affected by 15 pathogens (see Chapter 13, Table 13.1). Apparently, any portion of *Cannabis* plant can be impacted by diseases (Fig. 1.5).

Other first reports on *Cannabis* diseases include hemp canker caused by *Sclerotinia sclerotiorum* in Alberta, Canada (Bains *et al.*, 2000), southern blight of hemp caused by *Sclerotium rolfsii* in Sicily and southern Italy (Pane *et al.*, 2007), hemp root-knot caused by *Meloidogyne javanica* in China (Song *et al.*, 2017), hemp witches' broom disease associated with phytoplasma of elm yellows group (16SrV) in China (Zhao *et al.*, 2007), bacterial leaf spot of hemp caused by *Xanthomonas campestris* in Japan (Netsu *et al.*, 2014), charcoal rot of hemp caused by *Macrophomina phaseolina* in southern Spain (Casano *et al.*, 2018), hemp crown and root rot caused by *Pythium ultimum* in Indiana (Beckerman *et al.*, 2018), foliar blight of hemp

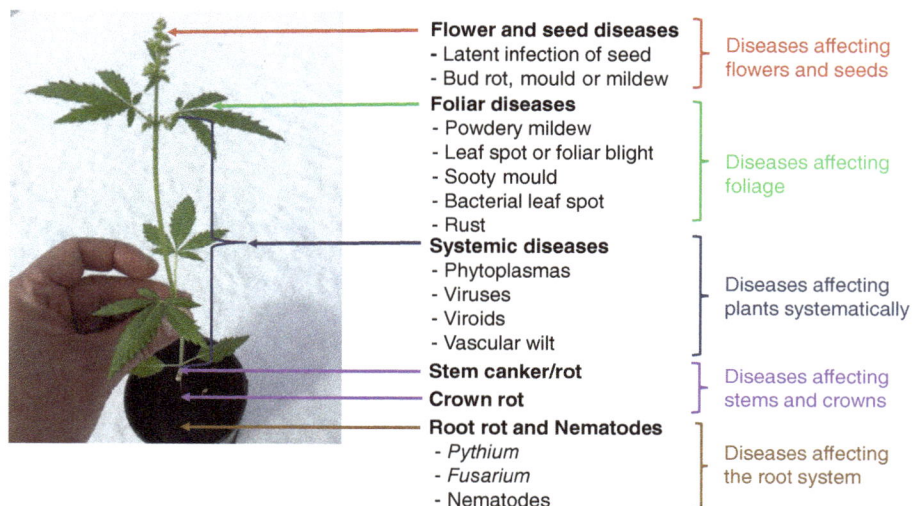

**Flower and seed diseases**
- Latent infection of seed
- Bud rot, mould or mildew

Diseases affecting flowers and seeds

**Foliar diseases**
- Powdery mildew
- Leaf spot or foliar blight
- Sooty mould
- Bacterial leaf spot
- Rust

Diseases affecting foliage

**Systemic diseases**
- Phytoplasmas
- Viruses
- Viroids
- Vascular wilt

Diseases affecting plants systematically

**Stem canker/rot**
**Crown rot**

Diseases affecting stems and crowns

**Root rot and Nematodes**
- *Pythium*
- *Fusarium*
- Nematodes

Diseases affecting the root system

**Fig. 1.5.** Representative diseases detected from each part of *Cannabis sativa* plants.

caused by *Exserohilum rostratum* (Thiessen and Schappe, 2019), hemp leaf spot caused by *Cercospora* cf. *flagellaris* in Kentucky (Doyle *et al.*, 2019) and hemp crown rot caused by *Sclerotinia minor* in California (Koike *et al.*, 2019). Besides these first reports, there is a wealth of hemp disease information in the databases of plant diagnostic labs where many hemp samples are submitted. In most cases, pathogens detected from hemp plants are common species and such information is often treated as routine diagnostic data without being written as a first report from a specific state or geographical area.

Some non-pathogenic microorganisms or endophytes are found in *Cannabis* plants. For example, endophytic fungi such as *Chaetomium*, *Trametes*, *Trichoderma*, *Penicillium* and *Fusarium* were found in the crown, stem and petiole tissues (Punja *et al.*, 2019). Endophytes of plants live within the tissues and organs without causing harmful diseases. Rather, they contribute to plant growth, enhance nutrient uptake, induce defence systems against pathogens and potentially modulate the production of secondary metabolites (Taghinasab and Jabaji, 2020). Understanding endophytic microorganisms and their partnerships with *Cannabis* plants at organismal, cellular and molecular levels can potentially improve plant fitness and yield.

## References

Amaducci, S., Zatta, A., Raffanini, M. and Venturi, G. (2008) Characterisation of hemp (*Cannabis sativa* L.) roots under different growing conditions. *Plant and Soil* 313, 227–235. doi: 10.1007/s11104-008-9695-0

Andre, C.M., Hausman, J. and Guerriero, G. (2016) *Cannabis sativa*: the plant of the thousand and one molecules. *Frontiers in Plant Science* 7, 19. doi: 10.3389/fpls.2016.00019

Bains, P.S., Bennypaul, H.S., Blade, S.F. and Weeks, C. (2000) First report of hemp canker caused by *Sclerotinia sclerotiorum* in Alberta, Canada. *Plant Disease* 84, 372. doi: 10.1094/PDIS.2000.84.3.372B

Beckerman, J., Stone, J., Ruhl, G. and Creswell, T. (2018) First report of *Pythium ultimum* crown and root rot of industrial hemp in the United States. *Plant Disease* 102, 2045. doi: 10.1094/PDIS-12-17-1999-PDN

Bouma, T.J., Nielsen, K.L. and Koutstaal, B. (2000) Sample preparation and scanning protocol for computerised analysis of root length and diameter. *Plant and Soil* 218, 185–196. doi: 10.1023/A:1014905104017

Brand, E.J. and Zhao, Z. (2017) Cannabis in Chinese medicine: are some traditional indications referenced in ancient literature related to cannabinoids? *Frontiers in Pharmacology* 8, 108. doi: 10.3389/fphar.2017.00108

Casano, S., Cotan, A.H., Delgado, M.M., García-Tejero, I.F., Saavedra, O.G. *et al.* (2018) First report of charcoal rot caused by *Macrophomina phaseolina* on hemp (*Cannabis sativa*) varieties cultivated in southern Spain. *Plant Disease* 102, 1665. doi: 10.1094/PDIS-02-18-0208-PDN

Congressional Research Service (2019) Defining hemp: a fact sheet. Available at https://crsreports.congress.gov/product/pdf/R/R44742 (accessed 29 May 2020).

Doyle, V.P., Tonry, H.T., Amsden, B., Beale, J., Dixon, E. *et al.* (2019) First report of *Cercospora* cf. *flagellaris* on industrial hemp (*Cannabis sativa*) in Kentucky. *Plant Disease* 103, 1784. doi: 10.1094/PDIS-01-19-0135-PDN

Edwards, K.J. and Whittington, G. (1992) Male and female plant selection in the cultivation of hemp, and variations in fossil cannabis pollen representation. *The Holocene* 2, 85–87. doi: 10.1177/095968369200200111

Erkelens, J.L. and Hazekamp, A. (2014) That which we call Indica, by another name would smell as sweet. *Cannabinoids* 9, 9–15.

Flachowsky, H., Schumann, E., Weber, W.E. and Peil, A. (2001) Application of AFLP for the detection of sex-specific markers in hemp. *Plant Breeding* 120, 305–309. doi: 10.1046/j.1439-0523.2001.00620.x

Hilig, K.W. (2005) Genetic evidence for speciation in *Cannabis* (Cannabaceae). *Genetic Resources and Crop Evolution* 52, 161–180.

Janischevsky, D.E. (1924) A form of hemp in wild areas of southeastern Russia. *N.G. Černyševskogo Universiteta* 2, 3–17.

Koike, S.T., Stanghellini, H., Mauzey, S.J. and Burkhardt, A. (2019) First report of Sclerotinia crown rot caused by *Sclerotinia minor* on hemp. *Plant Disease* 103, 1771. doi: 10.1094/PDIS-01-19-0088-PDN

Lindley, J. (1838) *Flora Medica, a botanical account of all the more important plants used in medicine, in different parts of the world*. Longman, Orme, Brown, Green, & Longmans, London. doi: 10.5962/bhl.title.19534

Linnaeus, C. (1753) *Species Plantarum*. Laurentius Salvius, Stockholm.

Netsu, O., Kijima, T. and Takikawa, Y. (2014) Bacterial leaf spot of hemp caused by *Xanthomonas campestris* pv. *cannabis* in Japan. *Journal of General Plant Pathology* 80, 164–168. doi: 10.1007/s10327-013-0497-8

Pane, A., Cosentino, S.L., Copani, V. and Cacciola, S.O. (2007) First report of southern blight caused by *Sclerotium rolfsii* on hemp (*Cannabis sativa*) in Sicily and southern Italy. *Plant Disease* 91, 636. doi: 10.1094/PDIS-91-5-0636A

Pollio, A. (2016) The name of *Cannabis*: a short guide for nonbotanists. *Cannabis and Cannabinoid Research* 1, 234–238. doi: 10.1089/can.2016.0027

Punja, Z.K., Collyer, D., Scott, C., Lung, S., Holmes, J. et al. (2019) Pathogens and molds affecting production and quality of *Cannabis sativa* L. *Frontiers in Plant Science* 10, 1120. doi: 10.3389/fpls.2019.01120

Sacilik, K., Öztürk, R. and Keskin, R. (2003) Some physical properties of hemp seed. *Biosystems Engineering* 86, 191–198. doi: 10.1016/S1537-5110(03)00130-2

Schluttenhofer, C. and Yuan, L. (2017) Challenges towards revitalizing hemp: a multifaceted crop. *Trends in Plant Science* 22, 917–929. doi: 10.1016/j.tplants.2017.08.004

Schoener, J.L. and Wang, S. (2019) First detection of a phytoplasma associated with witches' broom of industrial hemp in the United States. (Abstr.) *Phytopathology* 109, S2.1. doi: 10.1094/PHYTO-109-10-S2.1

Schoener, J.L., Wilhelm R., Rawson, R., Schmitz, P. and Wang, S. (2017) Five *Fusarium* species associated with root rot and sudden death of industrial hemp in Nevada. (Abstr.) *Phytopathology* 107, S5.98. doi: 10.1094/PHYTO-107-12-S5.1

Schoener, J.L., Wilhelm, R., Yazbek, M. and Wang, S. (2018) First detection of *Pythium aphanidermatum* crown rot of industrial hemp in Nevada. (Abstr.) *Phytopathology* 108, S1.187. doi: 10.194/PHYTO-108-10- S1.187

Schoener, J.L., Wilhelm, R. and Wang, S. (2019) Emerging diseases of *Cannabis* crops in Nevada. (Abstr.) NPDN Fifth National Meeting, April 15–18, 2019, Indianapolis, Indiana.

Schultes, R.E., Klein, W.M., Plowman, T. and Lockwood, T.E. (1974) *Cannabis*: an example of taxonomic neglect. *Botanical Museum Leaflets, Harvard University* 23, 337–367.

Song, Z.Q., Cheng, F.X., Zhang, D.Y., Liu, Y. and Chen, X.W. (2017) First Report of *Meloidogyne javanica* infecting hemp (*Cannabis sativa*) in China. *Plant Disease* 101, 842–843. doi: 10.1094/PDIS-10-16-1537-PDN

Spitzer-Rimon, B., Duchin, S., Bernstein, N. and Kamenetsky, R. (2019) Architecture and florogenesis in female *cannabis sativa* plants. *Frontiers in Plant Science* 10, 350. doi: 10.3389/fpls.2019.00350

Suriyong, S., Krittigamas, N., Pinmanee, S., Punyalue, A. and Vearasilp, S. (2015) Influence of storage conditions on change of hemp seed quality. *Agriculture and Agricultural Science Procedia* 5, 170–176. doi: 10.1016/j.aaspro.2015.08.026

Taghinasab, M. and Jabaji, S. (2020) Cannabis microbiome and the role of endophytes in modulating the production of secondary metabolites: an overview. *Microorganisms* 8, 355. doi: 10.3390/microorganisms8030355

Thiessen, L.D. and Schappe, T. (2019) First report of *Exserohilum rostratum* causing foliar blight of industrial hemp (*Cannabis sativa* L.). *Plant Disease* 103, 1414. doi: 10.1094/PDIS-08-18-1434-PDN

Törjék, O., Bucherna, N., Kiss, E., Homoki, H., Finta-Korpelová, Z. et al. (2002) Novel male-specific molecular markers (MADC5, MADC6) in hemp. *Euphytica* 127, 209–218. doi: 10.1023/A:1020204729122

USDA (2020) The PLANTS database. Natural Resources Conservation Service, US Department of Agriculture, Washington, DC. National Plant Data Team, Greensboro, North Carolina. Available at: http://plants.usda.gov (accessed 17 May 2020).

Vamerali, T., Saccomani, M., Bona, S., Mosca, G., Guarise, M. et al. (2003) A comparison of root characteristics in relation to nutrient and water stress in two maize hybrids. *Plant and Soil* 255, 157–167. doi: 10.1023/A:1026123129575

van Bakel, H., Stout, J.M., Cote, A.G., Tallon, C.M., Sharpe, A.G. et al. (2011) The draft genome and transcriptome of *Cannabis sativa*. *Genome Biology* 12, R102. doi: 10.1186/gb-2011-12-10-r102

Zhao, Y., Sun, Q., Davis, R.E., Lee, I.M. and Liu, Q. (2007) First report of witches'-broom disease in a *Cannabis* spp. in China and its association with a phytoplasma of elm yellows group (16SrV). *Plant Disease* 91, 227. doi: 10.1094/PDIS-91-2-0227C

# 2 The Concept of Plant Disease

## What is Plant Disease?

Plant disease is a condition where plant tissue or growth is damaged or altered by a pathogen or environmental factors. The initial tissue damage may be very subtle at microscopic level and invisible. When plant cells or tissue are killed or altered over a large area of the plant, we see the abnormality, also known as (aka) the symptom. If a symptom is associated with and demonstrated to be caused by a pathogen, the condition is defined as an infectious disease. For example, a hemp plant infected with a species of *Pythium* shows noticeable symptoms of brown colour, necrosis or rot at the crown (the basal part of the stem) (Fig. 2.1). In contrast, healthy crown tissue exhibits green colour externally and white colour internally (Fig. 2.2). This abnormality in colour and texture of crown tissue defines an unhealthy condition for the plant. In many cases, structures or growth of a specific pathogen are visible on symptomatic tissue. As seen in Fig 2.1, the whitish mycelium of a *Pythium* pathogen is growing on the stem surface and covers the crown area. Some pathogens, e.g. viruses, may only alter plant growth and development rather than killing the tissue. For example, hemp leaves may abnormally curl due to a virus infection (Fig. 2.3). If a symptom of the plant is not associated with a pathogen but, rather, is caused by an environmental factor, the condition is often called an abiotic disorder. Some plants have genetic defects and show various chimeric symptoms. This type of condition is called a genetic or physiological disorder. Abiotic and genetic disorders are non-infectious diseases and not transmissible from plant to plant.

## Plant Disease Triangle and Pyramid

A plant disease is a result of the interactions between the host plant, pathogen(s) and environmental conditions. When a pathogen reaches a susceptible plant under a favourable environmental condition, the pathogen may infect the plant and kill plant cells, eventually compromising the health of the entire plant. Some plant species are susceptible to certain pathogens and others are more resistant or even immune to specific pathogens. Plant pathogens are very diverse, including fungi, oomycetes, bacteria, viruses and nematodes. These microorganisms and nematodes commonly exist in the environment; some are even ubiquitous. However, only a small portion of them cause plant diseases. Environmental conditions are critical for disease development. High humidity and optimum temperature favour many fungal diseases, especially those such as powdery mildew, *Botrytis* blight and downy mildew. Hot and dry conditions are not favourable to fungal diseases, but they cause abiotic disorders and induce secondary infections by weak pathogens or insect pests. This interaction between the host plant, pathogen and environmental conditions is called a disease triangle, as shown in Fig. 2.4.

The occurrence of a disease requires a period of conducive environment and the time needed for the plant and pathogen to interact. When a host plant is present and lacks resistant genes for a pathogen, and that pathogen lands on a leaf surface under an inducible environmental condition, an initial infection may occur. If the pathogen is a fungus, it will germinate, penetrate plant tissue, grow and proliferate, and kill many cells inside plant tissue. This infectious process takes time. Unless a visible symptom, for example, a leaf spot, is expressed and visible to naked eyes, the early stages of infection are hardly noticeable. The time between initial infection and symptom appearance is defined as the incubation period. Depending on the type of disease, the incubation period can range from 2 weeks to 1 month. That said, an infectious disease does not occur instantaneously or overnight. Therefore, time is

DOI: 10.1079/9781789246070.0002

considered to be the fourth component of a disease, and the concept of disease triangle has evolved into the four-dimensional disease pyramid (Fig. 2.5),

**Fig. 2.1.** Plant tissue killed by a pathogen. The crown portion of hemp was infected by *Pythium aphanidermatum*, an oomycete species generally causing root and crown rot on many plant species. Note the white, cotton-like *Pythium* mycelia on the stem surface (right-hand specimen in the picture).

**Fig. 2.2.** A healthy hemp plant exhibiting normal crown texture in contrast to the infected plants in Fig. 2.1.

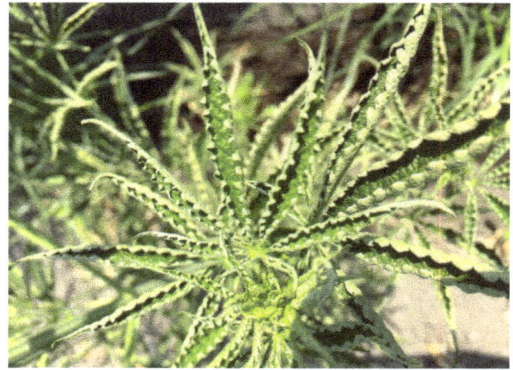

**Fig. 2.3.** A hemp plant exhibiting altered foliage (upward curling of leaves) due to *Beet curly top virus* infection. In this case, plant tissues are not killed but diverge from normal growth.

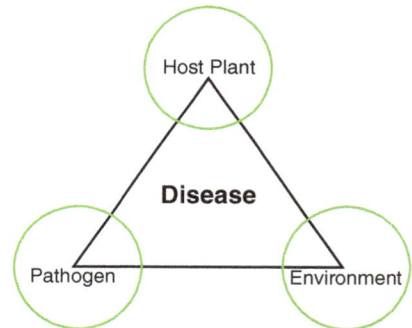

**Fig. 2.4.** Plant disease triangle. The interactions between the host plant, pathogen and environment define a plant disease.

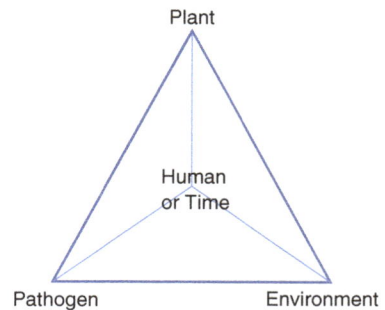

**Fig. 2.5.** Disease pyramid illustrating that time is needed for a disease to develop and that humans are actively involved in plant disease incidence, development and management.

which emphasizes that time is a critical and important factor for an infection to progress and for symptoms to be developed (Francl, 2001).

There is another disease pyramid proposed by some plant pathologists. Instead of adding time to the disease triangle, the human factor is added to the disease triangle, forming the disease pyramid (Fig. 2.5). In the modern agricultural system, humans play an important role in disease incidence, epidemiology and management. For example, humans decide what plant species or cultivars will be planted and how to cultivate specific crops under certain environmental conditions. Human activities may also be associated with the unintended spread of pathogens. Although the impact of human activities on plant health is significant, this model is more useful in illustrating the importance of humans manipulating the disease triangle so that a disease cannot occur. For example, humans can breed and use resistant varieties, modify environment conditions to be less conducive to diseases and take actions to mitigate risk of pathogen introduction and spread. Other human factors such as sanitation, crop rotation and the use of certified disease-free seeds are among the most common practices to break up relationships in the disease triangle and therefore prevent diseases. Understanding the relationships among host, pathogen and environment as well as time and human factors help us to manage plant diseases effectively.

## Types of Plant Diseases

Plant diseases can be classified into two types: infectious diseases and non-infectious diseases.

The infectious disease is defined as a condition caused by pathogenic microorganisms such as fungi, bacteria, viruses and nematodes. Plants infected by a pathogen can be contagious, which means the disease can spread from plant to plant. However, not all plant diseases are infectious or contagious. Many environmental factors, i.e. high soil salinity, chronic drought stress, poor nutrients and misuse of chemicals, can all cause visible damage to individual plants or crops. Plants affected may show various symptoms and the condition is referred to as an abiotic disorder. Since there is no pathogen involved, the disease is not transmitted from plant to plant. Understanding the difference between infectious diseases and abiotic disorders is important in plant problem diagnosis and management.

There are other ways to classify plant diseases. For examples, plant diseases can be classified into fungal diseases, bacterial diseases, viral diseases and nematode diseases. This classification is based on the type of pathogens causing the disease. Sometimes, a disease is called a soilborne disease or an airborne disease. This classification is based on the mode of disease transmission. Some diseases are monocyclic, which means the disease only has one infection cycle during the growing season. Some are polycyclic, meaning the disease has multiple infection cycles in a season. Knowing the disease type is one step forward in understanding the disease's biology as well as its aetiology. Table 2.1 lists some examples of *Cannabis* diseases that fit into each category. Note that some diseases can be assigned to different categories according to the cause, transmission mode and infection cycle.

**Table 2.1.** The basis of disease classification and common types of plant diseases

| Classification base | Type of diseases | Example |
| --- | --- | --- |
| Nature of contagiousness | Infectious diseases | Hemp stem canker |
| | Non-infectious diseases | Cannabis stem overgrowth |
| Pathogen type | Fungal diseases | Hemp *Alternaria* leaf spot |
| | Bacterial diseases | Hemp crown soft rot |
| | Viral diseases | Hemp leaf curl (*Beet curly top virus*) |
| | Nematode diseases | Hemp root-knot |
| | Oomycete diseases | Hemp *Pythium* crown rot |
| Transmission mode | Airborne diseases | Hemp powdery mildew |
| | Soilborne diseases | Hemp *Fusarium* root rot |
| | Vector-borne diseases | Hemp witches' broom (phytoplasma) |
| | Seed-borne diseases | Seedling damping-off |
| | Plant-borne diseases | Cannabis *Fusarium* wilt |
| Disease cycle | Monocyclic diseases | Hemp *Rhizoctonia* root rot |
| | Polycyclic diseases | Cannabis powdery mildew |

## Modes of Disease Transmission

When it comes to infectious diseases, some are soil-borne, some are airborne and some are seed-borne. Some viral diseases are transmitted by insects, mites, fungi or nematodes. Some diseases are spread by cuttings or other propagative materials. One disease may be spread by multiple means. The mode of transmission determines how a disease spreads in a field, locally, regionally, or even internationally. From a management perspective, understanding the transmission mechanisms helps growers take effective measures for disease control. Table 2.2 lists the most common means of disease transmission and their associations with disease types. The specific transmission mode for each hemp disease is discussed in Chapters 6 to 10.

## Plant Disease Cycles

Infectious diseases have a starting point. As illustrated in the disease triangle, a pathogen must reach the plant surface during favourable environmental conditions and initiate an interaction with plant cells to determine if an infection can occur. If the pathogen conquers the plant defence system, it starts to infect plant tissue and spread inside specific plant organ(s). The infection process may result in the death of leaves, stem, roots or the entire plant. If the pathogen is a spore-producing fungus, it then produces massive asexual spores such as conidia. Conidia are a common type of spore produced by

many fungi during the growing season. This type of spore can spread among plants during the season, causing reinfection on the same or a different plant. From the initial contact of a spore on plant tissue to final production of new spores on diseased plant organs constitutes a disease cycle. This cycle may be repeated many times in a season, consequently killing more plants. At the end of a season, a fungal pathogen may start to produce sexual spores or other structures that can survive in dead plant tissue or soil during the winter. These spores or structures become the primary source of infection for the coming year. A disease such as this having more than one infection cycles is called a polycyclic disease (Fig. 2.6).

Some diseases only have one infection cycle that starts with primary infection, kills plant organs, produces special spores or structures, and then survives in soil during the winter. These are called monocyclic diseases. Most soilborne diseases are monocyclic, as their causative agents usually do not produce spores and use them to reinfect plants during the same season.

However, not all plant diseases can be defined as monocyclic or polycyclic, because the disease biology is complex. For example, most viral diseases can be monocyclic if no vector is present but will be polycyclic if vectors are present. Nematode diseases can have more than one infection cycle if the growing season is long enough to allow nematode eggs to hatch and reinfect roots.

The concept of the disease cycle is important for disease management. For monocyclic diseases, one

**Table 2.2.** The common means of pathogen dissemination.

| Dissemination mode | Example of diseases or disease type |
|---|---|
| Soil | Soilborne diseases, plant parasitic nematodes |
| Seed | Seed-borne disease such as *Allium* crop white rot |
| Insects | Diseases caused by phytoplasmas and some viruses |
| Mites | Diseases caused by certain viruses |
| Nematodes | Disease caused by certain viruses |
| Plant cuttings | Vascular wilt diseases such as *Fusarium* wilt |
| Nursery stock | Many diseases can be disseminated through prerogative plant materials |
| Wind | Most airborne diseases such as leaf rust diseases |
| Rain splashes | Leaf spot or blight diseases such as alfalfa spring black stem and leaf spot disease |
| Irrigation/flooding | Oomycete-caused diseases such as *Phytophthora* root rot or blight |
| Field equipment, such as tractors | Soilborne diseases such as *Verticillium* wilt |
| Tools used in the field, such as pruning shears or knives | Vascular wilt disease, such as Canary Island date palm wilt |
| Boots or shoes | Plant parasitic nematodes, soilborne diseases |
| Hands or direct contact of plants | Certain plant viruses such as *Tobacco mosaic virus* |

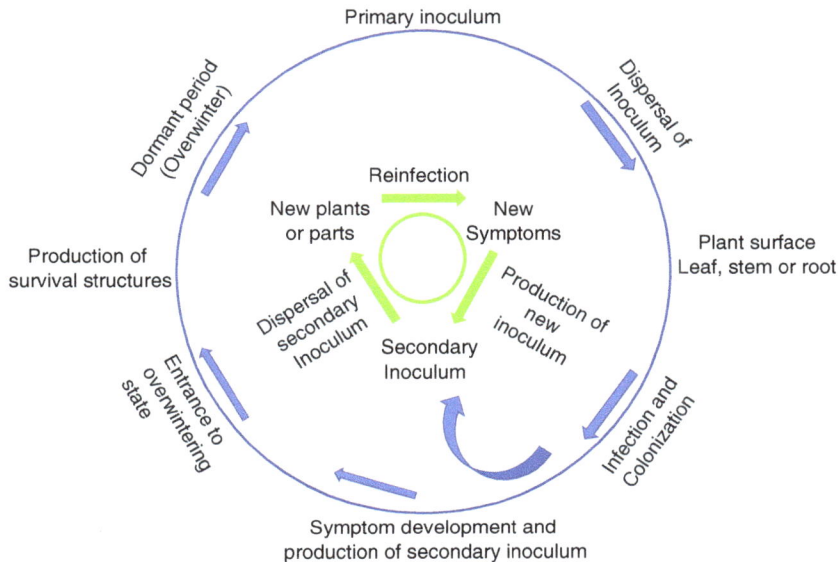

**Fig. 2.6.** Monocyclic disease (blue circle) and polycyclic disease (blue and green circle).

important control strategy is the elimination of the primary source of inoculum, such as fungal spores, sclerotia, mycelia, and other fungal structures. This can be achieved by removing all diseased plants or organs at the end of the season. Soil fumigants and other fungicides are good products to reduce primary inocula. In cases where a disease occurs, a single treatment may significantly suppress the disease. The control strategy for polycyclic plant diseases is more focused on mitigation of multiple-cycle infections by applying fungicides, when appropriate. Sometimes, periodic use of a fungicide during the growing season is necessary to control secondary inocula and prevent a crop from being destroyed by repeated infections.

## Plant-pathogenic Fungi and Plant Diseases They Cause

Fungi are a distinct group of organisms that are generally filamentous in structure, produce spores and grow and feed on either living or dead organic materials. Most fungal species live upon and decompose dead organic materials, but some species can cause diseases in plants. Fungi that only live on dead plant material are called saprophytes; and fungi that only live on live plant tissue and cause diseases are called obligate plant-pathogenic fungi. Some plant-pathogenic fungi can infect live plants but also survive on dead tissue and these fungi are called opportunistic plant pathogens. Fungal infections can cause a variety of symptoms in plants, including necrosis (tissue death), leaf spot, canker, blight, root rot, dieback, damping-off, rot, reduced growth, wilt and plant death.

## Morphology

The morphology of fungi is diverse and therefore used to describe many distinct fungal species. All fungi have living filaments that grow in a mass. An individual filament is called a hypha (Fig. 2.7) and the mass of hyphae is called mycelium (plural: mycelia) (Fig. 2.8). The fungal body spreads in all directions and forms a round colony by growing the hyphal tips (Fig. 2.9). The size and shape of a fungal body is determined by the environmental conditions and nutrient availability. Most fungi produce two kinds of spores: sexual and asexual. The sexual spores are produced for their survival during the winter or other adverse environmental conditions. The asexual spores are produced to aid in fungal dispersal. The shape and types of asexual spores vary greatly among fungal species. The most common asexual spore is the conidium (plural: conidia) (Fig. 2.10), which is often seen in infected

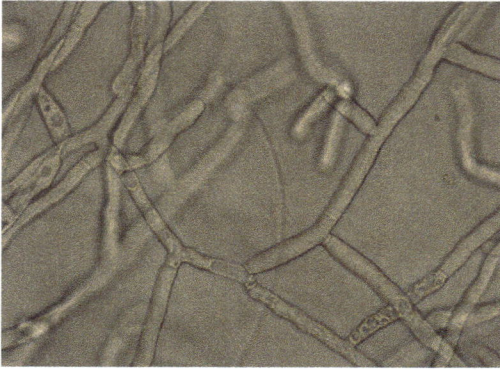

**Fig. 2.7.** Microscopic image of individual hyphae of *Rhizoctonia* species isolated from a hemp root. Note that hyphae are characteristically vertically branched.

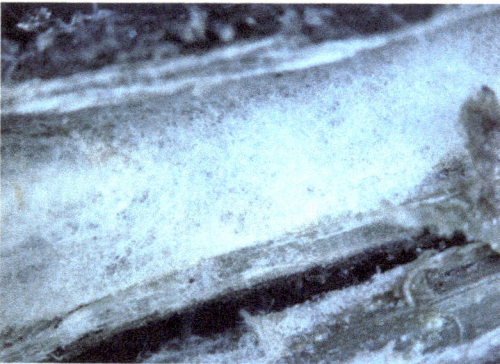

**Fig. 2.8.** Microscopic image of whitish mycelium growth inside the stem of a tomato plant (*Solanum lycopersicum*).

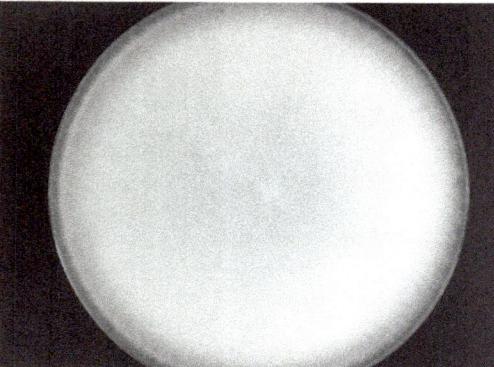

**Fig. 2.9.** A colony of *Fusarium oxysporum* isolated from a hemp plant.

plants during the growing season. Some fungi produce visible fruiting bodies, such as mushrooms. Some produce sclerotia (sing. sclerotium) (Fig. 2.11), which are tiny structures composed of massive, compact hyphal cells. Sclerotia can survive in the soil, dead plant tissue, or seed for many years, even under unfavourable environmental conditions. When environmental conditions are favourable and host plants are present, they germinate and infect plants.

## Life cycles

There are about 300,000 species of fungi that have been described. Among different fungal species, the life cycle can be quite different. The life cycles of

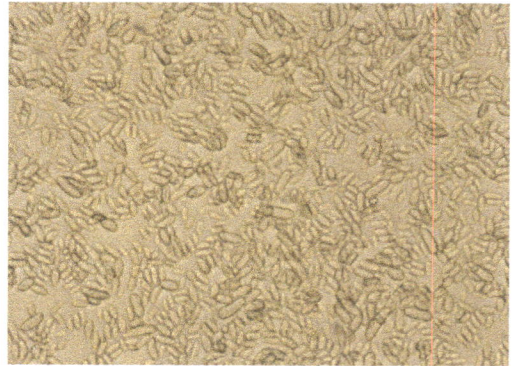

**Fig. 2.10.** Microscopic image showing massive microconidia of *Fusarium oxysporum* isolated from a cannabis plant.

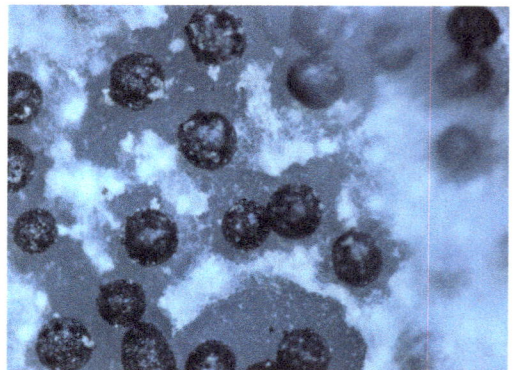

**Fig. 2.11.** Microscopic image showing smooth, shiny, round and black sclerotia produced by *Sclerotium cepivorum*.

plant-pathogenic fungi are better characterized as disease cycles (see Fig. 2.6), because they are causing agents of diseases. In general, spores of a pathogenic fungus are dispersed to the surface of a plant and initiate infection when temperature and humidity are optimal. Spores germinate, penetrate plant tissue, grow mycelium intercellularly and kill a part of or the whole plant. The fungus either survives in dead plant tissue or continues to produce asexual spores as secondary inoculum for further dispersal to new plants, resulting in a large number of plants infected during the growing season. After plants are killed, or at the end of the season, some fungi produce sexual spores, fruiting bodies or other types of structures such as sclerotia that can survive in soil, dead plants and plant debris.

### Plant diseases caused by fungi and oomycetes

Fungi cause many plant diseases, and almost all plant species are subject to fungal infection. Any part of a plant can be infected by a fungal pathogen. Fungi cause more plant diseases than any other type of pathogen. Oomycetes, especially species in *Phytophthora* and *Pythium* genera, can cause diseases similar to those caused by fungi. Predominantly, oomycetes cause root and crown rots, stem canker and damping-off.

#### Diseases caused by oomycetes

**PHYTOPHTHORA DISEASES** The *Phytophthora* species used to be classified as a group of pathogenic fungi, but they are different from true fungi. *Phytophthora* is a genus of pathogens that cause a variety of diseases on many plant species. They cause leaf spot, foliage dieback, stem canker (Fig. 2.12), bleeding canker (Fig. 2.13), crown and root rot (Fig. 2.14). Diseases caused by *Phytophthora* can be either airborne, like potato late blight caused by *Phytophthora infestans* (Nowicki *et al.*, 2011), or soilborne such as many root rot diseases (Erwin and Ribeiro, 1996). *Phytophthora* was listed as one of the pathogens infecting *C. sativa* in Korea (Cho and Shin, 2004; Farr and Rossman, 2019); however, formal reports of hemp and cannabis diseases caused by *Phytophthora* species are scarce at the time of writing. The management of *Phytophthora* diseases should be focused on integrated measures such as the use of healthy plant stock and resistant varieties, sanitation, crop rotation, and chemical treatment.

**Fig. 2.12.** Azalea (*Rhododendron* sp.) stem canker caused by *Phytophthora citricola*. Note that brown colour of lower stem indicates infection, and the canker may expand further to upper part of stem under favourable environmental conditions.

**PYTHIUM DISEASES** Like *Phytophthora*, *Pythium* is another genus of oomycetes that cause diseases in many plant species. Typical diseases caused by *Pythium* are root rot (Fig. 2.15), crown rot (Fig. 2.16), seedling damping-off (Fig. 2.17) and soft rot of vegetables. *Pythium* species generally grow fast, especially under high levels of moisture. For example, *P. aphanidermatum*, the pathogen causing hemp crown rot (Schoener *et al.*, 2018), completely colonized on a corn meal agar plate (9.0 cm in diameter) within 24 h. In the field, *Pythium* is often recognized as a white to grey mould on the surface of infected plant parts (Fig. 2.17). In marijuana plants, *P. dissotocum*, *P. myriotylum* and *P. aphanidermatum* were frequently found from plants exhibiting root rot symptoms (Punja and Rodriguez, 2018; Punja *et al.*, 2018). In hemp crops, *P. aphanidermatum* and *P. ultimum* were found to be associated with crown and root rot (Beckerman *et al.*, 2017, 2018; Schoener *et al.*, 2018). The characteristics and control of specific diseases caused by *Pythium* in hemp and marijuana plants are presented in Chapters 8 and 9.

**DOWNY MILDEW DISEASES** Downy mildews are foliage diseases caused by a number of downy mildew fungi, which are now classified as fungus-like oomycetes. Downy mildew pathogens include species in genera of *Bremia*, *Peronospora*, *Peronosclerospora*, *Plasmopara*, *Pseudoperonospora*, *Sclerophthora*, and *Sclerospora* (Agrios, 1997). Each genus has a limited host range, causing downy mildew only

**Fig. 2.13.** Maple (*Acer* sp.) bleeding canker caused by *Phytophthora cactorum*. Note the brown to dark sap oozing out from inside the bark. Internal bark tissue is decayed (not shown).

**Fig. 2.14.** Rosemary (*Salvia rosmarinus*) root rot caused by *Phytophthora nicotianae*. Note that some roots become black and the root system lacks new growth of healthy roots.

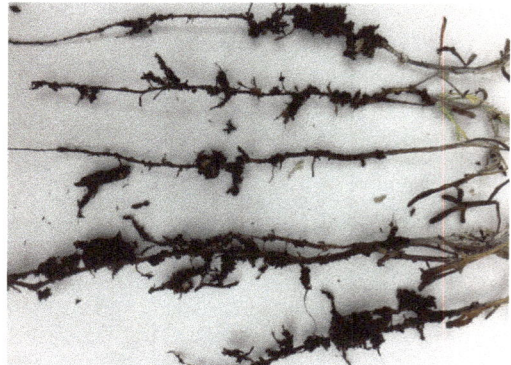

**Fig. 2.15.** Big sagebrush (*Artemisia tridentata*) root rot caused by *Pythium dissotocum*. Note the blackened taproots with very limited lateral roots.

on certain crops or plant species. They affect various vegetable crops, grape, alfalfa, maize, wheat, grasses, sunflower and others. The mouldy growth of mycelium and spores predominantly occurs on the lower surface of the leaf (Fig. 2.18). Cool temperatures and high humidity are favourable conditions for disease development. In hemp crops, downy mildew caused by *Pseudoperonospora* *cannabina* was observed in central Asia, and downy mildew caused by *Peronospora cannabina* was observed in China, Pakistan and Poland (Farr and Rossman, 2019). Infected hemp plants exhibit yellowing, discoloration and eventually death of leaves. The application of fungicides is effective in preventing and controlling downy mildew diseases in crops.

### Diseases caused by zygomycetes

Zygomycetes are generally saprophytes and can be easily found in the environment. Only a few species are recognized as plant pathogens or opportunistic parasites, causing soft rot of vegetables or fruits. Some species can cause human diseases such as zygomycosis, especially those in the orders of Mucorales and Entomophthorales (Ribes *et al.*, 2000). These fungi may cause opportunistic infections in immunocompromised human patients.

Three genera of zygomycetes are recognized as plant-pathogenic pathogens: *Choanephora*, *Rhizopus* and *Mucor*. The genus *Rhizopus* is the most common in this group. It includes at least eight species (Abe *et al.*, 2010). Species in *Rhizopus* infect fleshy organs of plants, causing soft rot. They grow rapidly and extensively on rotted tissue and produce long, aerial sporangiophores and black spherical sporangia (Fig. 2.19). One species called *R. oryzae* is frequently found on hemp seeds during the germination test (Fig. 2.20) (see Chapter 6).

### Diseases caused by ascomycetes and imperfect fungi

Ascomycetes and some asexual fungi (also known as imperfect fungi) cause some of most well-known

**Fig. 2.16.** Hemp (*Cannabis sativa*) crown rot caused by *Pythium aphanidermatum*. Note that the internal tissue of lower stem and stem-root junction area is decayed and brown in colour.

**Fig. 2.18.** Downy mildew on lettuce (*Lactuca sativa*) caused by *Bremia lactucae*. Note the discoloured leaf tissue and white mycelia growing on the underside of the leaf.

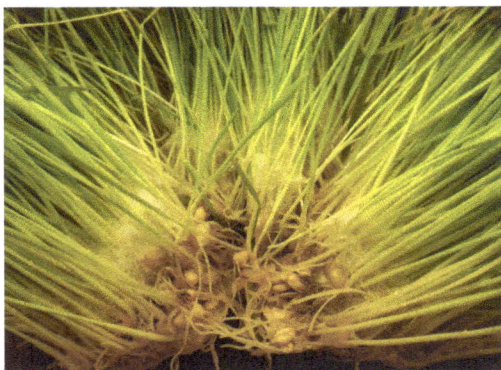

**Fig. 2.17.** Wheat (*Triticum* sp.) seedling damping off caused by *Pythium* sp. Note the whitish mould growing out from seedlings and brown lesions at the basal part of seedlings.

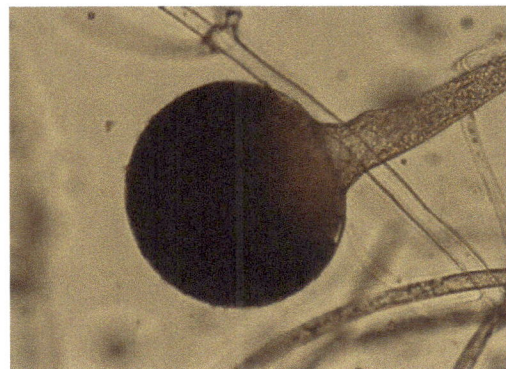

**Fig. 2.19.** Long sporangiophore and a round sporangium of *Rhizopus oryzae* isolated from hemp seeds. Note that sporangiospores are released from the sporangium.

**Fig. 2.20.** White, coarse mycelium of *Rhizopus oryzae* growing out from inside hemp seeds and taproots on a germination paper. Note that one taproot (second seed from the left) had a sunken lesion. Several seeds with *Rhizopus* growth failed to germinate.

plant diseases. Ascomycetes produce both sexual spores called ascospores and asexual spores known as conidia. Imperfect fungi only produce asexual spores in natural conditions, or their sexual stages have not been found. Both groups of fungi cause various symptoms such as leaf spot, blight, canker, fruit spot, fruit rot, anthracnose, stem and root rot, vascular wilt and soft rot. Examples of diseases caused by these fungi include sooty mould, various powdery mildews, grey mould, leaf spot, needle blight of conifers, anthracnose of grasses, apple scab, *Fusarium* wilt, *Verticillium* wilt, Dutch elm disease and onion white rot. Plate 2.1 shows some representative plant diseases caused by these two groups of fungi. Depending on the causative fungal species, each disease may be quite different in terms of symptom development and severity. Control of each specific disease requires an accurate diagnosis and an understanding of the disease biology.

### Diseases caused by basidiomycetes

Basidiomycetes produce sexual spores called basidiospores. Most basidiomycetes are fleshy fungi that produce mushrooms and puffballs, which either live on decaying organic matter or grow on the roots and trunk of trees, causing wood decay. However, certain fungi in two orders of basidiomycetes cause rust or smut diseases in crops, ornamental plants and trees. Common rust diseases include cereal rust, stem rust of wheat, cedar-apple rust, white pine blister rust, daylily rust and sagebrush rust. Smut diseases include corn smut, kernel smuts of small grains, loose smut of cereals and covered smut of wheat. Plate 2.2 shows some representative rust and smut diseases caused by basidiomycetes. In *Cannabis* plants, only *Puccinia cynodontis* was recorded in China (Farr and Rossman, 2019). Rusts and smuts are airborne diseases that are spread by strong winds and their spores can be dispersed several hundred kilometres. Infected seeds can also carry pathogens to new regions. The life cycle of rust fungi is complex and some species require multiple host plants to complete their life cycle. The management of rust and smut diseases relies on the use of resistant crop varieties. When a disease breaks out in a field, systemic fungicides can be used to control the damage. However, in most cases multiple applications may be needed during the season.

### Fungi reported from Cannabis sativa plants

Many fungal species have been reported from *Cannabis* plants based on either preliminary disease observations or complete diagnostic research (Farr and Rossman, 2019). Some of them are from old records. A complete list of these fungal species is provided in Table 2.3. The large number of reported fungal pathogens suggest that *Cannabis* plants are prone to fungal infections, but it does not mean that all listed fungal species will cause diseases in a *Cannabis* crop. Instead, growers may exercise cautions to prevent those diseases from occurring in a field or facility. The genera in bold in Table 2.3 are likely to be more problematic in *Cannabis* plants, as they are major pathogens in many types of crops.

## Plant-pathogenic Bacteria and Plant Diseases They Cause

Bacteria are microscopic single-celled organisms that are found ubiquitously in the environment. Plants

**Plate 2.1.** Examples of plant diseases caused by ascomycetes and imperfect fungi that affect the health of leaves, stem and underground tubers or bulbs. **(A)** Powdery mildew of marijuana plants (*Cannabis sativa*) caused by *Golovinomyces ambrosiae*. **(B)** Vascular wilt of a cannabis plant caused by *Fusarium oxysporum*. Note the internal vascular discoloration at the lower portion of the stem. **(C)** *Alternaria* leaf spot of cucumber (*Cucumis sativus*) caused by *Alternaria alternata*. Note that the dark colour indicates massive conidia spores. **(D)** *Verticillium* wilt of peppermint (*Mentha* sp.) caused by *Verticillium dahliae*. Note that the infected plant was collapsed. **(E)** Black scurf of potato (*Solanum tuberosum*) caused by *Rhizoctonia solani*. Note the raised, black patches that are actually compact masses of mycelia attached to the tuber skin. **(F)** White rot of onion (*Allium cepa*) caused by *Sclerotium cepivorum*. Note the white mould on the fresh onion.

**Plate 2.2.** Examples of rust and smut diseases caused by basidiomycetes. **(A)** Daylily (*Hemerocallis* sp.) rust caused by *Puccinia hemerocallidis*. Note the brown lesions and pustules. **(B)** Wheat (*Triticum aestivum*) rust caused by *Puccinia* sp. Note the opened pustules with orange uredospores inside and dark pustules containing teliospores. **(C)** Reed grass (*Calamagrostis* sp.) rust caused by *Puccinia* sp. Note the orange uredospores inside pustules. **(D)** Garlic (*Allium sativum*) rust caused by *P. allii*. Note the extensive orange pustules and chlorotic spots. **(E)** Sagebrush (*Artemisia tridentata*) rust caused by *P. tanaceti*. Note the brown uredospores dispersed on the leaf surface. **(F)** Wheat (*Triticum aestivum*) loose smut caused by *Ustilago tritici*. Note the kernels destroyed by the fungus, compared with healthy kernels on the right.

**Table 2.3.** Fungal species reported from *Cannabis sativa* and the geographical occurrence of associated diseases (Farr and Rossman, 2019).

| Fungal species[a] | Geographical occurrence of observed diseases |
|---|---|
| *Achlya aquatica* | India |
| *Achlya* sp. | USA: Alabama |
| *Acrosporium* sp | USA: Illinois |
| *Aecidium cannabis* | Russia |
| **Alternaria** alternata | India |
| *Alternaria porri* | France |
| *Alternaria tenuis* (*Alternaria alternata*) | Kyrgyzstan |
| **Ascochyta** boehmeriae | China |
| *Ascochyta prasadii* | China, India |
| *Ascochyta* sp. | China |
| *Aspergillus flavus* | USA: Maryland, Virginia, Wisconsin |
| *Aspergillus niger* | USA: Maryland, Wisconsin |
| *Aspergillus parasiticus* | USA: Virginia |
| *Botryosphaeria marconii* | USA: Maryland, Virginia |
| | France, Italy, Lithuania, Russia |
| **Botrytis** cinerea | USA: California, Oregon, Virginia |
| | Bulgaria, Canada |
| **Cercospora** cannabina | USA: Mississippi, Wisconsin |
| (*Pseudocercospora cannabina*) | Cambodia, China, India, Pakistan, Uganda, USSR |
| *Cercospora cannabis* | China, India, Japan |
| *Cercospora* sp. | USA: Kentucky, |
| | Cambodia |
| *Chaetomium succineum* | India |
| *Cladosporium tenuissimum* | India |
| **Colletotrichum** dematium | China |
| *Colletotrichum lini* | China |
| *Corticium solani* (*Rhizoctonia solani*) | Greece |
| *Cylindrosporium* sp. | USA: Maryland |
| | Korea |
| *Dendrophoma marconii* | Chile |
| *Diaporthe arctii* var. *achilleae* | Italy |
| *Diaporthe ganjae* | USA: Illinois |
| *Didymella arcuata* | Germany |
| *Diplodiella ramentacea* | Poland |
| *Epicoccum purpurascens* (*Epicoccum nigrum*) | China, India |
| *Exserohilum rostratum* | USA: North Carolina |
| **Fusarium** avenaceum var. *herbarum* | Poland |
| *Fusarium brachygibbosum* | USA: California |
| *Fusarium equiseti* | USA: California |
| | Canada |
| *Fusarium javanicum* var. *radicicola* (*Fusarium radicicola*) | Poland |
| *Fusarium lateritium* | Poland |
| *Fusarium oxysporum* | USA: California |
| | Canada, China, Poland |
| *Fusarium oxysporum* f. sp. *cannabis* | USA: California |
| *Fusarium solani* | California, Canada, Poland |
| *Fusarium* sp. | USA: Illinois, Indiana, Virginia, Wisconsin |
| | Canada, Mexico, USSR |
| *Gibberella saubinetii* (*Fusarium sulphureum*) | USA: Indiana, Virginia |
| *Glomus mosseae* | USA: Illinois |
| **Golovinomyces** cichoracearum | Canada |
| *Golovinomyces spadiceus* | USA: Kentucky |
| *Helicomina cannabis* | India |

*Continued*

**Table 2.3.** Continued.

| Fungal species[a] | Geographical occurrence of observed diseases |
|---|---|
| *Hypomyces cancri* | USA: Maryland |
| *Leptosphaeria cannabina* | Italy |
| *Leptosphaeria woroninii* | Romania |
| *Leptosphaerulina trifolii* | India |
| *Leveillula taurica* | France, Turkey, USSR |
| *Leveillula taurica* f. sp. *cannabis* | Asia, USSR |
| *Macrophomina phaseoli* - (*Macrophomina phaseolina*) | USA: Illinois<br>China, Cyprus |
| *Macrophomina phaseolina* | USA: Illinois<br>Iran, Spain |
| *Micropeltopsis cannabis* | France |
| *Mucor* sp. | USA: Wisconsin |
| *Mycosphaerella cannabis* - (*Neodidymelliopsis cannabis*) | China |
| *Myrothecium roridum* (*Paramyrothecium roridum*) | India |
| *Oidium* sp. | South Africa |
| *Ophiobolus anguillides* | USA: Minnesota |
| *Orbilia luteola* | France |
| *Papularia sphaerosperma* (*Arthrinium phaeospermum*) | India |
| *Pellicularia rolfsii* (*Athelia rolfsii*) | Korea |
| *Penicillium copticola* | Canada |
| *Penicillium olsonii* | Canada |
| *Penicillium* sp. | USA: Wisconsin |
| *Periconia byssoides* | India |
| *Peronoplasmopara cannabina* (*Pseudoperonospora cannabina*) | Central Asia |
| **Peronospora** *cannabina* (*Pseudoperonospora cannabina*) | China, Pakistan, Poland |
| **Phoma** *cannabis* (*Neodidymelliopsis cannabis*) | Asia, Europe, USA |
| *Phoma herbarum* (*Phoma herbarum* var. *herbarum*) | China |
| *Phoma nebulosa* (*Phomatodes nebulosa*) | Netherlands |
| *Phoma* sp. | Poland |
| *Phomopsis cannabina* | India |
| *Phomopsis ganjae* (*Diaporthe ganjae*) | USA: Illinois |
| *Phomopsis* sp. | USA: Illinois |
| **Phyllosticta** *cannabis* (*Neodidymelliopsis cannabis*) | USA: Wisconsin<br>Bulgaria, China, India |
| *Phyllosticta straminella* (*Macrophoma straminella*) | China, Korea |
| **Phymatotrichum** *omnivorum* (*Phymatotrichopsis omnivora*) | USA: Arizona, Texas |
| *Phymatotrichum* sp. | Mexico |
| **Phytophthora** sp. | Korea |
| *Pithomyces chartarum* (*Pseudopithomyces chartarum*) | India |
| *Pleosphaerulina cannabina* | USSR |
| *Podosphaera macularis* | Switzerland |
| *Pseudocercospora cannabina* | China, India, Korea, Poland |
| **Pseudoperonospora** *cannabina* | Andorra, Austria, China, France, Hungary, Japan, Korea, Kyrgyzstan, Latvia, Poland, Portugal, Romania, Russia, Spain, Switzerland |
| **Puccinia** *cynodontis* | China |
| **Pythium** *aphanidermatum* | USA: California, Indiana<br>Canada |
| *Pythium dissotocum* | Canada |
| *Pythium myriotylum* | Canada |
| *Pythium ultimum* (*Globisporangium ultimum*) | USA: Indiana |
| *Ramularia collo-cygni* | Austria |

*Continued*

**Table 2.3.** Continued.

| Fungal species[a] | Geographical occurrence of observed diseases |
|---|---|
| **Rhizoctonia** sp. | Netherlands |
| **Sclerotinia** minor | USA: California |
| Sclerotinia sclerotiorum | USA: Montana |
| | Canada, China, France |
| **Sclerotium** bataticola (Macrophomina phaseolina) | Bulgaria |
| Sclerotium rolfsii (Athelia rolfsii) | USA: South Carolina, Texas |
| | Italy |
| **Septoria** cannabina (Septoria neocannabina) | Canada, Romania |
| Septoria cannabis | USA: Florida, Iowa, Minnesota, South Dakota, Texas |
| | Bulgaria, Canada, China, India, Korea, Poland, Romania |
| Sphaerotheca macularis (Podosphaera macularis) | USA: Illinois |
| Stemphylium cannabinum | Bulgaria |
| Thanatephorus cucumeris (Rhizoctonia solani) | China |
| Trichothecium roseum | India |
| Uredo kriegeriana | Germany, former USSR |
| Verticillium albo-atrum | China |

[a]Genera in bold likely to be more problematic in *Cannabis* plants.

are associated with many different types of bacteria. Some bacteria are beneficial to plant growth, some are neutral, but some are harmful and cause diseases in plants. There are thousands of bacterial species named in scientific literature, but only a small portion of bacterial species cause plant diseases under certain environmental conditions. Those species causing plant diseases are called plant-pathogenic bacteria. There are about 29 bacterial genera that contain plant-pathogenic species (Kado, 2010) and the most important genera are *Agrobacterium*, 'Candidatus Liberibacter', *Ralstonia*, *Xanthomonas*, *Xylella*, *Pseudomonas*, *Dickeya*, *Erwinia*, *Pantoea*, *Spiroplasma* and 'Candidatus Phytoplasma'.

The plant-pathogenic bacteria can be classified into two major groups by the structure of the bacterial cell walls. One group of bacteria has thick cell walls and the other has thin cell walls. When using Gram stain, bacteria with thick cell walls are stained purple (called Gram-positive) and bacteria with thin walls are not stained (Gram-negative). Most plant-pathogenic bacteria are Gram-negative. Some in the class Actinobacteria are Gram-positive, such as *Clavibacter* and *Streptomyces*.

Although most plant-pathogenic bacteria found on plant tissue are culturable on synthetic nutrient media, some cannot be cultured *in vitro*. These non-culturable organisms are classified as being in the Mollicutes class of bacteria. Because these bacteria only inhabit live plant tissue, especially the phloem vessels (Doi *et al.*, 1967), they are often called fastidious bacteria or obligate parasites. Two genera, *Spiroplasma* and 'Candidatus Phytoplasma', are in this group and have been found from diseased hemp plants (Schoener and Wang, 2019).

## Morphology

Most plant-pathogenic bacteria are bacilliform, with some strains being more elongated or filamentous, or even slightly club-shaped. Many bacteria have thin, whip-like protrusions called flagella. Some species have only one flagellum and others have flagella across the entire surface. Mollicutes such as phytoplasmas have variable morphology ranging from round or ovoid to elongated.

Plant-pathogenic bacteria are 1–3 µm in size and are only visible under a microscope. When the bacterium is grown on a culture medium, hundreds to thousands of bacterial cells can form a colony visible to the naked eye (Fig. 2.21). In some genera or species, the growth of bacteria may emit a strong odour when cultured in plates or infecting plant tissue. Therefore, a strong odour associated with soft rot may indicate a bacterial infection.

## Life cycle

Plant pathogenic bacteria reproduce and develop within the host plant, on the surface of the plant, on plant debris, or in the soil, if the nutrients to support bacterial growth are available and environmental conditions are favourable. The rod-shaped bacteria reproduce through binary fission, in which one bacterium splits into two, those two bacteria become four, and so on. Binary fission occurs when the cytoplasmic membrane grows inwards towards the centre of the cell, forming a membranous partition and dividing the cytoplasm in half. When the formation of the cell walls between the two is complete, the two cells split apart. While the rate of reproduction depends on the environmental conditions, bacteria can multiply exponentially in a brief period. For example, a single cell of *Escherichia coli* can grow into a visible colony 1–3 mm in diameter on a lysogeny broth (LB) agar plate overnight at 37°C. This rate of cell growth is approximately equal to doubling of cell numbers every 30 min, which means that a single cell can grow into a colony containing $10^7$–$10^8$ cells in 12 h (Lodish *et al.*, 2000).

Most plant-pathogenic bacteria are facultative parasites, which means that the bacteria can survive without associating with a host plant. Therefore, they can be easily cultured on a standard nutrient medium. Meanwhile, fastidious bacteria are only associated with live plant tissue and cannot be cultured on a nutrient medium unless a specialized medium is developed. Some fastidious plant-pathogenic bacteria are xylem-limited, such as *Xylella fastidiosa* (Wells *et al.*, 1987) which causes bacterial leaf scorch and Pierce's disease of grapevine. Some are phloem-limited, like phytoplasmas (Hogenhout *et al.*, 2008), which cause diseases

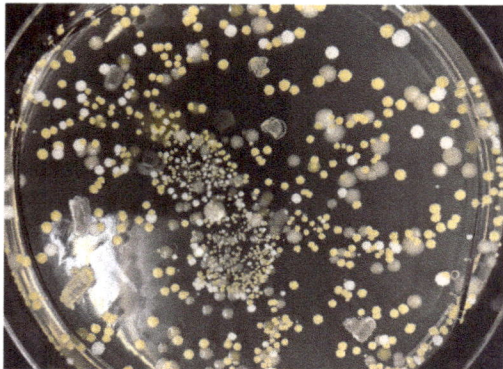

**Fig. 2.21.** Bacterial growth on an LB agar plate showing multiple types of bacterial colonies isolated from diseased hemp root. Note that the colony sizes are about 1–3 mm in diameter and that there are at least two types of bacterial colonies in the plate.

on many monocot and dicot plant species. Both *Xylella* and phytoplasmas are transmitted by insect vectors. For *X. fastidiosa*, the most common vectors are xylem-feeding insects such as leafhoppers and spittle bugs (Purcell, 1997). Since phytoplasmas live inside phloem tissue of plants, only phloem-feeding insects can effectively acquire a substantial amount of phytoplasma cells and then transmit them to new host plants via feeding. However, not all phloem-feeding insects are vectors; only some leafhoppers, four families of planthoppers and two genera of psyllids are confirmed to be vectors.

Most plant-pathogenic bacteria do not produce special structures for survival in adverse conditions. They can survive on leaves, flowers, fruits, stems and roots for a short period of time as epiphytes. Some can survive in soils as soil inhabitants or saprophytes. These bacteria can infect new plants whenever they are in contact with a host plant surface under favourable environmental conditions.

### Diseases caused by bacteria

When plant-pathogenic bacteria infect plants, they cause various types of symptoms such as leaf spot, yellowing, chlorosis, canker, gall, soft rot, vascular wilt, witches' broom and eventual death of plants. The initial symptom might be water-soaked lesions on the plant tissue due to the enzymatic digestion of the plant cells by the bacteria. Depending on the plant tissue affected and the type of plant-pathogenic bacteria involved, the symptoms of infected plants vary significantly. Table 2.4 lists some common genera of plant-pathogenic bacteria that cause disease on many plant species. However, it is unknown if bacterial species in some of these genera will cause hemp or cannabis diseases under laboratory, greenhouse or field conditions. Some species of bacteria have been found in hemp fields causing significant damage to the crops. For example, phytoplasmas were the major pathogen detected from hemp plants exhibiting severe witches' broom and leaf proliferation symptoms (Schoener and Wang, 2019; Wang, 2019) (see Chapter 10). Others were only reported from hemp crops years ago (Netsu *et al.*, 2014), or were merely found on hemp plants as endophytes (Scott *et al.*, 2018).

#### *Agrobacterium diseases*

Bacteria in genus *Agrobacterium* cause crown gall (close to the soil line or the stem base), tumour-like

**Table 2.4.** Common plant-pathogenic bacterium genera that may impact hemp growth.

| Bacterial genus | Disease symptoms | Reports from *Cannabis* plants |
|---|---|---|
| *Agrobacterium* | Crown gall, gall, hairy root | Wahby *et al.*, 2013 |
| '*Candidatus* Liberibacter' | Citrus greening, yellowing, growth decline | Not known |
| *Ralstonia* | Bacterial wilt | Not known |
| *Xanthomonas* | Leaf spot, leaf streak, leaf blight, dieback, canker | Netsu *et al.*, 2014; Jacobs *et al.*, 2015 |
| *Xylella* | Bacterial leaf scorch such as Pierce's disease of grape, leaf scorch on shading trees such as oak leaf scorch | Not known |
| *Pseudomonas* | Leaf spot, blight, knot, node, gall, canker | Gardan *et al.*, 1999; Bull and Rubio, 2011 |
| *Dickeya* | Soft rot | Not known |
| *Erwinia* | Fire blight, leaf spot, soft rot, canker, vascular wilt | Not known |
| *Pantoea* | Soft rot, blight, dieback | Scott *et al.*, 2018 |
| *Spiroplasma* | Stunting, little leaf, chlorosis, over-branching, witches' broom, | Schoener and Wang, 2019 |
| '*Candidatus* Phytoplasma' | Yellowing, dwarf, small leaf, chlorosis, crowded buds or proliferation, witches' broom, chronic decline | Wang, 2019; Schoener and Wang, 2019 |

abnormal growth on the aerial part of stems, or hairy root disease. The most common species are *A. tumefaciens* (crown gall), *A. vitis* (aerial parts) and *A. rhizogenes* (hairy root). The diseases affect woody and herbaceous plants in about 140 genera of more than 60 families, but it is mostly found on pome and stone fruit trees and grapes. The initial symptom is a small, round, whitish and soft gall on the stem and roots near the soil line. The gall may continue to grow into a very large size with the outer tissue becoming dark brown. Galls are generally irregularly shaped and usually surround the stem (Fig. 2.22) or root.

There is no report demonstrating that *Agrobacterium* can cause diseases in hemp crops, even though *A. tumefaciens* was listed as one of the pathogens for hemp (https://en.wikipedia.org/wiki/List_of_hemp_diseases). One research team successfully infected hemp plants with *A. rhizogenes* while they were developing a lab protocol for transformation of *C. sativa* plants (Wahby *et al.*, 2013). Although the purpose of the study was to overcome the difficulty of hemp plant transformation, the research did confirm that hemp was susceptible to several wild-type strains of *A. rhizogenes* in a lab condition. The susceptibility of hemp to *Agrobacterium* may signal a potential problem for hemp production. When hemp crops are widely cultivated in diverse types of farmland, diseases caused by *Agrobacterium* may start to show up in the field. Therefore, *Agrobacterium*-caused diseases need to be monitored during hemp production. Inspecting seedlings, cuttings and other propagative materials for signs of *Agrobacterium*-caused diseases and destroying

**Fig. 2.22.** Galls on a grapevine caused by *Agrobacterium vitis*. Note the cracked bark tissue and tumour-like overgrowth in the aerial part (stem) of the plant.

infected plants are crucial to prevent the spread of the bacteria to new cultivation sites. Biological control of crown gall has been practised for years by soaking germinated seeds or dipping rootstocks in a suspension of a biocontrol agent such as the No. 84 strain of *Agrobacterium radiobacter* (Moore and Warren, 1979). The website of the US Environmental Protection Agency (EPA) offers a factsheet on the *Agrobacterium radiobacter* Strain K84 (114201) product.

### '*Candidatus* Liberibacter' diseases

The genus '*Candidatus* Liberibacter' contains several species that cause serious plant diseases. Most notably, three species cause citrus greening disease, also known as 'huanglongbing', meaning 'yellow dragon disease' in Chinese (Li *et al.*, 2006). 'C. Liberibacter americanus' is a species causing citrus greening in South America, 'C. Liberibacter africanus' causes citrus greening in Africa and 'C. Liberibacter asiaticus' causes citrus greening in Asia. Infected citrus plants exhibit leaf chlorosis (Fig. 2.23), stunting, underdeveloped leaves, reduced branch size and gradual decline of growth vigour.

'*Candidatus* Liberibacter' also causes zebra chip disease in potatoes, but the strain associated with zebra chip is different from species causing citrus greening disease (Abad *et al.*, 2009; Lin *et al.*, 2009). One species named '*Candidatus* Liberibacter psyllauous' was found to infect solanaceous plants such as tomato and potato, causing 'psyllid yellows' (Hansen *et al.*, 2008). All these reported '*Candidatus* Liberibacter' species or strains are Gram-negative, unculturable and phloem-inhabiting bacteria and they are transmitted by specific psyllids. In phloem tissue, '*Candidatus* Liberibacter' coexists and interacts with many other bacterial species; interestingly, other phloem-inhabiting bacteria support the survival of '*Candidatus* Liberibacter' (Fujiwara *et al.*, 2018). It is speculated that '*Candidatus* Liberibacter' colonizes the host and then constructs a microbial community that favours disease development. There has been no case of '*Candidatus* Liberibacter' infection reported in *Cannabis* plants. However, there is a possibility that '*Candidatus* Liberibacter' may be easily transmitted to the phloem of *Cannabis* plants from solanaceous or other crops through the feeding of bacteria-carrying psyllids. If this happens, the colonization of '*Candidatus* Liberibacter' inside the phloem and subsequent development of a disease may depend on the native microbial community and the nutrients in the phloem of *Cannabis* plants.

### Ralstonia diseases

*Ralstonia* is a genus in the Ralstoniaceae family of the Betaproteobacteria class. There are only two species in this genus causing plant diseases: *Ralstonia solanacearum* and *R. syzygii*. The most important species is *R. solanacearum*, which causes diseases in over 50 plant families. This bacterium is soilborne, colonizing the xylem and causing vascular wilt (aka bacterial wilt). Cross-sections of the infected stem may ooze a slimy bacterial exudate. *R. solanacearum* is differentiated into five races according to host range, which are equivalent to pathovars in other plant-pathogenic bacteria such as *Pseudomonas* and *Xanthomonas*. Some races are further differentiated into different biovars based on biochemical characteristics. For example, *R. solanacearum* race 1 biovar 1 is a strain that infects tomatoes and other vegetable and ornamental plants; while *R. solanacearum* race 3 biovar 2 mainly infects certain solanaceous crops and geranium plants, causing brown rot of potato and systemic wilt of tomato, eggplant and geranium. *R. solanacearum* also infects banana plants, causing Moko disease (aka blood disease). For instance, a strain characterized as *Ralstonia solanacearum* race 2 biovar 1 was found to infect banana, causing yellowing, wilting and collapsing of plants in Malaysia (Zulperi and Sijam, 2014). *R. syzygii*, which causes Sumatra disease of clove, was considered to be a separate species in the genus *Ralstonia*. However, Remenant *et al.* (2011) suggested that *R. solanacearum*, *R. syzygii* and the strain infecting banana all fall into the Phylotype IV subgroup of the *R. solanacearum* species complex, based on their DNA sequences and DNA–DNA hybridization data, despite being phenotypically different. Thus, the *Ralstonia* species infecting plants can be treated as a single species complex – all are soilborne, colonize plant xylem vessels and cause wilt diseases. Still, the host specificity in *R. solanacearum* is poorly understood and the classification of *R. solanacearum* strains is complex (Peeters *et al.*, 2013). The fact that there are diverse strains in the *R. solanacearum* species complex and that they infect both monocot and dicot plants raises the question of whether any strain of this bacterium may evolve to infect *Cannabis* plants.

**Fig. 2.23.** Chlorosis of citrus (*Citrus* sp.) leaves caused by '*Candidatus* Liberibacter asiaticus'.

## Xylella diseases

The *Xylella* genus belongs to the Xanthomonadaceae family, but *Xylella* bacteria behave quite differently from *Xanthomonas*, another major plant-pathogenic genus in the same family (see next section). *Xylella* bacteria are fastidious, meaning they are difficult to culture on synthetic media. *X. fastidiosa* is probably the only known species in this genus that causes diseases in many plant species. This bacterium infects plants and multiplies in the xylem, causing various types of leaf scorch diseases, symptomatically similar to acute or chronic drought stress. The species is speculated to have multiple strains that infect specific plant species (Hopkins and Purcell, 2002). For example, grape strains cause Pierce's disease (grapevine leaf scorch) (Fig. 2.24), alfalfa dwarf and almond leaf scorch, while peach-plum strains cause phony peach, plum leaf scald, citrus variegated chlorosis and coffee leaf scorch. There are some uncharacterized strains that cause various leaf scorch diseases on shade trees, fruit trees and ornamental plants. *X. fastidiosa* is transmitted by xylem-feeding insects such as the glassy-winged sharpshooter (*Homalodisca coagulata*) (Hopkins and Purcell, 2002), but infected plant materials may also transmit the bacteria though grafting or other propagation methods. *X. fastidiosa* has not been reported in *Cannabis* crops. It is not known whether this bacterium can survive in the xylem of

**Fig. 2.24.** Pierce's disease of grapevine showing a typical leaf scorch symptom on a leaf infected by *Xylella fastidiosa*. Note that reddish and brown lesions between veins indicate water deficit resulting from clogged xylem vessels. This symptom (with reddish colour) is distinct from leaf marginal scorch caused by drought stress (often grey to brown in colour).

hemp plants and cause any disease when it is introduced through xylem-feeding insects.

## Xanthomonas diseases

*Xanthomonas* bacteria cause bacterial spots on the leaves, stems and fruits of certain crops. For example, *X. cucurbitae* causes leaf spots on cucurbits such as cucumber, pumpkin, winter squash and gourds. Infected plant tissue exhibits water-soaked lesions of variable sizes and may coalesce to a large area, resembling a blight symptom. Like some fungal pathogens and other bacterial pathogens, *Xanthomonas* bacteria may develop host specificity, and thus one species may have multiple pathovars (pv.) or a type of subspecies (subsp.) that only or predominantly infects a specific host species. For example, *X. citri* subsp. *citri* causes fruit canker on citrus trees and *X. oryzae* pv. *oryzae* causes bacterial blight in rice crops. *Xanthomonas* bacteria can be transmitted via seeds, splashing rainwater, propagative plants, or pruning tools.

In *Cannabis* crops, *Xanthomonas* was reported to cause a serious disease on hemp in former Yugoslavia (Šutić and Dowson, 1959). This disease caused severe wilting of hemp leaves and was found in multiple areas of Serbia. Later, *Xanthomonas* was also isolated from hemp plants in Romania (Severin, 1978). In 1982, bacterial leaf spot caused by *Xanthomonas campestris* was found from hemp in Tochigi Prefecture, Japan (Takikawa *et al.*, 1984). This disease caused necrotic lesions ranging from 1 to 2 mm in size with a yellow halo 2–3 mm wide. Further studies of the pathological, physiological and genetic properties of this bacterium found that it belonged to *Xanthomonas campestris* pv. *cannabis*, a cannabis-specific strain that was also reported in Romania (Severin, 1978; Netsu *et al.*, 2014). In some plant-pathogenic bacteria species, *hrp* genes are essential for pathogenicity. When these genes are expressed, a set of proteins (type III secretion system) function collectively to determine if an infection is to occur (Hirano *et al.*, 1999). To gain insight into the evolution and pathogenicity of *Xanthomonas* in *Cannabis* plants, Jacobs *et al.* (2015) sequenced and compared the genomes of both strains isolated from Japan and Romania. They found that both strains lacked the Hrp Type III secretion system and any of the known Type III effectors when compared with other *Xanthomonas* species. This study demonstrated that *Xanthomonas campestris* pv. *cannabis* evolved significantly in pathogenicity, which gains more insight into the molecular mechanism of the pathogenicity of *Xanthomonas* to *Cannabis* plants.

### Pseudomonas diseases

*Pseudomonas* is a genus of bacteria causing stem canker, leaf spot, leaf blight and gall, depending on the host species. One of the most common species, *Pseudomonas syringae*, has over 50 pathovars that infect specific host plants (Kado, 2010). One pathovar of *P. syringae* was designated as *P. syringae* pv. *cannabina* and it infects hemp plants, causing bacteriosis. *P. cannabina* is another species that can infect hemp, causing leaf blight. Bull *et al.* (2010) suggested that *P. syringae* pv. *alisalensis* strains, originally isolated from broccoli and other non-*Cannabis* hosts, belonged to *P. cannabina* based on DNA sequences in five gene fragments used in multi-locus sequence typing, but they were not pathogenic on *C. sativa*. *P. cannabina* strains, originally isolated from *C. sativa*, were not pathogenic on broccoli and other hosts. Thus, the *P. cannabina* species appeared to have two distinct pathovars: one that infected *C. sativa* and hence named *P. cannabina* pv. *cannabina*; and one that did not infect *Cannabis*, named *P. cannabina* pv. *alisalensis* (Bull *et al.*, 2010). The species of *P. cannabina* (ex Šutić and Dowson 1959) sp. nov. was first described in 1999 based on its relation to the generic name of the host plant, *Cannabis sativa* (Gardan *et al.*, 1999). This bacterium is Gram-negative and rod-shaped with dimensions of 1.1–3.0 μm wide × 3.0–4.0 μm long. It has one to four polar flagella that make the bacterial cell motile. Colonies are grey in colour on culture media. The bacterium infects *C. sativa* plants in nature and can also infect green bean (*Phaseolus vulgaris* L.) via inoculation. *Pseudomonas* bacteria were also found inside hemp tissue as endophytes (Scott *et al.*, 2018). Endophytic *Pseudomonas* could function as a growth promoter and play a role in microbial competition, which could be potentially useful in the control of plant diseases caused by other pathogenic species or strains.

### Dickeya diseases

*Dickeya* is a relatively new genus in the Enterobacteriaceae family. It was created by reclassifying 75 strains of *Pectobacterium chrysanthemi* and the type strains of *Brenneria paradisiaca* (CFBP 4178T) into a new genus, *Dickeya*, based on phenotypic characteristics, serology and 16S rRNA gene-based phylogenetic analyses (Samson *et al.*, 2005). Bacteria in this genus usually cause soft rot, a condition defined as disintegration and disruption of normal tissue and production of cellular fluids and strong odours. Infected tissue feels soft, looks 'dissolved', and smells offensive. *Dickeya* species can also infect the stem or leaves on some plant species, causing water-soaked lesions. For *Cannabis* crops, there have been no reports on diseases caused by *Dickeya* species.

### Erwinia diseases

*Erwinia amylovora*, the most important species in *Erwinia* genus, notoriously causes fire blight. It is a rod-shaped bacterium with peritrichous flagella (a tail-like projection all over its surface). Fire blight is a very contagious disease and causes a characteristic 'fire-scorched" appearance on leaves, fruits, twigs and branches (Fig. 2.25). While it is a major bacterial disease, it has not been reported from *Cannabis* crops. The disease is destructive to some susceptible varieties of pear and apple trees. In the active disease-developing stage, water-soaked lesions and bacterial oozing appear on affected branches. Cool and humid conditions favour disease development. Control of fire blight requires integrated disease management practices. These include: (i) all blighted twigs, branches and cankers should be cut at about 5 inches (12–13 cm) below the visible infection point and burned; (ii) chemical control with appropriate bactericides to protect trees from new

**Fig. 2.25.** Fire blight of apple (*Malus domestica*) caused by *Erwinia amylovora*. Note the blackened leaves, as if 'fire-burned', and dark lesions on the stem. Fresh bacterial fluids are usually present on the surface of infected tissue during the disease outbreak.

infection; and (iii) severely infected or dead trees should be removed from the field and destroyed.

### Pantoea diseases

*Pantoea* is another genus in the Enterobacteriaceae family and they are yellow-pigmented, rod-shaped and Gram-negative bacteria. Like *Dickeya*, *Pantoea* mainly causes soft rot. The genus *Pantoea* contains many species associated with plants, soil, water, insects, animals and humans (Walterson and Stavrinides, 2015). Some species in the genus are major plant pathogens, causing soft rot (Gitaitis and Gay, 1997; Walcott *et al.* 2002), internal fruit rot (Kido *et al.*, 2008), wilt (Pataky, 2004) and galls (Manulis and Barash, 2003) in a variety of plants. Some may cause human infections. For example, *Pantoea agglomerans*, could cause opportunistic infections in humans, especially those who are immunocompromised (Dutkiewicz *et al.*, 2016). In *Cannabis* plants, *Pantoea* species were only found inside hemp tissue as endophytes rather than as parasites (Scott *et al.*, 2018). Depending on the species, it is speculated that *Pantoea* may have a beneficial effect on plant health such as outcompeting pathogenic organisms, producing antibiotics and inducing plant systemic resistance (Walterson and Stavrinides, 2015).

### Spiroplasma diseases

Both *Spiroplasma* and 'Candidatus Phytoplasma' are mollicutes. Unlike phytoplasmas, which have many members and cause serious diseases on many plant species, *Spiroplasma* is only known to contain three plant-pathogenic species: (i) *S. citri*, causing citrus stubborn disease; (ii) *S. kunkelii*, causing corn stunt and horseradish brittle root; and (iii) *S. phoeniceum*, causing periwinkle yellows disease (Kado, 2010). The typical symptom caused by *Spiroplasma* is yellowing, like phytoplasma-caused diseases. *Spiroplasma* spp. are vectored by phloem-feeding insects such as leafhoppers. *S. citri* DNA was detected from hemp plants exhibiting yellowing, stunting and leaf proliferation symptoms (Schoener and Wang, 2019). This pathogen may play an important role in the outbreak of hemp witches' broom disease in certain varieties.

### 'Candidatus Phytoplasma' diseases

Phytoplasmas cause apple proliferation, pear decline, peach X-disease, aster yellow, ash yellow and decline, and hemp witches' broom (Zhao *et al.*, 2007; Raj *et al.*, 2008; Sichani *et al.*, 2011; Wang, 2019; Feng *et al.*, 2019), and other destructive plant diseases. The main symptoms caused by phytoplasma infections include stunting, uniform discoloration of leaves (mainly yellow), abnormally small leaves, shoot proliferation, the loss of pigment in flowers, shortening of internodes, reduced yield and dieback. Figure 2.26 shows the typical symptoms on hemp plants that were found to be infected with a phytoplasma species in the clover proliferation phytoplasma group (Schoener and Wang, 2019). The infection caused many hemp plants to become unproductive, resulting in no value for harvest (see Chapter 10).

## Plant Viruses and Viroids and Plant Diseases They Cause

Viruses are infectious, submicroscopic and intracellular pathogens composed of coat proteins and

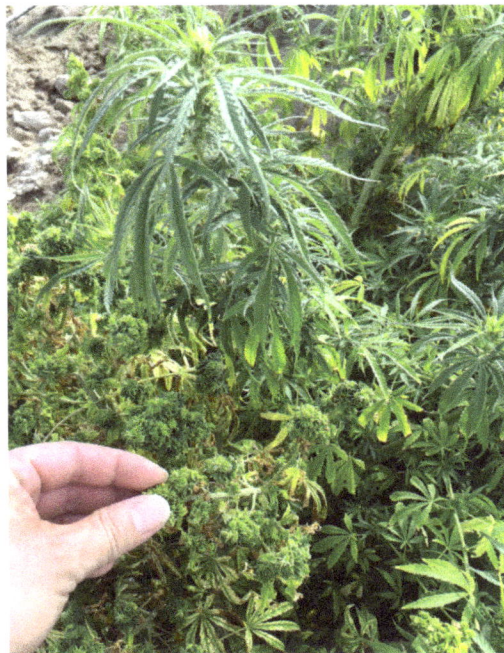

**Fig. 2.26.** Witches' broom symptom on hemp plants (*Cannabis sativa*). Note the characteristics of over-branching, clusters of underdeveloped leaves and shortened internodes. Plants are usually affected wholly, by exhibiting stunting and smaller size, but the one shown in this photo appeared to be partially infected, with only certain branches showing witches' brooms.

nucleic acid. They are obligate parasites living inside plant cells. A viroid is a simpler organism consisting only of nucleic acid. It does not form a particle like a virus, but it infects plant cells. Both viruses and viroids infect plants and hijack the host's nucleic acid replication system to reproduce. Because of this, the normal physiology of a plant is affected, thus resulting in a wide range of symptoms, including mosaic, dwarfing, yellowing, ringspot, leaf roll, deformation and death of plants. Symptoms caused by viruses are generally systemic and they can be seen on any part of the plant, but mostly on leaves. Symptoms caused by viruses can be confused with herbicide damage or other abiotic factors. Therefore, a test is often needed to confirm a viral infection. Many plant species are susceptible to virus/viroid infections and symptoms are often not treatable.

## Morphology

Individual virus particles are extremely small and not visible under a standard light microscope. The size of virus particles is measured in nanometres and they must be magnified over 10,000 times under an electron microscope to become visible. A typical plant virus particle is rod-shaped (Fig. 2.27) or spherical (Fig. 2.28). The rod-shaped *Tobacco mosaic virus* is 300 nm × 15 nm in size and spherical viruses can be 20 nm or less in diameter. Some plant viruses are bacilliform or filamentous. Most plant viruses have only two types of macromolecules: nucleic acid and proteins. The proteins form a coat that encloses the nucleic acid. The nucleic acid is either ribonucleic acid (RNA) or deoxyribonucleic acid (DNA). Most plant viruses have single-stranded RNA (ssRNA). Some have double-stranded DNA (dsDNA) or double-stranded RNA (dsRNA). Unlike the virus, a viroid is simply an infectious, short, circular, single-stranded RNA molecule. Since viroids do not have coat proteins, they do not have a defined shape. The first recognized viroid was the *Potato spindle tuber viroid* (PSTVd) (Diener, 1971). PSTVd only has 356–359 nucleotides but can form a high degree of secondary structure. Most viroids known today are only a few hundred nucleotides long and are the smallest infectious agents.

## Life cycles and transmission

Plant viruses complete their life cycles inside the cells of a host plant. Viruses use the host cell's replication system to replicate its RNA or DNA and

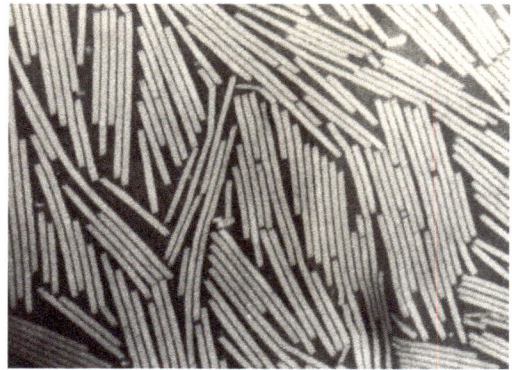

Fig. 2.27. Electron micrograph of *Tobacco mosaic virus* (TMV) purified from diseased banana leaves (Wang *et al.*, 1998). Note that each particle is 300 nm long and some particles are connected to appear as 600 nm or 900 nm.

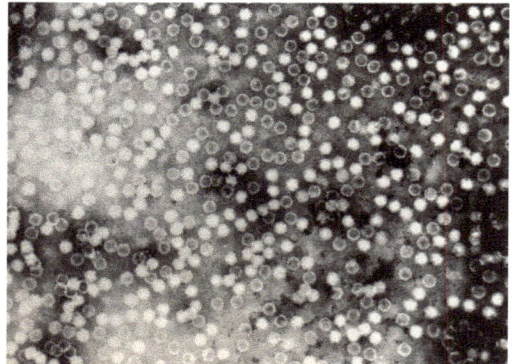

Fig. 2.28. Electron micrograph of purified *Tobacco ringspot virus* (TRSV). Note that all particles are nearly identical.

synthesize the coat protein. Replicated new RNA or DNA and newly synthesized coat protein are assembled together to form a new virus particle. In each host plant cell, hundreds to millions of virus particles can be reproduced. The infected plant can spread viruses to healthy plants in a number of ways. Some plant viruses are easily transmitted from plant to plant via direct contact or other mechanical means. For example, the *Tobacco mosaic virus* is easily transmitted through contact of contaminated hands, soil, tools and infected leaves. However, many viruses are not transmitted mechanically. Some viruses are passed to the next generation of plants through the seed, grafting, pollen, or vegetative propagative materials such as cuttings, tubers, bulbs, corms and runners. Certain viruses are transmitted by dodder (*Cuscuta* spp.), a plant-parasitic plant.

Many plant viruses are transmitted by biological vectors such as insects and mites, including aphids (Fig. 2.29), whiteflies, leafhoppers, planthoppers, thrips, beetles, mealybugs, beetles, eriophyid mites and flat mites (Table 2.5.). Transmission of plant viruses by insect vectors involves specific biological interactions between vectors and viruses (Dietzgen *et al.*, 2016). There are four types of interactions: (i) non-persistent transmission (viruses are retained in the distal tip of the insect stylet); (ii) non-circulative, semi-persistent transmission (viruses are bound to chitin lining the gut, but do not enter tissues) (iii) circulative, non-propagative transmission (viruses do not replicate inside the insect vector but pass multiple insect organs to reach the salivary glands); and (iv) circulative, propagative transmission (viruses replicate and systemically infect insect organs). The four types of vector–virus interactions explain how insect-borne viruses have developed a high specificity in terms of virus acquisition, retention and release to the plant. Disruption of any phase of these interactions could result in an effective control of diseases caused by vector-borne viruses.

**Fig. 2.29.** Aphids feeding on the leaf of a Bradford pear (*Pyrus calleryana*). Note that aphid mouthparts contain a stylet bundle used to feed on plants. Initially, the aphid may temporarily penetrate the epidermal and/or mesophyll cells of any plant it lands on to search for a host plant and find ideal feeding sites. During this feeding process, aphids may acquire a virus from an infected plant and release the virus into another plant. Many viruses are transmitted in a non-persistent or semi-persistent way through the probing and feeding process. Once an aphid finds an appropriate host plant and feeding site, it settles and feeds on the plant tissue by inserting its stylet into the deeper layers until it reaches a phloem sieve tube where many phloem-limited viruses are located. Phloem-feeding insect vectors can transmit viruses in a circulative manner (the virus reaches the salivary glands).

There are a number of viruses transmitted by migratory ectoparasitic nematodes, such as dagger nematodes (Fig. 2.30), needle nematodes and stubby-root nematodes (Wang *et al.*, 1998, 2002; Brown and MacFarlane, 2001). These nematodes acquire viruses by feeding on the virus-infected plant root and absorbing virus particles at the lumen of the stylet and oesophagus (Fig. 2.31). Viruses bound to the inner surface of the lumen can be released into a healthy plant by subsequent feeding on its roots, thus transmitting the viruses from plant to plant. The transmission mechanism is similar to that of non-circulative transmission of viruses by insect vectors. Virus-vector nematodes do not move further than a few feet naturally, so viruliferous nematodes (virus-carrying nematodes) do not spread viruses from one field to another in the same season. However, long-distance transmission can occur through the soil containing viruliferous nematodes. The most troublesome issue of virus transmission is that the viruliferous nematodes (mostly adults) serve as a virus reservoir for subsequent crop infection, as nematodes can retain viruses for their entire lives.

**Table 2.5.** Common biological vectors transmitting viruses or viroids from plant to plant.

| Kingdom | Phylum | Class/Order | Common vectors |
|---------|--------|-------------|----------------|
| Animalia | Arthropoda | Insecta | Aphids, leafhoppers, planthoppers, whiteflies, thrips, mealybugs, heteropterans/bugs, beetles, weevils |
| | | Arachnida | Eriophyid mites, spider mites, flat mites |
| | Nematoda | Dorylamida | Dagger nematodes (*Xiphinema*), needle nematodes (*Longidorus* and *Paralongidorus*) |
| | | Triplonchida | Stubby-root nematodes (*Trichodorus* and *Paratrichodorus*) |
| Fungi | Chytridiomycota | Spizellomycetales | Species in *Olpidium* |
| Protista | Cercozoa | Plasmodiophorida | Species in *Polymyxa* and *Spongospora* |

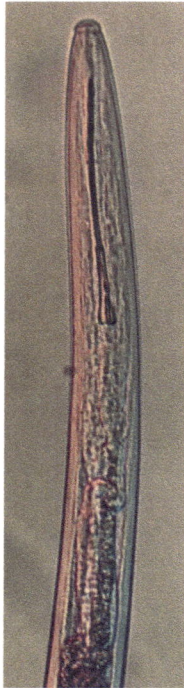

**Fig. 2.30.** The anterior end (head region) of American dagger nematode (*Xiphinema americanum*) showing its feeding apparatus that contains a long stylet, the oesophagus tract and the oesophageal bulb. Note that virus particles are preferably retained inside the lumen of the lower part of the stylet (aka stylet extension), the entire oesophagus tract and the triradiate lumen of the oesophageal bulb (see Fig. 2.31). (Reprinted from Wang, S., Gergerich, R.C., Wickizer, S.L. and Kim, K. S. (2002) Localization of transmissible and nontransmissible viruses in the vector nematode *Xiphinema americanum*. *Phytopathology* 92, 646–653.)

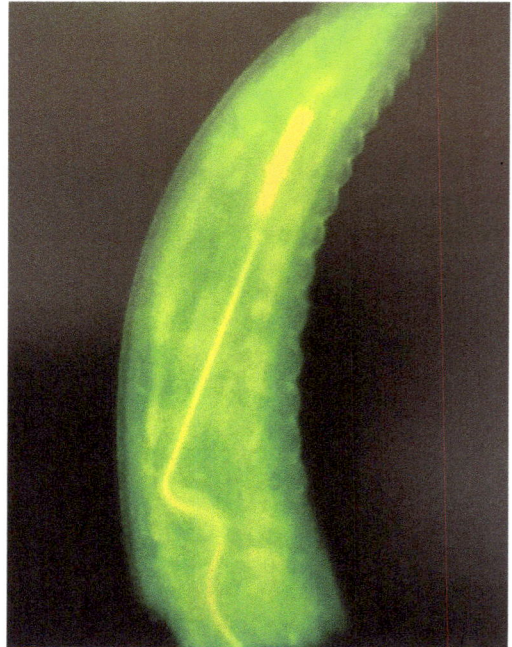

**Fig. 2.31.** Fluorescent labelling of *Tobacco ringspot virus* (TRSV) inside the nematode *Xiphinema americanum* showing exact virus retention sites inside the lumen of stylet and oesophagus (also see Fig 2.30 for structure detail). Note that strong yellow fluorescence in the stylet (straight line) and oesophagus tract (curved line) indicates the presence of TRSV particles (Wang *et al.*, 1998).

Therefore, a nematode-transmitted virus disease behaves as a soilborne disease in nature. There is a high degree of specificity in nematode transmission of viruses. Dagger and needle nematodes only transmit viruses in the *Nepovirus* genus, such as the *Tobacco ringspot virus* and the *Tomato ringspot virus*; stubby-root nematodes only transmit viruses in the *Tobravirus* genus, such as the *Tobacco rattle virus*. Although there are approximately 3500 known plant-parasitic nematode species, fewer than 1% of them are virus vectors. Additionally, not all nepoviruses or tobraviruses are transmitted by nematode

vectors. In the genus *Nepovirus*, only 13 out of 38 members are confirmed to be transmitted by dagger or needle nematodes (Brown and MacFarlane, 2001). *Cherry leaf roll virus*, a member of *Nepovirus*, has not been found to be vectored by a nematode species.

A few soilborne fungi and protists also transmit certain plant viruses. Similar to nematode transmission of viruses, these fungi and protists inhabit the soil and the diseases they transmit may behave as soilborne diseases. There are quite a few viruses that have been proven to be transmitted by fungi and protists. These viruses are in the genera *Tombusvirus*, *Carmovirus*, *Necrovirus*, *Furovirus*, *Pecluvirus*, *Benyvirus*, *Pomovirus*, *Bymovirus*, *Ophiovirus* and *Varicosavirus* (Hull, 2014).

In summary, plant viruses can be transmitted in several ways. Infected seeds can directly carry viruses to the subsequent crop. Vegetative propagation of infected stock plants effectively passes viruses to the next generation. Insect vectors can efficiently transmit viruses from plant to plant during the growing season. Virus-carrying nematodes, fungi or protists can serve as virus reservoirs of subsequent crop infections.

## Plant diseases caused by plant-pathogenic viruses

Viruses attack all types of plants. They can damage any or all parts of a plant. Some viruses can cause catastrophic crop loss in yield and quality. The most diagnostic symptoms caused by plant viruses are leaf mosaic, deformation, stunting and poor growth (Fig 2.32). Symptoms caused by different viruses may look similar. One plant can be infected by two or more viruses. However, a plant infected by a virus may not necessarily show a disease symptom during the entire infection period, i.e. the viral symptom may disappear under certain environmental

**Fig. 2.32.** Tobacco (*Nicotiana tabacum*) leaf mosaic caused by *Tobacco mosaic virus*. Note the randomly distributed green islands, leaf distortion and asymmetry.

conditions, such as high temperatures. A weak or mild strain of the virus may cause very subtle symptoms or unnoticeable damage to the plant. Although virus particles can be translocated systemically inside the plant organs, not all plant tissues exhibit viral symptoms. Unlike fungal or bacterial infections that cause canker, rot or other tissue damage, viral infections usually cause abnormal plant growth, resulting in losses of productivity and economic value.

Viroids infect a number of plant species including both dicotyledonous and monocotyledonous plants. There are 31 species of viroids described and they are classified into eight genera in two families (Hull, 2014) (Table 2.6). Symptoms on plants caused by viroid infections include stunting, mottling, necrosis, or leaf malformation; these symptoms are generally not distinguishable from those caused by virus infections. Some viroids infect plants but do not induce any visible symptoms. The first recognized and most widely studied viroid is the *Potato spindle tuber viroid* (PSTVd) (Diener, 1971). PSTVd is an important plant pathogen that poses a threat to potato production worldwide. Potato plants infected by PSTVd exhibit symptoms of stunting, small and twisted leaves, and small and spindle-shaped tubers. PSTVd is transmitted by contact. Mechanical injury and some insect feeding activities also spread the viroid throughout a field. This viroid is also transmissible by infected seed potato tubers and true potato seeds. Management of PSTVd relies on cultural practices to exclude infected seed stock from planting. Certification of seed potatoes by periodically monitoring and testing for PSTVd is an essential programme for providing high-quality and viroid-free seed potatoes.

Virus and viroid diseases may become a serious issue as *Cannabis* crop production continues to increase. A new viroid was reported from hop

**Table 2.6.** A list of typical viroids and their classification (Hull, 2014).

| Family | Genus | Representative species |
|---|---|---|
| Pospiviroidae | *Pospiviroid* | Potato spindle tuber viroid |
| | *Hostuviroid* | Hop stunt viroid |
| | *Cocadviroid* | Coconut cadang-cadang viroid |
| | *Apscaviroid* | Apple scar skin viroid |
| | *Coleviroid* | Coleus blumei viroid |
| Avsunviroidae | *Avsunviroid* | Avocado sunblotch viroid |
| | *Pelamoviroid* | Chrysanthemum chlorotic mottle viroid |
| | *Elaviroid* | Eggplant latent viroid |

plants and was considered as the first member of a new viroid group (Puchta *et al.*, 1988). Because this viroid did not cause visible symptoms on plants, it was tentatively named as *Hop latent viroid* (HpLVd). HpLVd is a circular RNA containing 256 nucleotides and can infect hop plants through mechanical inoculation. Recently, HpLVd was detected from multiple cultivars of *C. sativa* plants in California (Bektaş *et al.*, 2019; Warren *et al.*, 2019;). The infected plants exhibited stunting, leaf chlorosis and malformation, brittle stems, outwardly horizontal plant structure and yield reduction. To confirm if the detected viroid RNA sequence was related to disease symptoms, infectious RNA constructs were inoculated into healthy plants and resulted in stunting, malformation and chlorosis of leaves in those plants. This study demonstrated that HpLVd was a cause of this specific disease occurring in some cannabis production areas of California (Warren *et al.*, 2019).

There is also a virus called *Hop latent virus* (HpLV) in the literature. It is a single-stranded positive-sense RNA virus (Jo *et al.*, 2015). HpLV consists of filamentous RNA particles and contains six proteins (Hataya *et al.*, 2000). It can be transmitted mechanically or through aphids in a nonpersistent manner (Adams and Barbara, 1982). HpLV was described as a member of the genus *Carlavirus* in the family Betaflexiviridae. Other members of this genus include the *Hop mosaic virus* and the *American hop latent virus*. These three viruses infect various hop (*Humulus lupulus* L.) cultivars.

The *Hop latent virus* can infect *Cannabis* plants and hemp has been used as propagation host plants for HpLV. Using hemp plants to study hemp latent virus, researchers found that spherical virus particles with a diameter of around 34 nm were present in hemp plants (Ziegler *et al.*, 2012). RNA sequence data suggested it was a hemp cryptic virus, a putative new member of the genus *Partitivirus*.

The cryptic viruses, or cryptoviruses, are in the family *Partitiviridae* that is taxonomically reorganized to contain the *Alphapartitivirus*, *Betapartitivirus*, *Gammapartitivirus* and *Deltapartitivirus* genera (Nibert *et al.*, 2014). These viruses mainly infect fungi, plants and protozoans; and those infecting plants can spread viruses via pollen or seeds. Plants infected by cryptoviruses may not show any visible symptoms, making this virus hard to detect. In some cases, they may only induce very mild symptoms. Because few symptoms are induced, the economic importance of cryptoviruses is not well known.

Hemp mosaic and leaf curling have been seen in hemp fields in recent years. It is not known which virus is causing leaf mosaic in hemp, but many plants in a field show a mix of viral symptoms such as leaf mosaic (Fig. 2.33) and leaf curling (see Fig. 2.3). In California, Nevada and Colorado, affected hemp plants were found to be infected with *Beet curly top virus* (see Chapter 10). As more research is involved in diagnosing viral agents in *Cannabis* crops, more viruses may be detected.

Confirmation of a virus infection requires serological or molecular tests. Most viruses can be detected in a laboratory using either immunostrip or DAS-ELISA testing. Control of plant viruses requires a system integrated with quarantine, inspection and certification, which is the best way to exclude viruses from a crop. When a destructive virus is detected, eradication is necessary and often achievable. Rogueing has been used successfully to eliminate many infected plants from a tree crop. There is no chemical product that directly kills viruses in plants; however, insecticides are widely used to protect plants or crops from further spread of viral diseases by insects. Virus-resistant varieties are available for some major or special crops.

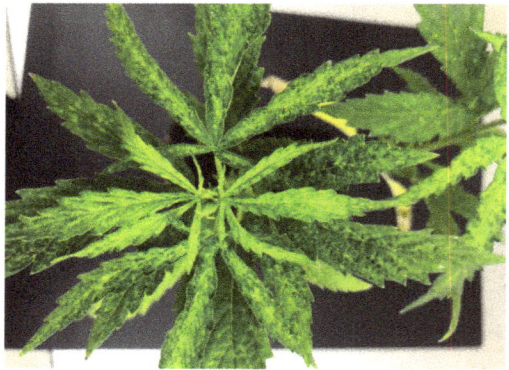

**Fig. 2.33.** A hemp plant showing leaf mosaic and deformation typical of a virus infection. Note that the characteristic dark green islands (DGIs, dark green surrounded by light green or yellow) are sporadically distributed in leaves and are associated with mild leaf distortion. DGIs are a symptom commonly shown in plants infected by a mosaic virus. In DGIs, green leaf cells are not infected by a virus, but the surrounding cells are infected by the virus and exhibit a yellow or light green colour. The cause of DGIs is posttranscriptional gene silencing (Moore *et al.*, 2001).

## Plant-parasitic Nematodes

Nematodes are microscopic, non-segmented round-worms (Fig. 2.34). They are probably the most abundant multicellular animals on earth. Most nematodes are free-living in soil or in fresh or salt water without harming plants or other organisms. A small number of nematode species can cause mild to significant damage to agricultural crops, horticultural plants and forestry. These species possess a specialized feeding structure called a stylet. Most plant-parasitic nematodes have a stylet and use it to penetrate plant cells and draw nutrients from the host plants. Plant-parasitic nematodes can attack leaves, stem and roots, but most of them cause root damage. Many plant-parasitic nematodes spend part of their lives freely in the soil, with the highest density around the roots of host plants. Plant-parasitic nematodes may be spread by host plants, soil, field equipment, boots, or flood irrigation. Common symptoms of nematode infection include root galls, prolific root branching, root lesions, leaf yellowing, wilting, poor plant growth and yield reduction. It has been estimated that nematodes cause about 12.3% yield loss, equivalent to a loss of approximately US$157 billion worldwide (Singh *et al.*, 2015).

### Morphology

The body of the plant-parasitic nematode is a hollow tube extending from the mouth through the oesophagus, intestine, rectum and anus (Fig. 2.35). It is transparent and its internal organs are visible when observed under a light microscope. The body is covered by a layer of colourless cuticle, and the body cavity is full of fluid. The nematode's cuticle is usually marked by striations or other markings and it is shed during the moulting process. The nematode moults several times when it goes through juvenile stages to become an adult. There are muscles in the nematode that enable it to move. The nematode has a complete digestive tract that includes the mouth, stylet, oesophagus, intestine and anus. It has a simple reproductive system that contains testis and spicule (male) or ovary and vulva/vagina (female). Reproduction of the plant-parasitic nematode is through laying eggs and may require mating between males and females. Many species of nematodes are parthenogenetic, meaning that males are not required for reproduction. Some species even do not have males found in nature. Most nematodes are vermiform during their life cycles, but some sedentary endoparasitic nematodes develop from vermiform juvenile stages into pear-shaped females (Fig. 2.36) or round cysts. However, males are always vermiform and may be found inside root tissue or soil. The size of plant-parasitic nematodes ranges from 250 µm to 12 mm in length and 15µm to 1mm in width. It is hard to see individual nematodes without a microscope. However, a population of nematodes suspended in water in a clear glass bottle or beaker will be visible under a light.

**Fig. 2.34.** 'Worm'-like stem nematodes (*Ditylenchus dipsaci*) inside the stem tissue of an alfalfa plant causing severe stunting and poor growth. Note that a large number of nematodes were reproduced inside plant tissue and intertwined. This nematode lives freely in soil and infects stem or underground bulbs when susceptible host plants are present.

### Life cycle

Most plant-parasitic nematodes lay eggs, go through four juvenile stages, and become adult males or females (Fig. 2.37). Females of ectoparasitic nematodes lay eggs in the soil, root-knot nematodes lay eggs in a gelatinous egg sac and cyst nematodes produce eggs inside a cyst (a naturally matured and dead female). Eggs can survive in the soil for a period of time with or without host plants and they continue to develop until a first-stage juvenile is present. This first-stage juvenile moults for the first time while still within the egg and becomes a second-stage juvenile when it emerges from the egg. This process is called hatching (Fig 2.38). When a susceptible host plant is present and environmental conditions (especially temperature) are optimal, eggs hatch into second-stage juveniles and then moult three more times to become adult males or females. Nematode sex is established after the final moult. Depending on the

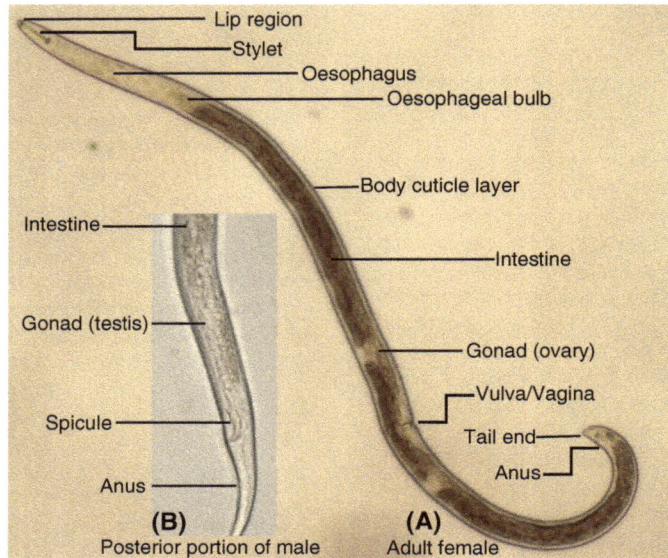

**Fig. 2.35.** The general anatomy of the plant-parasitic nematode. **(A)** An adult spiral nematode female (*Helicotylenchus* sp.) showing general morphological structures used in nematode identification. Note that each genus of nematodes can be identified based on the shape of the lip region (mouth), size or shape of stylet, distinct or weak median oesophageal bulb, overlapping or non-overlapping oesophagus with intestine, single or double branched ovaries, location (middle or posterior) of vulva and tail shapes. A well-fed nematode may look darker in the intestine region, while a hungry nematode appears to be more transparent. **(B)** Posterior portion of stem nematode (*Ditylenchus dipsaci*) to show the spicule, a needle-like structure used in mating to open the vulva of a female nematode. Some males have a bursa, a male structure used to grasp a female during copulation.

species, females may produce eggs either sexually or asexually. In favourable environmental conditions, the nematode can complete its life cycle from egg to adult within a month. At the infectious stages (both juvenile stages and adults), plant-parasitic nematodes must feed upon susceptible hosts in order to survive. Certain plant-parasitic species may die if a suitable host plant is not available.

### Plant diseases caused by plant-parasitic nematodes

Plant-parasitic nematodes can be classified into four types based on their feeding behaviours and parasitic relationships with host plants (Table 2.7). Most nematodes are free-living in the soil, migrating to plant root surfaces and using their stylet to uptake nutrients for a short period of time. The nematodes in this group are called migratory ectoparasites. The second group are called migratory endoparasites, which enter the plant tissue such as the root,

stem and even seed. They move, feed and reproduce inside the plant tissue, causing substantial damage to the plant. The third group are sedentary endoparasites, which enter into the plant tissue and stay at a fixed site feeding on the giant cells they induce. The fourth type are called semi-endoparasites. These nematodes feed on a fixed location of a root with most of their body exposed outside of the root. From an economic importance standpoint, the sedentary endoparasites such as root-knot nematodes and cyst nematodes are economically devastating groups of pathogens. These nematodes are able to alter the host's metabolism by establishing a permanent feeding site within the plant roots. Migratory endoparasites, such as lesion nematodes and stem nematodes, also cause significant damage to certain crops. Most nematode species mainly infect roots, causing mild to moderate damage to crops. Some species infect leaves, flowers, stems, bulbs or seeds. Because nematode infections draw nutrients from plants and ultimately impair the root system, affected plants may show symptoms of stunting, wilting, leaf yellowing, or poor growth, similar to symptoms caused by nutrient deficiency and/or drought stress. Other symptoms caused by the nematodes include root necrosis, root proliferation or stubby-root, root-knot or enlargement, root-tip gall, short internode, stem lesions, leaf distortion and seed galls.

**Fig. 2.36.** A young root-knot nematode female (*Meloidogyne* sp.) removed from the inside of a root-knot of cucumber (*Cucumis sativus*) had pale to white, pear-shaped body with an elongated neck. The second-stage juvenile (infectious stage) penetrates root tissue and establishes a feeding site where the nematode stays and grows (becoming sedentary). The feeding site is a group of giant cells in the root induced by secretions of the nematode and serves as a nutrient sink for the nematode to ingest nutrients from the plant. Parasitism by the nematode induces surrounding cells to divide rapidly and enlarge, resulting in visible knots in the root system. Large knots may contain many females, each with a gelatinous sac in which eggs are deposited.

### Migratory ectoparasitic nematodes

Ectoparasitic nematodes use a stylet to penetrate the cells of plant roots and uptake nutrients from those cells, while the nematode body remains outside of the plant. Some species, such as needle nematodes and sting nematodes, have extremely long stylets and they can draw nutrients from the deep layers of root cells. The feeding process may be short but frequent and it can cause mild damage to roots. When high populations of nematodes exist in the rhizosphere (soil and space in the vicinity of plant roots) and frequently feed on roots, the plant may exhibit symptom of poor growth. Mechanical damage to roots caused by nematode feeding makes plants more susceptible to secondary bacterial or fungal infections, resulting in more severe symptoms such as root rot and collapse of plants. Certain genera of ectoparasitic nematodes also transmit viruses to the plants they feed on (Brown and MacFarlane, 2001). Some ectoparasitic nematodes can induce root tip gall or stubby-root symptoms. For example, *Trichodorus nanjingensis* caused a severe stubby-root disease on apple trees in an orchard (Wang *et al.*, 1994). Under greenhouse conditions, the nematode caused root-tip gall in cucumber (*Cucumis sativus*) plants (Fig. 2.39).

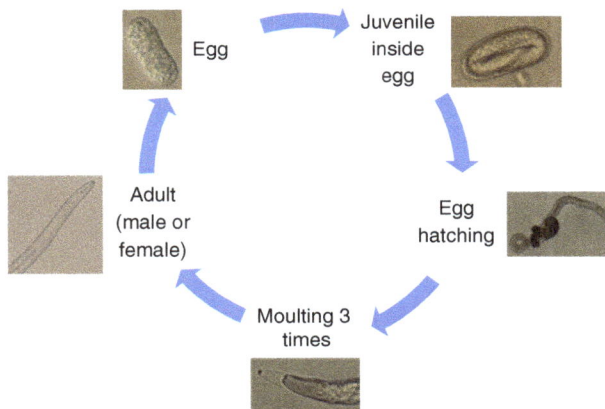

**Fig. 2.37.** A typical life cycle of the plant-parasitic nematode. Note that a nematode has four juvenile stages ($J_1$, $J_2$, $J_3$ and $J_4$) before becoming a male or female adult.

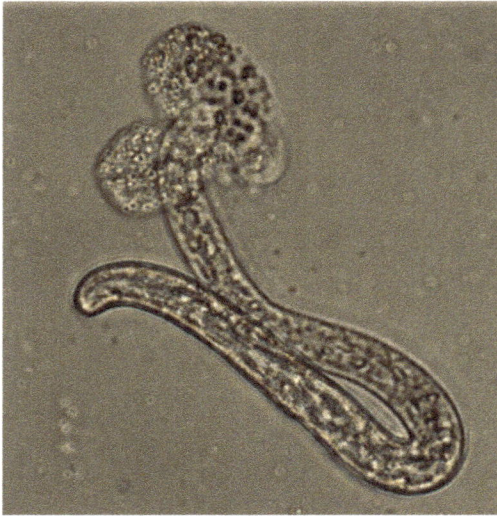

**Fig. 2.38.** A second-stage juvenile (J₂) hatching from the egg. Note that J₁ develops inside the egg and moults to become J₂ before emerging from the egg into the soil. J₂ is an infectious stage and migrates toward a root for feeding.

## Migratory endoparasitic nematodes

Migratory endoparasitic nematodes enter into root or stem tissues and migrate progressively when feeding on plant cells. Because of migration, feeding and massive reproduction, the internal root or stem tissue is completely disrupted. Once plant cells are killed, the nematodes move to healthy tissue, where they cause more lesions. Although nematodes feed, moult and reproduce mostly inside the plant tissue, they do move back to the soil to search for new roots, especially when the current root tissue is completely rotten.

**LESION NEMATODES (PRATYLENCHUS SPP.)** Lesion nematodes are important migratory endoparasites (Fig. 2.40). Their economic impact might be just behind that of root-knot nematodes and cyst nematodes. The genus contains diverse species that can invade many crops. The nematode penetrates root tissue, moves internally through the roots, feeds on root cells and ultimately causes severe disruption to the integrity of internal root tissue. The damage is often visible as dark lesions in the root. In addition

**Table 2.7.** Important nematode genera and their parasitic relationships with host plants.

| Genus | Common name | Type of parasitism | Adult female shape | Plant part affected |
|---|---|---|---|---|
| *Anquina* | Seed and leaf gall nematode | Ectoparasites on leaf but endoparasites in seed | Vermiform | Seed and leaves |
| *Aphelenchoides* | Foliar nematodes | Migratory endoparasites | Vermiform | Leaves |
| *Bursaphelenchus* | Pine wood nematode | Migratory endoparasites | Vermiform | Stem (pine tree) |
| *Ditylenchus* | Stem and bulb nematodes | Migratory endoparasites | Vermiform | Stem and bulb |
| *Pratylenchus* | Lesion nematodes | Migratory endoparasites | Vermiform | Root |
| *Radopholus* | Burrowing nematodes | Migratory endoparasites | Vermiform | Root |
| *Hirschmanniella* | Rice root nematode | Migratory endoparasites | Vermiform | Root |
| *Meloidogyne* | Root-knot nematodes | Sedentary endoparasites | Swollen to pear shape | Root |
| *Heterodera* | Cyst nematodes | Sedentary endoparasites | Swollen to lemon shape | Root |
| *Globodera* | Cyst nematodes | Sedentary endoparasites | Swollen to lemon or round shape | Root |
| *Tylenchulus* | Citrus nematodes | Semi-endoparasites | Posterior end swollen to kidney shape | Root |
| *Rotylenchulus* | Reniform nematodes | Semi-endoparasites | Swollen to kidney shape | Root |
| *Xiphinema*[a] | Dagger nematodes | Migratory ectoparasites | Vermiform | Root |
| *Longidorus*[a] | Needle nematodes | Migratory ectoparasites | Vermiform | Root |
| *Trichodorus*[a] | Stubby-root nematodes | Migratory ectoparasites | Vermiform | Root |
| *Belonolaimus* | Sting nematodes | Migratory ectoparasites | Vermiform | Root |
| *Helicotylenchus* | Spiral nematodes | Migratory ectoparasites | Vermiform | Root |
| *Hoplolaimus* | Lance nematodes | Migratory ectoparasites | Vermiform | Root |
| *Hempcycliophora* | Sheath nematodes | Migratory ectoparasites | Vermiform | Root |
| *Tylenchorhynchus* | Stunt nematodes | Migratory ectoparasites | Vermiform | Root |
| *Criconemoides* | Ring nematodes | Migratory ectoparasites | Vermiform | Root |
| *Paratylenchus* | Pin nematodes | Migratory ectoparasites | Vermiform | Root |

[a]These nematodes also transmit plant viruses.

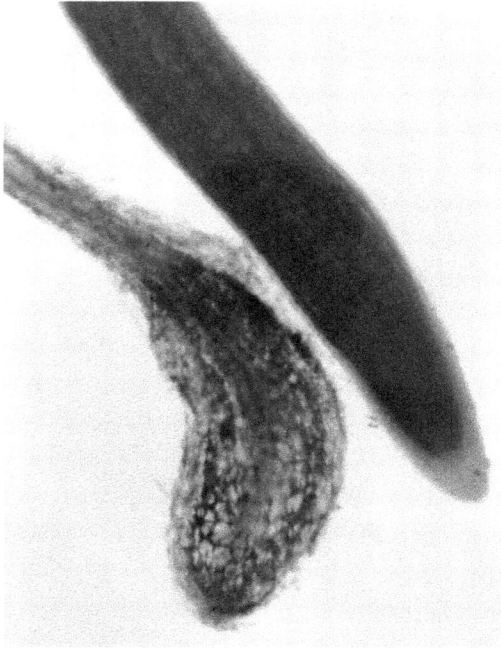

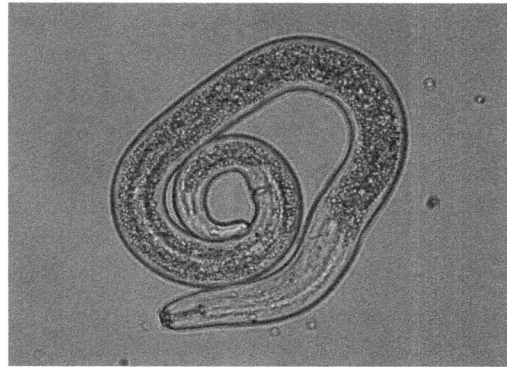

**Fig. 2.40.** Lesion nematode (*Pratylenchus* sp.) female showing a strong 'muscular' appearance and actively coiled. Note that the nematode has a robust stylet to facilitate penetration and feeding. The key identification characteristics include a broad and flatted head (lip) region, a large stylet basal knob, an oesophageal basal bulb overlapping with the intestine lateroventrally, a bluntly round tail and a posteriorly positioned vulva.

**Fig. 2.39.** Root-tip gall of cucumber plants (*Cucumis sativus*) induced by the stubby-root nematode, *Trichodorus nanjingensis*, after 30 days of inoculation under greenhouse conditions. Note the normal root tip on the right. The gall induced by the nematode is about double the size of the normal root in cross-section and is composed of enlarged, less structured and nutrient-rich cells that serve as a nutrient sink for the nematode (Wang and Chiu, 1997).

to direct damage to the roots, nematode intrusion disrupts the existing defensive structures of a root system and makes it more prone to fungal and bacterial infections. The severity of damage depends on the nematode population levels in the soil. When a susceptible crop is planted, lesion nematode population can reach very high levels in the field, causing significant damage to the crop. Lesion nematodes have very wide host ranges and can attack more than 400 crop species, including both dicots and monocots. This means that crop rotation may have limited success as a control measure, especially when non-host crops are not an option for the field. Lesion nematodes survive in a wide range of climates. Because of the wide host range and broad geographical distribution, the nematode may pose a potential threat to *Cannabis* crops. An effective

practice to control lesion nematodes is to apply soil fumigants (nematicides) before planting. This control strategy aims to reduce the nematode population to a level below the economic damage threshold.

**STEM AND BULB NEMATODES (*DITYLENCHUS* SPP.)** Stem nematodes, as a group of nematode species in the *Ditylenchus* genus, occur worldwide but are most prevalent in the temperate zone. They are one of the most destructive plant-parasitic nematode groups. One species classified as *Ditylenchus dipsaci* causes severe bulb rot in *Allium* crops such as garlic and onions. A field infested with this nematode is no longer suitable for planting *Allium* species unless the nematode is eradicated from the field. The initial symptoms on infected onion and garlic crops include erratic stands and collapsing of foliage. In the late stages of disease development, onion and garlic bulbs rot, desiccate, shrink and become light in weight. The severity of damage to bulbs depends on the density of nematodes in the soil. Highly infected crops are seldom harvested. Early detection and eradiation of the nematode from garlic and onion production areas helps prevent nematode spread. Once the nematode is established in a field, various nematicides and soil fumigants can be used to reduce the nematode population. Crop rotation with non-host crops has also been proven to be effective.

Stem nematodes also cause serious damage to alfalfa crops (Fig. 2.41). Infected plants show enlarged stems and nodes and have very short internodes. These nematodes infect stem tissue, then feed, moult and reproduce internally as migratory endoparasites. Nematodes move to fresh areas of stem to continue feeding and reproducing after the originally infected tissue is completely destroyed. All stages of juveniles except J1 (still inside egg) and adults can feed on stem tissue, and the density of nematodes inside the tissue can reach several thousand per gram of tissue. As stem tissue is macerated, leaves turn yellow and plants become stunted. The entire crop may fail if the field is infected with a high population of nematodes.

### Semi-endoparasitic nematodes

There are two genera of nematodes that feed on plant roots as semi-endoparasites: *Tylenchulus* and *Rotylenchulus* (see Table 2.7). The most important

**Fig. 2.41.** Typical symptom on alfalfa plant caused by the stem nematode (*Ditylenchus dipsaci*). Note that the stem was swollen with short internode. Inside the stem were thousands of nematodes at different developmental stages (see Fig. 2.34).

species include the reniform nematode (*Rotylenchulus reniformis*) and the citrus nematode (*Tylenchulus semipenetrans*). Unlike root-knot nematodes where the whole nematode body lives inside the root tissue during its feeding, semi-endoparasitic nematodes only have their head region embedded in the root while the rest of the body protrudes from the root surface. Because of this, they are classified as semi-endoparasites. However, this type of nematode behaves quite differently from other endoparasites such as lesion nematodes or root-knot nematodes. Initially, an immature female penetrates the outer surface of the root and enters the deep cortical layers. Once an ideal site is identified, it becomes sedentary and permanently feeds on a group of specialized cells or nurse cells. As the nematode grows, the posterior portion of its body swells and is exposed to the outside of the root while its anterior portion (head and neck) remains embedded inside the root tissue (the cortex). A mature female body outside of a root is usually swollen into a kidney-like shape and produces eggs that are embedded in a gelatinous matrix secreted by the nematode through its excretory pore. Because of enlarged females and their egg masses attached to the root, an infected root is often called a dirty root. This means when roots are washed under running water to remove soil, they still look dirty due to the clay soil or debris stuck to the gelatinous egg masses. Dirty-root appearance is one of the diagnostic characteristics for semi-endoparasitic nematodes. Symptoms of semi-endoparasitic nematode parasitism are similar to those caused by other nematodes. Infected plants are typically underdeveloped. At high nematode population levels, severe root damage and above-ground symptoms may occur. The life cycle may take 4–8 weeks, depending on the crop species and environmental conditions. The citrus nematode has a narrow host range, mainly infecting trifoliate orange, grapevines, persimmon, lilac and olive, and it has not been reported from herbaceous plants (Inserra *et al.*, 1994). In contrast, the reniform nematode has a broad host range, infecting more than 300 plant species, including cotton, soybean, pineapple and vegetable crops (Robinson *et al.*, 1997).

### Sedentary endoparasitic nematodes

The most destructive nematodes are sedentary endoparasitic nematodes. The most well-known nematodes in this group are root-knot nematodes (*Meloidogyne* spp.) and cyst nematodes (*Heterodera* spp. and *Globodera* spp.). Unlike migratory ectoparasites and migratory endoparasites, nematodes in this group invade plant roots by their second-stage juveniles (J2) and stimulate the formation of giant cells or large feeding cells. Giant cells are special feeding cells for the root-knot nematodes. When the nematode initially penetrates a root cell, it injects secretory proteins into the cell. The secretory proteins are produced by the nematode's oesophageal gland cells and can stimulate plant cell growth. Because of this, affected cells grow bigger and divide repeatedly but without new cell wall formation. Therefore, a parasitized cell becomes a giant cell containing many nuclei. Cells surrounding the parasitized cells may also become giant due to the diffusion of nematode secretions. The large feeding cells induced by cyst nematodes are formed in a different way. They are formed by several neighbouring cells combining into a single cell called a syncytium through the breakdown of neighbouring cell walls. Thus, the syncytium is an enlarged and multinuclear cell with abundant cytoplasm, endoplasmic reticulum, ribosomes, plastids and mitochondria. Both giant cells and syncytia serve as a nutrient sink providing large amounts of nutrients to nematode growth and development while the plant suffers nutrient loss. This mechanism of parasitizing on giant cells or large feeding cells enables the nematode to remain sedentary inside the root for the rest of its development.

**ROOT-KNOT NEMATODES (*MELOIDOGYNE* SPP.)** Root-knot nematodes infect a broad range of host plants and they are the most common nematode pathogens causing diseases in both agricultural and horticultural crops. The characteristic symptoms are the small to giant root knots visible to the naked eye when the root system is pulled out from the soil (Fig. 2.42). Inside each knot, there may be one to several nematode females each with an egg mass of at least 100 eggs. Root-knot nematodes severely affect root function, so infected plants may exhibit wilting during the daytime due to an insufficient uptake of water from the soil. It is not uncommon to see plants infected by root-knot nematodes that are also infected by fungal or oomycete pathogens, a condition called a disease complex. Control of root-knot nematodes is difficult and expensive. Exclusion of the nematode from the crop production area is the best management practice. When the nematode is established in a field, nematicides are often used to reduce the

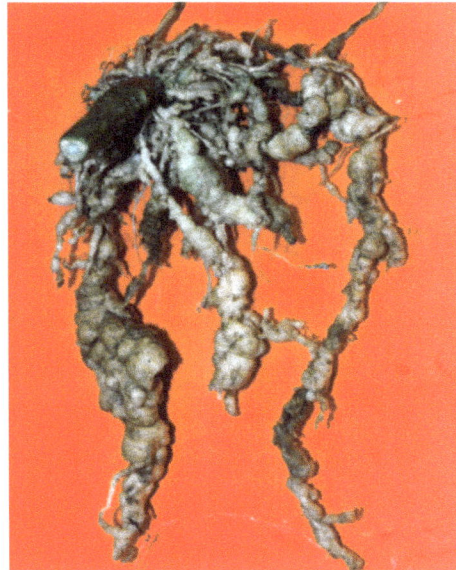

**Fig. 2.42.** Root knots on a cucumber plant (*Cucumis sativus*). Note that the root is abnormally enlarged and that a single root has numerous knots of different sizes. Inside a large knot there are usually multiple nematode females. Matured females lay eggs in a gelatinous egg sac. Males may occasionally be found associated with a root, but most root-knot nematodes reproduce by parthenogenesis (male not required).

nematode population so that a crop can be grown without a significant impact on its yield.

**CYST NEMATODES (*HETERODERA* SPP. AND *GLOBODERA* SPP.)** Cyst nematodes do not have as broad a host range as root-knot nematodes do. However, some species are critical pathogens on certain crops. For example, the soybean cyst nematode (*Heterodera glycines*) is one of the most widespread and destructive pathogens for soybean and is the only species that has caused damage to soybean in all major soybean-producing countries (Wang and Riggs, 1999). Another destructive cyst nematode is the pale cyst nematode (*Globodera pallida*). This nematode is a major pest to potato crops but also affects other plant species in the Solanaceae family, including tomatoes and eggplants. There are several other species in both the *Heterodera* and *Globodera* genera that can cause significant economic loss in some crops (Table 2.8). Because the syncytium (large feeding cell) induced by cyst

**Table 2.8.** Important cyst nematode species and their primary host plants.

| Scientific name | Common name | Major host plants |
|---|---|---|
| *Globodera pallida* | Pale cyst nematode | Potato, tomato, eggplant |
| *Globodera rostochiensis* | Golden nematode | Potato, tomato, eggplant, solanaceous weeds |
| *Heterodera glycines* | Soybean cyst nematode | Soybean |
| *Heterodera sacchari* | Sugarcane cyst nematode | Sugarcane |
| *Heterodera latipons* | Mediterranean cereal cyst nematode | Wheat, barley, oat |
| *Heterodera ciceri* | Chickpea cyst nematode | Lentil, chickpea |
| *Heterodera cajani* | Pigeonpea cyst nematode | Pigeonpea |
| *Heterodera avenae* | Cereal cyst nematode | Wheat, barley, oats, rye |
| *Heterodera carotae* | Carrot cyst nematode | Carrot |
| *Heterodera humuli* | Hop cyst nematode | Hop |
| *Heterodera schachtii* | Sugar beet nematode | Over 200 plant species |
| *Heterodera trifolii* | Clover cyst nematode | Clover and other plant species |

nematode infections functions as a nutrient and metabolic sink into which the nutrients are diverted and accumulated, infected plants have decreased or disturbed nutrient concentrations in the roots, leaves and other parts of the plant. Such a physiological disturbance causes plant symptoms similar to nutrition deficiency syndrome. Infected plants are generally smaller, have a decreased dry weight and have much lower root volume than normal plants. In the field, patches of poor growth and/or foliage yellowing may be seen in areas with a high nematode population density. However, the most obvious and definitive diagnostic symptoms are the tiny but visible lemon-shaped females and cysts attached to the root when the plant is pulled out from the ground (Fig. 2.43). If left uncontrolled, cyst nematode infections can cause up to an 80% reduction in crop yield. Early detection of cyst nematodes is the key to preventing introduction and minimizing treatment costs but the use of resistant varieties and nematicides is an effective way to manage existing nematode infections.

## Plant-parasitic Plants

Certain plant species parasitize on higher plants and hence are called parasitic plants. Parasitic plants are classified as hemiparasites or holoparasites. Hemiparasites contain chlorophyll (an essential green pigment for photosynthesis) and can use sunlight to synthesize food from carbon dioxide and water, but they still connect to the xylem of host plants via haustoria to obtain water and nutrients. In contrast, holoparasites lack chlorophyll and must obtain water and nutrients from the xylem and phloem of host plants. Therefore, all

**Fig. 2.43.** Yellow and brownish soybean cyst nematodes detached from soybean roots. Note that in the early stage of infection, nematodes are completely embedded inside the root, but later the enlarged females protrude from the root with their head region attached to the feeding sites. Young females are generally white and gradually developed into mature females with a light-yellow colour. A female lays 200–400 eggs inside her body, hardens her cuticle body wall and eventually turns a brown or dark colour, becoming a cyst. Unlike root-knot nematodes, cyst nematodes are exclusively sexually reproduced and the mobile vermiform males are required for females to produce eggs. (Picture courtesy of Dr Weimin Ye.)

holoparasites are obligate parasites and completely depend on a host plant to complete their life cycle. Parasitic plants can cause wooden host plants to decline over time (Fig. 2.44) or make a patch of crop collapse (Fig. 2.45). Although there are about 4500 species of flowering plants known to be parasitic, only a few cause significant damage to host plants in nature. The most common and economically impacting parasitic plants are mistletoes and

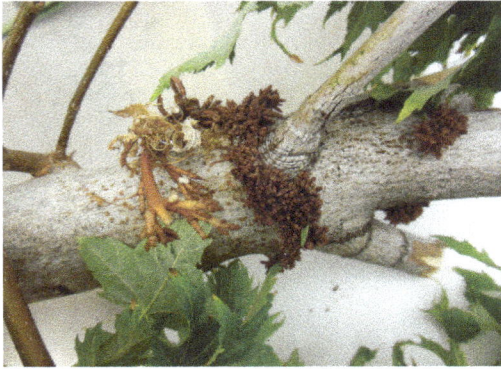

**Fig. 2.44.** A species of mistletoe parasites on a maple tree (*Acer* sp.) causing growth vigour decline and branch dieback. Note that the parasitism makes the tree more prone to branch canker disease caused by the fungus *Cytospora* sp. (internal canker symptom not shown).

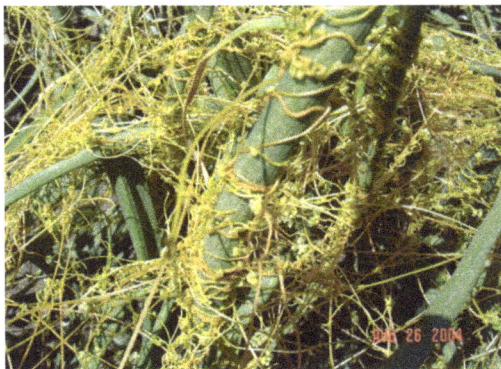

**Fig. 2.45.** Onion plants (*Allium cepa*) parasitized by dodders (*Cuscuta* sp.).

dodders (*Cuscuta* spp.), both of which parasitize on the stem of host plants. Mistletoes are a type of obligate hemiparasitic plants in the order Santalales and are commonly associated with trees or woody plants. Dodders, an obligate holoparasite, can parasitize hundreds of different hosts in diverse families (Nickrent and Musselman, 2004). Some crops or plant species may suffer more damage from dodder parasitism than other plant species. For example, alfalfa, clover, onions, chives, tomatoes, carrots, spinach, flax, hemp, cotton, potatoes and beets were found to be damaged or endangered mostly by dodder infestations in several pratological ecosystems of the south-eastern region of

Romania (Tanase *et al.*, 2012). Parasitic plants are mainly dispersed via contaminated crop seeds, especially for dodders. Therefore, to reduce the risk of dodders infecting a hemp crop, hemp seeds must be cleaned to eliminate the chance of contamination with dodder seeds. Another source of contamination is the nursery stock and soil associated with potted plants. Control of a dodder infestation relies on these preventive measures.

## General Considerations for Disease Control

### Prevention

Controlling plant diseases takes time and money and sometimes the methods used may not be effective, especially when a disease is incorrectly or not diagnosed. Preventing a disease in a hemp field or marijuana cultivation facility is the first strategy in combating any potential diseases or pests. Government agencies have adopted various quarantines against certain invasive and/or destructive diseases and have provided funding to states for conducting pest and disease surveys. This way, once a disease or a pest is detected, it can be easily eradicated with minimum costs. This prevention strategy using quarantine and regulations has helped safeguard US agriculture by preventing the introduction of dangerous diseases and pests. This strategy should also be adopted for *Cannabis* production, which would prevent some, if not all, diseases from occurring in a field or facility.

### Resistant varieties

Crop varieties resistant to specific or multiple diseases have been developed in many crop species and used by many growers. When available, resistant varieties offer an easier and much more effective method of controlling diseases than any other single measure. Most resistant varieties are bred using traditional breeding or genetic engineering. However, a resistant cultivar may exist naturally and thus be collected by growers or horticulturists for cultivation. Such a cultivar generally has a complex defence system that includes both physical or structural barriers against pathogens and a molecular interface between plant cells and pathogens. These special traits confer on the plants an immunity, resistance, or high levels of tolerance to a specific disease. In the *Cannabis* industry, cultivators may select their own cultivars for production, some of which may be disease resistant.

## Disease-free seeds and mother plants

Most hemp crops are started with seeds, although cuttings are also used in greenhouse and outdoor production. Seeds carry pathogens without showing visible symptoms, aka latent infection (see Chapter 6). Infected or contaminated seeds may result in poor germination, damping off of seedlings and crop failure. Using certified seeds may help minimize the risk of seed-borne diseases. Seed certification requires both field disease inspections and seed testing. A seed-producing hemp field without visible disease symptoms during the growing season is much less likely to have seed-borne diseases than a crop with rampant disease problems. When both field infections and seed viability tests meet the certification standards, the seed lots are certified. If no certified seeds are available, a common practice is treating seeds with an appropriate disinfectant to eliminate the pathogens. Many cannabis production crops are started with cuttings from mother plants. Therefore, maintaining disease-free mother plants is the key to preventing diseases passing into the new crop.

## Crop rotations

Rotating crops in a field is a very effective method in disease management, especially for those diseases that have a narrow host range. However, certain pathogens have broad host ranges and can render rotation ineffective. To know which crop should be planted after one crop, growers need to keep records of diseases associated with each crop planted in the field. Such records will help determine if the pathogens found in the previous crops will impact the following crop. The purpose of rotation is to mitigate disease pressure and reduce pathogen inoculum build-up during years of crop production. Hemp is a new crop and susceptible to *Fusarium* and other soilborne diseases. Therefore, growing hemp in the same field continuously may result in disease occurrence. A more aggressive rotation scheme may be necessary for hemp crops than for other crops.

## Weed and arthropod control

Many viruses infect unwanted weeds and are carried by insects and mites. Certain fungal pathogens may survive on weeds during the winter. These infected weeds and insects serve as pathogen reservoirs and are the primary source for infection. During the growing seasons, insects not only infest hemp crops, causing mechanical damage, but also spread viral diseases throughout the field. A virus infection is generally not reversible with treatment and is much more difficult to treat than an insect infestation. Therefore, control of weeds and arthropods should be incorporated into routine cultural practices in hemp production.

## Sanitation

Sanitation is an especially important cultural practice for indoor cannabis production. It also applies to outdoor hemp production. Soils or media used for cultivation should be pasteurized; and pots and tools should be sterilized. Frequent sanitation procedures should be implemented during the entire cultivation period. For the hydroponic or aeroponic cultivation of cannabis, the water source and fertilizer solution should be tested periodically for *Pythium* or *Phytophthora* pathogens, as these water-mould oomycetes survive, grow and spread in water to cause root and crown rot.

## Chemical treatment

Chemical-based products such as insecticides and fungicides are effective in plant pest and disease control. They are often used as a last resort when cultural or preventive measures do not yield ideal results. However, many fungicides and insecticides are not labelled for use in *Cannabis* plants. This lack of effective chemical treatment frustrates growers dealing with *Cannabis* plant diseases and pests. There are some alternative products that have been assigned by each US state for use on *Cannabis* crops. As more legal processes move on to define hemp as a regular crop, some fungicides and pesticides may become available for use in hemp crops.

## Integrated pest management

No single method alone can achieve effective long-term disease or pest control without harming the environment. The best way to minimize ecosystem damage is to integrate multiple approaches, including cultural control, biological control, mechanical and physical controls, developing resistant varieties and chemical control. This is defined as integrated pest management (IPM), in which a pest is defined as any weed, vertebrate, nematode, fungus, bacterium, virus, or other organism harming the ecosystem. The IPM programme is guided by scientific research and embraces several major components, including disease and pest diagnosis, monitoring pathogen/pest population and epidemiology, predicting disease occurrence and outbreak, guiding management and assessing the effectiveness of pest management. The principles and practices of IPM programmes can and will be used to manage hemp pests and diseases.

## Clinical diagnosis

The treatment of plant diseases and the practice of IPM principles rely on timely and accurate diagnosis. It is almost impossible to treat a plant disease correctly without knowing what causes the problem. That said, a clinical diagnosis of a plant problem is the foundation for any prescribed treatment, like medical diagnosis is for human disease treatment. While this chapter has overviewed the basic concept of plant diseases using examples of diseases associated with hemp and other crops, the next chapter will illustrate the science and art of plant diagnostics and how to use them in *Cannabis* crop disease diagnosis.

## References

Abad, J.A., Bandla, M., French-Monar, R.D., Liefting, L.W. and Clover, G.R.G. (2009) First report of the detection of 'Candidatus Liberibacter' species in zebra chip disease-infected potato plants in the United States. *Plant Disease*, 93, 108. doi: 10.1094/PDIS-93-1-0108C

Abe, A., Asano, K. and Sone, T. (2010) A molecular phylogeny-based taxonomy of the genus *Rhizopus*. *Bioscience, Biotechnology, and Biochemistry* 74, 1325–1331. doi: 10.1271/bbb.90718

Adams, A.N. and Barbara, D.J. (1982) Host range, purification and some properties of two carlaviruses from hop (*Humulus lupulus*): hop latent and American hop latent. *Annals of Applied Biology* 101, 483–494. doi: 10.1111/j.1744-7348.1982.tb00849.x

Agrios, G.N. (1997) *Plant Pathology*, 4th edn. Academic Press, San Diego, California.

Beckerman, J., Nisonson, H., Albright, N. and Creswell, T. (2017) First report of *Pythium aphanidermatum* crown and root rot of industrial hemp in the United States. *Plant Disease* 101, 1038–1039. doi: 10.1094/PDIS-09-16-1249-PDN

Beckerman, J., Stone, J., Ruhl, G. and Creswell, T. (2018) First report of *Pythium ultimum* crown and root rot of industrial hemp in the United States. *Plant Disease* 102, 2045. doi: 10.1094/PDIS-12-17-1999-PDN

Bektaş, A., Hardwick, K.M., Waterman, K. and Kristof, J. (2019) Occurrence of hop latent viroid in *Cannabis sativa* with symptoms of cannabis stunting disease in California. *Plant Disease* 103, 2699. doi: 10.1094/PDIS-03-19-0459-PDN

Brown, D.J.F. and MacFarlane, S. A. (2001) 'Worms' that transmit viruses. *Biologist* 48, 35–40.

Bull, C.T. and Rubio, I. (2011) First report of bacterial blight of crucifers caused by *Pseudomonas cannabina* pv. *alisalensis* in Australia. *Plant Disease* 8, 1027. doi: 10.1094/PDIS-11-10-0804

Bull, C.T., Manceau, C., Lydon, J., Kong, H., Vinatzer, B.A. et al. (2010) *Pseudomonas cannabina* pv. *cannabina* pv. nov., and *Pseudomonas cannabina* pv. *alisalensis* (Cintas Koike and Bull, 2000) comb. nov., are members of the emended species *Pseudomonas cannabina* (ex Šutič & Dowson 1959) Gardan, Shafik, Belouin, Brosch, Grimont & Grimont 1999. *Systematic and Applied Microbiology* 33, 105-115. doi: 10.1016/j.syapm.2010.02.001

Cho, W.D. and Shin, H.D. (2004) *List of Plant Diseases in Korea*, 4th edn. Korean Society of Plant Pathology, Seoul, Republic of Korea.

Diener, T.O. (1971) Potato spindle tuber 'virus': IV. A replicating, low molecular weight RNA. *Virology* 45, 411–428. doi: 10.1016/0042-6822(71)90342-4

Dietzgen, R.G., Mann, K.S. and Johnson, K.N. (2016) Plant virus–insect vector interactions: current and potential future research directions. *Viruses* 8, 303. doi: 10.3390/v8110303

Doi, Y., Teranaka, M., Yora, K. and Asuyama, H. (1967) Mycoplasma-or PLT group-like microorganisms found in the phloem elements of plants infected with mulberry dwarf, potato witches' broom, aster yellows or paulownia Witches' broom. *Annals of the Phytopathological Society of Japan* 33, 259–266.

Dutkiewicz, J., Mackiewicz, B., Kinga Lemieszek, M., Golec, M. et al. (2016) *Pantoea agglomerans*: a mysterious bacterium of evil and good. Part III. Deleterious effects: infections of humans, animals and plants. *Annals of Agricultural and Environmental Medicine* 23, 197–205. doi: 10.5604/12321966.1203878

Erwin, D.C. and Ribeiro, O.K. (1996) *Phytophthora Diseases Worldwide*. APS Press, St Paul, Minnesota.

Farr, D.F. and Rossman, A.Y. (2019) Fungal Databases. Available at https://nt.ars-grin.gov/fungaldatabases (accessed 28 October 2019).

Feng, X., Kyotani, M., Dubrovsky, S. and Fabritius, A.L. (2019) First report of *Candidatus* Phytoplasma trifolii associated with witches' broom disease in *Cannabis sativa* in Nevada, USA. *Plant Disease* 7, 103. doi: 10.1094/PDIS-01-19-0098-PDN

Francl, L.J. (2001) *The Disease Triangle: a plant pathological paradigm revisited*. Plant Health Instructor, Teaching Notes, American Phytopathological Society, St Paul, Minnesota. doi: 10.1094/PHI-T-2001-0517-01

Fujiwara, K., Iwanami, T. and Fujikawa, T. (2018) Alterations of *Candidatus* Liberibacter asiaticus-associated microbiota decrease survival of *Ca.* L. asiaticus in *in vitro* assays. *Frontiers in Microbiology* 9, 3089. doi: 10.3389/fmicb.2018.03089

Gardan, L., Shafik, H., Belouin, S., Broch, R., Grimont, F. et al. (1999) DNA relatedness among the pathovars of *Pseudomonas syringae* and description of *Pseudomonas tremae* sp. nov. and *Pseudornonas cannabina* sp. nov. (ex Sutic and Dowson 1959). *International Journal of Systematic Bacteriology* 49, 469–478. doi: 10.1099/00207713-49-2-469

Gitaitis, R.D. and Gay, J.D. (1997) First report of a leaf blight, seed stalk rot, and bulb decay of onion by *Pantoea ananas* in Georgia. *Plant Disease* 81, 1096. doi: 10.1094/PDIS.1997.81.9.1096C

Hansen, A.K., Trumble, J.T., Stouthamer, R. and Paine, T.D. (2008) A new Huanglongbing species, 'Candidatus Liberibacter psyllaurous,' found to infect tomato and potato, is vectored by the psyllid *Bactericera cockerelli* (Sulc). *Applied and Environmental Microbiology* 74, 5862–5865. doi: 10.1128/AEM.01268-08

Hataya, T., Uchino, K., Arimoto, R., Suda, N., Sano, T. et al. (2000) Molecular characterization of hop latent virus and phylogenetic relationships among viruses closely related to carlaviruses. *Archives of Virology* 145, 2503–2524. doi: 10.1007/s007050070005

Hirano, S.S., Charkowski, A.O., Collmer, A., Willis, D.K. and Upper, C.D. (1999) Role of the Hrp type III protein secretion system in growth of *Pseudomonas syringae* pv. *syringae* B728a on host plants in the field. *Proceedings of the National Academy of Sciences of the United States of America* 96, 9851–9856. doi: 10.1073/pnas.96.17.9851

Hogenhout, S.A., Oshima, K., Ammar, E., Kakizawa, S., Kingdom, H.N. et al. (2008) Phytoplasmas: bacteria that manipulate plants and insects. *Molecular Plant Pathology* 9, 403–423. doi: 10.1111/j.1364-3703.2008.00472.x

Hopkins, D.L. and Purcell, A.H. (2002) *Xylella fastidiosa*: cause of Pierce's disease of grapevine and other emergent diseases. *Plant Disease* 86, 1056–1066. doi: 10.1094/PDIS.2002.86.10.1056

Hull, R. (2014) *Plant Virology*, 5th edn. Academic Press, London.

Inserra, R.N., Duncan, L.W., O'Bannon, J.H. and Fuller, S.A. (1994) Citrus nematode biotypes and resistant citrus rootstocks in Florida. Available at https://ufdc.ufl.edu/IR00004655/00001 (accessed 30 November 2019).

Jacobs, J.M., Pesce, C., Lefeuvre, P. and Koebnik, R. (2015) Comparative genomics of a cannabis pathogen reveals insight into the evolution of pathogenicity in *Xanthomonas*. *Frontiers in Plant Science* 6, 431. doi: 10.3389/fpls.2015.00431

Jo, Y., Choi, H. and Cho, W.K. (2015) Complete genome sequence of a hop latent virus infecting hop plants. *Genome Announcements* 3, e00302-15. doi: 10.1128/genomeA.00302-15

Kado, C.I. (2010) *Plant Bacteriology*. APS Press, St Paul, Minnesota.

Kido, K., Adachi, R., Hasegawa, M., Yano, K., Hikichi, Y. et al. (2008) Internal fruit rot of netted melon caused by *Pantoea ananatis* (=*Erwinia ananas*) in Japan. *Journal of Genera Plant Pathology* 74, 302–312. doi: 10.1007/s10327-008-0107-3

Li, W., Hartung, J.S. and Levy, L. (2006) Quantitative real-time PCR for detection and identification of *Candidatus* Liberibacter species associated with citrus Huanglongbing. *Journal of Microbiological Methods* 66, 104–115. doi: 10.1016/j.mimet.2005.10.018

Lin, H., Doddapaneni, H., Munyaneza, J. E., Civerolo, E. L., Sengoda, V.G. et al. (2009) Molecular characterization and phylogenetic analysis of 16S rRNA from a new 'Candidatus Liberibacter' strain associated with zebra chip disease of potato (*Solanum tuberosum* L.) and the potato psyllid (*Bactericera cockerelli* Sulc). *Journal of Plant Pathology* 91, 215–219. doi: 10.4454/jpp.v91i1.646

Lodish, H., Berk, A., Zipursky, S.L., Matsudaira, P., Baltimore, D. et al. (2000) Growth of microorganisms in culture. In Lodish, H. et al. (eds) *Molecular Cell Biology*, 4th edn. W.H. Freeman and Co., New York. Available at: https://www.ncbi.nlm.nih.gov/books/NBK21593/ (accessed 30 November 2019).

Manulis, S. and Barash, I. (2003) *Pantoea agglomerans* pvs. *gypsophilae* and *betae*, recently evolved pathogens? *Molecular Plant Pathology* 4, 307–314. doi: 10.1046/j.1364-3703.2003.00178.x

Moore, C.J., Sutherland, P.W., Forster, R.L., Gardner, R.C. and MacDiamid, R.M. (2001) Dark green islands in plant virus infection are the result of post-transcriptional gene silencing. *Molecular Plant-Microbe Interactions* 14, 939–946. doi: 10.1094/MPMI.2001.14.8.939

Moore, L.W. and Warren, G. (1979) *Agrobacterium radiobacter* strain 84 and biological control of crown gall. *Annual Review of Phytopathology* 17, 163–179. doi: 10.1146/annurev.py.17.090179.001115

Netsu, O., Kijima, T. and Takikawa, Y. (2014) Bacterial leaf spot of hemp caused by *Xanthomonas campestris* pv. *cannabis* in Japan. *Journal of General Plant Pathology* 80, 164–168. doi: 10.1007/s10327-013-0497-8

Nibert, M.L., Ghabrial, S.A., Maiss, E., Lesker, T., Vainio, E.J. et al. (2014) Taxonomic reorganization of family Partitiviridae and other recent progress in partitivirus research. *Virus Research* 188, 128–141. doi: 10.1016/j.virusres.2014.04.007

Nickrent, D.L. and Musselman, L.J. (2004) *Introduction to parasitic flowering plants* (updated 2016). Plant Health Instructor, Teaching Notes, American Phytopathological Society, St Paul, Minnesota. doi: 10.1094/PHI-I-2004-0330-01

Nowicki, M., Foolad, M.R., Nowakowska, M. and Kozik, E.U. (2011) Potato and tomato late blight caused by *Phytophthora infestans*: An overview of pathology and resistance breeding. *Plant Disease* 96, 4–17. doi: 10.1094/PDIS-05-11-0458

Pataky, J.K. (2004) *Stewart's wilt of corn*. Plant Health Instructor, Teaching Notes, American Phytopathological Society, St Paul, Minnesota. doi: 10.1094/PHI-I-2004-0113-01

Peeters, N., Guidot, A., Vailleau, F. and Valls, M. (2013) *Ralstonia solanacearum*, a widespread bacterial plant pathogen in the post-genomic era. *Molecular Plant Pathology* 14, 651–662. doi: 10.1111/mpp.12038

Puchta, H., Ramm, K. and Sänger, H.L. (1988) The molecular structure of hop latent viroid (HLV), a new viroid occurring worldwide in hops. *Nucleic Acids Research* 16, 4197–4216. doi: 10.1093/nar/16.10.4197

Punja, Z.K. and Rodriguez, G. (2018) *Fusarium* and *Pythium* species infecting roots of hydroponically grown marijuana (*Cannabis sativa* L.) plants. *Canadian Journal of Plant Pathology* 40, 498–513. doi: 10.1080/07060661.2018.1535466

Punja, Z.K., Scott, C. and Chen, S. (2018) Root and crown rot pathogens causing wilt symptoms on field-grown marijuana (*Cannabis sativa* L.) plants. *Canadian Journal of Plant Pathology* 40, 528–541. doi: 10.1080/07060661.2018.1535470

Purcell, A.H. (1997) *Xylella fastidiosa*, a regional problem or global threat? *Journal of Plant Pathology* 79, 99–105.

Raj, S.K., Snehi, S.K., Khan, M.S. and Kumar, S. (2008) *Candidatus* phytoplasma asteris (group 16SrI) associated with a witches'-broom disease of *Cannabis sativa* in India. *Plant Pathology* 57, 1173. doi: 10.1111/j.1365-3059.2008.01920.x

Remenant, B., de Cambiaire, J.C., Cellier, G., Jacobs, J.M., Mangenot, S. *et al.* (2011) *Ralstonia syzygii*, the blood disease bacterium and some asian *R. solanacearum* strains form a single genomic species despite divergent lifestyles. *PLoS ONE* 6, e24356. doi: 10.1371/journal.pone.0024356

Ribes, J.A., Vanover-Sams, C.L. and Baker, D. J. (2000) Zygomycetes in human disease. *Clinical Microbiology Reviews* 13, 236–301. doi: 10.1128/cmr.13.2.236-301.2000

Robinson, A.F., Inserra, R.N., Caswell-Chen, E.P., Vovlas, N. and Troccoli, A. (1997) *Rotylenchulus* species: identification, distribution, host ranges, and crop plant resistance. *Nematropica* 27, 127–180.

Samson, R., Legendre, J.B., Christen, R., Fischer-Le Saux, M., Achouak, W. *et al.* (2005) Transfer of *Pectobacterium chrysanthemi* (Burkholder *et al.* 1953) Brenner *et al.* 1973 and *Brenneria paradisiaca* to the genus *Dickeya* gen. nov. as *Dickeya chrysanthemi* comb. nov. and *Dickeya paradisiaca* comb. nov. and delineation of four novel species, *Dickeya dadantii* sp. nov., *Dickeya dianthicola* sp. nov., *Dickeya dieffenbachiae* sp. nov. and *Dickeya zeae* sp. nov. *International Journal of Systematic and Evolutionary Microbiology* 55, 1415–1427. doi: 10.1099/ijs.0.02791-0

Schoener, J.L. and Wang, S. (2019) First detection of a phytoplasma associated with witches' broom of industrial hemp in the United States. (Abstr.) *Phytopathology* 109(S2), 110. doi: 10.1094/PHYTO-109-10-S2.1

Schoener, J.L., Wilhelm, R., Yazbek, M. and Wang, S. (2018) First detection of *Pythium aphanidermatum* crown rot of industrial hemp in Nevada. (Abstr.) *Phytopathology* 108(S1), 187. doi: 10.194/PHYTO-108-10- S1.187

Scott, M., Rani, M., Samsatly, J., Charron, J.B. and Jabaji, S. (2018) Endophytes of industrial hemp (*Cannabis sativa* L.) cultivars: identification of culturable bacteria and fungi in leaves, petioles, and seeds. *Canadian Journal of Microbiology* 64, 664–680. doi: 10.1139/cjm-2018-0108

Severin, V. (1978) Ein neues Bakterium an Hanf – *Xanthomonas campestris* pathovar *cannabis*. *Archives of Phytopathology and Plant Protection* 14, 7–15. doi: 10.1080/03235407809436295

Sichani, F.V., Bahar, M. and Zirak, L. (2011) Characterization of Stolbur (16SrXII) group phytoplasamas associated with *Cannabis sativa* witches'-broom disease in Iran. *Plant Pathology Journal* 10, 161–167. doi: h10.3923/ppj.2011.161.167

Singh, S., Singh, B. and Singh, A.P. (2015) Nematodes: a threat to sustainability of agriculture. *Procedia Environmental Sciences* 29, 215–216. doi: 10.1016/j.proenv.2015.07.270

Šutić, D. and Dowson, W.J. (1959) An investigation of a serious disease of hemp (*Cannabis sativa* L.) in Jugoslavia. *Journal of Phytopathology* 34, 307–314. doi: 10.1111/j.1439-0434.1959.tb01818.x

Takikawa, Y., Kijima, T., Yamashita, S., Doi, Y., Tsuyumu, S. *et al.* (1984) Bacterial leaf spot disease of hemp caused by *Xanthomonas campestris* pv. *cannabis*. *Annals of the Phytopathological Society of Japan* 50, 141.

Tanase, M., Stanciu, M.C., Moise, C. and Gheorghe, M. (2012) Ecological and economic impact of dodder species (*Cuscuta* spp. Convolvulaceae) on pratological ecosystems. *Journal of Horticulture, Forestry and Biotechnology*, 16, 93–97.

Wahby, I., Caba, J.M. and Ligero, F. (2013) *Agrobacterium* infection of hemp (*Cannabis sativa* L.): establishment of hairy root cultures. *Journal of Plant Interactions* 8, 312–320. doi: 10.1080/17429145.2012.746399

Walcott, R.R., Gitaitis, R.D., Castro, A.C., Sanders, F.H. and Diaz-Perez, J.C. (2002) Natural infestation of onion seed by *Pantoea ananatis*, causal agent of center rot. *Plant Disease* 86, 106–111. doi: 10.1094/PDIS.2002.86.2.106

Walterson, A.M. and Stavrinides, J. (2015) *Pantoea*: insights into a highly versatile and diverse genus within the Enterobacteriaceae. *FEMS Microbiology Reviews* 39, 968–984. doi: 10.1093/femsre/fuv027

Wang, S. (2019) Common marijuana and hemp diseases. *Cannabis Business Times*, 2019, 29–38.

Wang, S. and Chiu, W.F. (1997) Preliminary observations of the pathogenicity of *Trichodorus nanjingensis* to cucumber roots. *International Journal of Nematology* 7, 76–79.

---

Wang, S. and Gergerich, R.C. (1998) Immunofluorescent localization of tobacco ringspot virus in the vector nematode *Xiphinema americanum*. *Phytopathology* 88, 885–889. doi: 10.1094/PHYTO.2002.92.6.646

Wang, S. and Riggs, R. D. (1999) Nematode problems on soybean worldwide. In: Kauffman, H.E. (ed.) *Proceedings of World Soybean Research Conference VI*. Superior Printing, Champaign, Illinois, pp. 307–315.

Wang, S., Miao, Z. and Zhang, J. (1994) Incidence of apple stubby root disease and species identification of pathogenic nematodes. *Acta Phytopathologica Sinica* 24, 62–66.

Wang, S., Li, Y., Fan, Z., Li, H., Chiu, W.F. *et al.* (1998) Root knot nematode and tobacco mosaic virus complex involved in banana yellow bunchy-top disease. *International Journal of Nematology* 8, 5–12.

Wang, S., Gergerich, R.C., Wickizer, S.L. and Kim, K.S. (2002) Localization of transmissible and nontransmissible viruses in the vector nematode *Xiphinema americanum*. *Phytopathology* 92, 646–653. doi: 10.1094/PHYTO.2002.92.6.646

Warren, J.G., Mercado, J. and Grace, D. (2019) Occurrence of hop latent viroid causing disease in *Cannabis sativa* in California. *Plant Disease* 103, 2699. doi: 10.1094/PDIS-03-19-0530-PDN

Wells, J.M., Raju, B.C., Hung, H.Y., Weisberg, W.G., Mandelco-Paul, L. *et al.* (1987) *Xylella fastidiosa* gen. nov., sp. nov: gramnegative, xylem-limited, fastidious plant bacteria related to *Xanthomonas* spp. *International Journal of Systematic Bacteriology* 37, 136–143.

Zhao, Y., Sun, Q., Davis, R.E., Lee, I.M. and Liu, Q. (2007) First report of witches'-broom disease in a *Cannabis* spp. in China and its association with a phytoplasma of elm yellows group (16SrV). *Plant Disease* 91, 227. doi: 10.1094/PDIS-91-2-0227C

Ziegler, A., Matoušek, J., Steger, G. and Schubert, J. (2012) Complete sequence of a cryptic virus from hemp (*Cannabis sativa*). *Archives of Virology* 157, 383–385. doi: 10.1007/s00705-011-1168-8

Zulperi, D. and Sijam, K. (2014) First report of *Ralstonia solanacearum* race 2 biovar 1 causing Moko disease of banana in Malaysia. *Plant Disease* 98, 275. doi: 10.1094/PDIS-03-13-0321-PDN

# 3 The Art of Plant Diagnostics

## The Definition of Disease Diagnosis

### Diagnosis

Diagnosis is an investigative process to identify biological or non-biological factors that are responsible for a problem affecting a plant. There are many infectious diseases, abiotic disorders and arthropod (insects and mites) damage associated with a crop, so it is not always easy to determine the cause of a disease by observing the symptoms alone. Often, an extensive research may be carried out to find a responsible factor accurately. This is especially true for diseases that have never been described. The complexity of diagnosis is determined by the nature of the problem and the level of diagnosis to be achieved. In most cases, a diagnosis involves field observations, lab analyses and further identification of responsible pathogenic organisms.

### Identification

In *Glossary of Plant-Pathological Terms* (Shurtleff and Averre, 1997b), identification is defined as the study of the characters of an organism to determine its name. For example, identification of a fungus is a process of studying the characteristics of a given specimen so that a genus or species name can be given. Similarly, insect identification is a study classifying a specimen to an appropriate taxonomic level. In most cases, the purpose of identification is to give a name to a specific organism, e.g. virus, fungus, bacterium, nematode, plant, insect, or animal. Identification is part of the diagnostic process when an organism in question is suspected to be the cause of a disease and the identity of such an organism needs to be determined. In the world of plant diagnostics, many use the term 'disease identification'. The fact is that the disease is to be diagnosed so that a specific cause is found; the disease diagnosis is much more complex and broader in scope than simply the identification of an organism.

## Detection

Detection is a process to find something concealed – a target microorganism in plants, or a specific compound such as genetically modified organism (GMO) ingredients in food. The process is more interested in finding something of concern from a subject than in resolving a problem of a subject. For example, a lab testing a thousand citrus tissue samples for the citrus greening pathogen, '*Candidatus* Liberibacter asiaticus', aims to detect and map the occurrence of this bacterium in a region. Such an activity monitors the potential spread of the bacterium and is not intended to diagnose or resolve any disease associated with the plants the samples are taken from. However, detection is frequently used to verify the presence of a certain pathogen suspected of causing a disease. This is true when multiple aetiological hypotheses are formulated for a disease. For example, when a hemp sample exhibiting leaf distortion, mosaic and proliferative leaf growth is submitted to a lab, the lab diagnostician may formulate three diagnostic hypotheses: (i) herbicide injury; (ii) virus infection; and (iii) phytoplasma infection. A test is then performed to verify the presence of such a compound or organism in the plant tissue. Often, one of these tests turns out positive and can be used to explain the symptoms perfectly. Sometimes, no specific compounds or organisms are detected and additional hypotheses need to be proposed. Even when positive detection for an organism is confirmed, the cause of the disease may still be a mystery when the found organism is not always associated with symptomatic plants. Diagnosing a disease through a series of detections is time consuming, costly and not always successful. A universal detection may be more helpful. However, neither specific nor universal detection can lead to a full diagnosis, as a detection reports only whether a targeted organism is present or absent.

DOI: 10.1079/9781789246070.0003

## Testing and analysis

Related to detection and identification, a test is a process, method, or examination procedure designed to evaluate a condition or to demonstrate a scientific hypothesis. In plant and medical diagnostics, a test is often used to find the evidence of a disease or the cause of a condition. The method used for a test is generally validated and established for clinical diagnosis and the results can be presented as positive, negative, inconclusive, or a certain value. In plant diagnostics, plant pathologists prescribe and perform specific tests to support their own diagnosis. A test result is generally reviewed and then used to rule out or explain a disease. Analysis is a much broader term to describe a process or method used in studying the nature, features or relationships of things. In plant diagnostics, analysis can be used in the field or in a laboratory. For example, we analyse a disease pattern in the field, analyse plant tissue for the presence of pathogens or nutrient deficiencies, and analyse data to make a diagnosis. Such an analytical process is essential to plant diagnosis.

## The truth of plant diagnostics

A plant disease may require a number of tests to detect suspected pathogens or chemicals, followed by the identification of causative organisms. The integrated processes of testing, detection and identification are often required for a full diagnosis (Fig. 3.1). A diagnosis reporting only a test result, for example this hemp sample was examined and tested negative for *Phytophthora* infection, is not a full diagnosis, but rather a test result. Similarly, a diagnosis only reporting an identification result, for example a few aphids found from the sample were identified as *Neotoxoptera formosana* (onion aphid), is not a full diagnosis either. Clients receiving these types of reports may be disappointed and may call back for further diagnosis. Remember, when growers pull symptomatic plants from the field and submit them to a lab, they want to know what causes the symptoms so that they can treat the problem correctly. The determination of a cause is the main focus and should be concluded in a final diagnosis. With each diagnosis, a list of management recommendations should be provided to the client. For some reasons, if a definitive diagnosis based on lab procedures cannot be achieved, a professional opinion on the potential causes is helpful.

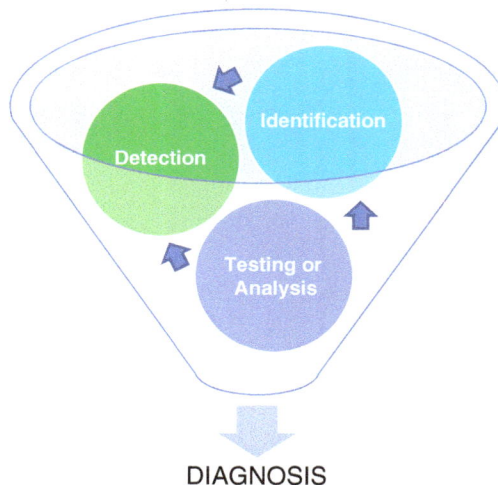

**Fig. 3.1.** A flow chart of plant diagnostics. Note that testing and analysis are essential to pathogen detection and also support organism identification. Identification is a part of the detection process. A full diagnosis embraces all processes of testing, analysis, detection and identification.

Detection, identification, testing, analysis and diagnosis of plant diseases have all been used in verbal communications and literature, but each term has a different meaning and scope of work. Differentiating between and appropriately using these terms will help growers and lab diagnosticians communicate. Growers who submit samples to a lab should ask for diagnosis rather than a specific test or identification, unless the sample is intended for a specific testing/identification programme. Lab diagnosticians should focus on determining causative agents or factors instead of reporting testing data that do not provide any clue to the cause of a plant problem.

## The Traits of a Diagnostician

### Knowledge and skills

Anyone who tries to diagnose a plant problem is considered to be a diagnostician. Hemp growers and cannabis cultivators are more likely to diagnose various problems themselves in the field before they send plant samples to a professional lab. Extension personnel mostly perform field diagnoses (field diagnostician) and lab personnel mostly perform lab diagnoses (lab diagnostician). A good diagnostician, working in either the field or the lab, generally has a solid knowledge of plant biology and physiology, understands agronomy practices (including indoor crop production such as greenhouse, hydroponic and aeroponic systems) and is trained in pest identification and disease diagnosis. A major in plant pathology and/or entomology is a great asset for plant diagnostics. When training in plant pathology, one should try to obtain research and diagnostic experience in all subjects of plant pathogens, namely mycology, bacteriology, virology and nematology. The ability to recognize diseases caused by all types of pathogens and to identify any encountered pathogen is a plus. This is particularly useful in diagnosing a disease complex caused by more than one type of pathogen. That said, a good diagnostician is competent in diagnosing plant diseases caused by all types of pathogens.

### Inquiring and open mind

In addition to the necessary foundational knowledge and skills, a plant diagnostician should have an inquiring mind. A person with an inquiring mind is interested in researching and exploring a subject or situation. Often, diagnosticians need to engage with clients to learn more information about the sample so that an accurate diagnosis can be made. Without this trait, a diagnostician may not look into a sample thoroughly and seriously, which may lead to an inaccurate diagnosis. An open mind is another important trait for a plant diagnostician. This trait motivates a diagnostician to wait until all the facts or evidence are received before forming an opinion or making a judgement. It requires listening to the client's description of disease problems, gathering all necessary information and thinking about all types of possibilities. Diagnosticians should take time to examine the plants or samples, perform clinical procedures to find abnormalities associated with the plants and run a series of tests to confirm or rule out certain causes. Until all steps of field and lab procedures are completed, a diagnostic statement should not be issued. On the other hand, a quick diagnosis based on a disease image posted online or in a disease compendium may not be reliable. Fig. 3.2 shows an example of a plant problem reported by a client and diagnosed by a plant diagnostician with an open and inquiring mind.

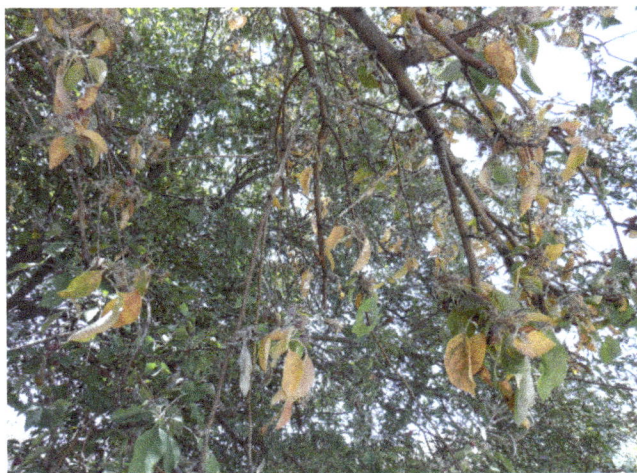

**Fig. 3.2.** A branch of crab apple tree (*Malus sylvestris*) damaged by the evaporation of fresh paint during the summer. Note that the entire foliage of the tree was infected by a powdery mildew disease, and beside the garage there was a branch extended 4–5 feet (1.2–1.5 m) above the roof. All leaves of that branch had recently started to turn yellow and die. By asking about the tree's previous health history and the timing of symptom appearance, the client confirmed that the symptom started right after the garage-wall painting project. The diagnostician observed that only the branch exposed to the painting showed leaf damage and therefore concluded that this was due to paint damage.

## Logical thinking

Thinking logically is another essential trait of a plant diagnostician. When a client drops off plant samples at a diagnostic lab, they often talk about what is going wrong with their plants or crops, intending to give as much information as possible. This is great communication between clients and lab diagnosticians, as all the information provided is valuable for diagnosis. When we digest a wealth of information, we need to use logical thinking to find a clue. For example, a client complained that almost 90% of her front- and backyard lawn had died very recently and claimed to have a melting-out disease in her yard (Fig. 3.3). She mentioned that the lawn had been perfectly healthy and green a week ago. For this case, one needs to think how a healthy lawn completely died in just a week and realizes that the damage could not have been caused by a common lawn disease. With this in mind, the diagnostician visited the site and found several empty large fertilizer bags in the yard. When the client confirmed that those bags of fertilizer were applied to the lawn a week ago (approximately five times of overdose per label), the diagnostician concluded that the cause was acute fertilizer burn. Logical thinking is applied in all areas of plant diagnostics, but especially when differentiating between infectious diseases and abiotic disorders. When diagnosing an infectious disease, the timing is a critical factor. Infectious diseases take time to develop symptoms (see Fig. 2.5). Sudden death of a plant or crop is likely to be an abiotic disorder.

## Research-oriented

In plant diagnostics, a research-oriented diagnostician often studies the nature and scope of the problem, develops a hypothesis, applies scientific methods to investigate the factors and analyses and interprets the findings to achieve a maximum understanding of the problem. This is a great attribute for a successful plant diagnostician. Since a plant disease is a problematic case, it often requires strong critical thinking and creativity in order to uncover all aspects of the disease. For example, a hemp crop exhibiting high mortality caused by an unknown disease (Fig. 3.4) has to be studied to find all the potential causes. In this case, a simple diagnostic procedure based on single or even multiple tissue samples may not be sufficient to figure out the true problem. Instead, systemic sampling and comprehensive research on the disease aetiology are needed (see Chapter 9). The use of research principles in disease diagnosis is the best way to find all-inclusive causes.

## Detective

A particularly useful trait for a diagnostician is to act like a detective. Plant diseases are caused by

**Fig. 3.3.** A home-yard lawn killed by over-fertilization. Note that some grass on the edge survived. In this case, although there were certain lawn-pathogenic fungi found in the dead grass, they were not the cause of this sudden lawn death.

**Fig. 3.4.** A hemp field exhibiting a large number of plants killed by *Fusarium* root rot. Note that five *Fusarium* species (*F. oxysporum*, *F. solani*, *F. redolens*, *F. tricinctum* and *F. equiseti*) were found to be associated with this disease (Schoener *et al.*, 2017).

many microorganisms that we do not see and most symptoms caused by them are subtle or invisible. A detective mindset will drive diagnosticians to observe the plant closely and carefully, paying special attention to the details of early stages of infection or very tiny arthropod pests. An orange spot, a slight discoloration, or a fuzzy-looking stem may indicate a disease or pest problem. These mild and often overlooked symptoms can be an indication of a pest infestation or a serious disease developing inside plant tissue. Fig. 3.5 shows a hemp leaf infested with hemp russet mites (*Aculops cannibicola*). In this case, affected plants only show very mild leaf chlorosis. In early infestation, the mite population may be low, making it easily overlooked if the leaf is not observed under a dissecting microscope with great attention.

### Patience

Diagnosis is a process requiring patience. For most cases, it is hard to find a shortcut. Rushing to a diagnostic conclusion without analysing all components and performing all necessary procedures may lead to an incomplete or wrong diagnosis. A quick visual diagnosis often works well for some obvious

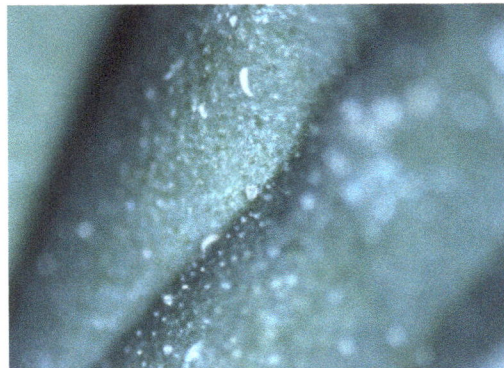

**Fig. 3.5.** Hemp russet mites (*Aculops cannibicola*) feeding on a hemp leaf. Note that this mite belongs to the Eriophyidae family and is often called eriophyid mite. Its body is sausage-shaped, very small (about 0.2 mm in length) and only visible under a microscope.

common diseases such as powdery mildew; however, this will not work when a disease is caused by multiple pathogens or when a disease requires a more indepth diagnosis. Often, plant samples are submitted with little background information and plant diagnosticians need to collect more information before they can proceed to sample processing. Sometimes, a field observation is needed to determine the nature of a problem. Once a plant sample is placed under laboratory procedures and testing, it may take anywhere from a few days to two weeks to complete, depending on the complexity of the case. When a sample is plated to isolate suspected pathogens, the plate should be incubated a few more days even after a fast-growing pathogen has already been identified from the plate. Fig. 3.6 shows dual infections of hemp stem tissue by both *Pythium* and *Fusarium* species in a culture plate. Rushing to read the plate in the first 48 h might have led to a *Pythium* crown rot diagnosis only, but by waiting a few more days, the slow-growing *Fusarium* was also recovered from the tissue, suggesting a disease complex (see Chapter 13). As a lab diagnostician, being patient and allowing certain lab procedures to continue over a period of time can increase the chance of new findings. Growers or clients should be patient too, to allow a lab more days to conclude a diagnosis.

### Accessibility to diagnostic labs

Plant pathology is a lab-based profession. When it comes to clinical diagnosis, a lab is the ultimate centre for a diagnostician to conduct observations,

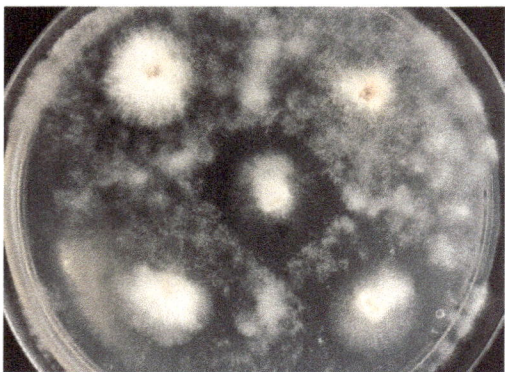

**Fig. 3.6.** A potato dextrose agar plate showing dual infections of hemp stem tissue by *Pythium* and *Fusarium* species. Note that there were five pieces of hemp tissue, all of which were infected by *Pythium* and *Fusarium*. In the plate, *Pythium* colonies grew out earlier and faster from the tissue and spread over the entire plate with a greyish and loose appearance; *Fusarium* grew later and more slowly from the tissue and was easily differentiated from *Pythium* by its white and dense colonies.

testing and experiments. Without a lab, a diagnosis is likely to be simply an opinion or guess. Only an evidence-backed diagnosis can achieve the highest level of reliability, which can be defended legally. Many growers and crop consultants perform initial field diagnosis and then submit meaningful samples to a laboratory for further diagnosis. If a crop has multiple issues with both infectious diseases and abiotic disorders, samples should be submitted to multiple labs with different expertise. For example, a fungal disease sample can be submitted to a general plant diagnostic clinic, but samples with nutrient issues should be submitted to a plant nutrition testing lab.

In plant diagnostics, there are four types of testing labs that can be used to determine the cause of a crop problem. The most common one is the plant disease and pest diagnostic lab. Every state in the USA has at least one lab dedicated to serving growers and the general public. These labs have expertise in plant disease diagnosis and insect/mite identification. The second type of lab is the plant nutrition lab. Some land-grant universities have this type of lab to provide nutritional tests for growers. Crops with nutritional issues should be sampled for testing to ensure all nutritional needs are met. The third type of lab is the soil lab. Many growers prefer to have the soil tested before planting a hemp crop. Soil testing includes the analysis of various nutrient contents, compositions, physical characteristics and acidity or pH level. A pre-planting soil test can determine if a chosen field is suitable for hemp crop production. The fourth type of lab is the chemistry lab. Most states have a regulatory chemistry lab to perform fertilizer and pesticide testing. When it is suspected that a crop has a chemical injury, samples should be collected and submitted to a chemistry lab for analysis. That said, growers and field diagnosticians should have connections to one or more labs. A lab confirmation is especially important when a suspected disease cannot be determined based on symptoms and a lab-based diagnosis is more accurate than a field-based or visual diagnosis.

## The Types of Diagnosis
### Full diagnosis

Diagnosis of a plant problem takes time and costs money. A full diagnosis of a specific disease can essentially become a research project, especially when it must meet the rule of Koch's postulates. The rule sets four criteria to demonstrate a causative relationship between a microbe and a disease.

**Step 1.** A microorganism must be found or isolated from all plants suffering from the disease, but not from healthy plants. This criterion is to demonstrate an obvious association between the microorganism and the host plant, and such an association should always be used in plant diagnostics.

**Step 2.** The microorganism must be isolated from a diseased plant and grown in a pure culture. Some groups of pathogens such as viruses cannot be cultured, but they can be extracted and purified from host plants. This step is mostly performed by diagnosticians to isolate causative agents from symptomatic plants.

**Step 3.** The cultured microorganism should cause the same disease or similar symptoms as originally observed when inoculated into a healthy plant of the same species and cultivars.

**Step 4.** The microorganism must be re-isolated from the inoculated and diseased host plants and it must be identified as the same microorganism that initially caused the disease. This step must be performed if the third step (inoculation) is conducted.

Koch's postulates are mainly used in research laboratories to demonstrate a new causative agent or a new host for a specific microorganism. Data from the inoculation tests are used to support a first report of a plant disease in a new host or geographical area. Figure 3.7 illustrates a full diagnosis of basal petiole rot and leaf blight of *Chamaerops humilis* caused by *Phytophthora nicotianae* based on Koch's postulates criteria (Bomberger *et al.*, 2016). A diagnosis meeting the Koch's postulates criteria is considered a full diagnosis for an infectious disease and has the highest level of reliability.

Some advanced clinical labs have both the capability and capacity to perform a full diagnosis. They are generally well equipped and have a lead scientist supervising each stage of diagnostic research. However, performing a full diagnosis does not necessarily serve the purpose of a routine plant sample. For example, a grower submits a garlic leaf sample with tiny orange spots and wants to know if the orange stuff is actually a rust fungus. In this case, microscopic examination to determine whether it is a rust fungus (*Puccinia allii*) or an abiotic disorder should serve the purpose well. The grower only needs this information to assess if the crop is at risk of rust disease and if a treatment is needed.

Unless a disease appears to be new to a crop, most diagnostic labs may not pursue steps 3 and 4 as they are time consuming and require a greenhouse facility. If a disease is found from a tree or other perennial plants, the inoculation test may take from several months to a year. With a diagnostic lab loaded daily with submitted plant samples, it is not practical to leave a sample in the queue for too long. Most labs require a short turnaround time (generally within 2 weeks) to meet the client's expectations.

### General diagnosis

In a typical diagnostic laboratory, most plant samples are diagnosed without performing all the steps prescribed in Koch's postulates. Rather, diagnosticians perform routine visual and microscopic examinations, analyse sample and field information, plate tissue on growth media, screen for potential pathogens and conclude a diagnosis. Because samples received in a clinic vary significantly in nature and importance, not all samples are diagnosed the same way. For example, a sample with an abiotic disorder is diagnosed differently than a sample with an infectious disease. A regulatory sample is diagnosed in a more in-depth manner than, say, a house insect specimen a homeowner is curious about. However, each sample has its own purpose and specific diagnostic flow determined by a diagnostician.

A general diagnosis by an experienced plant diagnostician can achieve a high level of reliability and is usually accepted by growers and clients. However, in the case of plant samples submitted in poor quality and/or without providing complete or relevant information, a diagnosis can only be indicative. In this situation, diagnosticians usually find only partial evidence of one or more causal agents. Sometimes, when major components of causative agents cannot be found from a sample or field, only an exclusion diagnosis can be made. By excluding some diseases based on the available information, growers can focus on other potential causes. To avoid a disease or problem being partially diagnosed, hemp growers should take several representative plants (not plant parts) with the root system intact. Additional information such as pictures of the disease pattern in the field and recent cultural practices are helpful for diagnosticians to determine if it belongs to an infectious disease or an abiotic disorder. A high-quality sample with more information always increases the chance of a timely, accurate and reliable diagnosis.

**Fig. 3.7.** An example of diagnosing an infectious plant disease using the four steps of Koch's postulates. Note that the leaf basal rot and leaf blight were observed on the diseased plants and also on the inoculated test plants. The complete diagnostic process confirms that *Phytophthora nicotianae* is a causative agent for the disease.

### Testing-only diagnosis

Many plant or soil samples in plant diagnostic labs only go through specific tests to find targeted pathogens. For example, samples for agricultural pest survey programmes are tested for some targeted pathogens using established testing protocols. This type of diagnosis involves science-based testing and has a high level of reliability. The diagnostic result is generally presented as positive/negative or present/absent. In most cases, samples submitted in this category are not for finding the cause of any diseases associated with the plants but, rather, screening for important pathogens.

### Identification-only diagnosis

A plant diagnostic lab receives many insect and mite specimens as well as mushroom and weed samples. For these samples, diagnosticians either conduct routine morphological examinations to determine the species or genus, or conduct further molecular analysis, depending on the objective of the specimen. When a taxonomic rank or a scientific name is determined, the identification is complete. Many diagnostic labs perform identification-only diagnosis for submitted organisms. However, if the organism is submitted because of concern about a problem in a crop, diagnosticians may perform a general diagnosis to determine if this organism is the cause of the problem.

### On-site field diagnosis

A field diagnosis is mainly performed by crop consultants, extension agents and arborists. Growers often conduct a preliminary diagnosis in the field when a disease first occurs in a crop. Some diseases can be correctly diagnosed in the field by observing disease patterns, assessing past and current cultural practices and examining plants closely for symptoms and pathogen signs. If a field assessment cannot determine the cause of a problem, representative samples are taken and submitted to a lab for further diagnosis. Field assessments provide valuable information on what a disease looks like and what factors may be involved. Data obtained from the field often complement lab-based diagnosis. For example, when a plant sample submitted is not sufficient for a diagnostician to determine the cause, a site visit may help towards a better understanding of the nature of the problem. This is especially true when a submitted sample does not contain the causative agents, or when a sample is irrelevant to the problem (Fig. 3.8). Although field diagnosis is commonly practised in plant health, the reliability

of a diagnosis based solely on field assessment may vary significantly from person to person.

## The Levels of Plant Diagnostics

Many plant problems can be diagnosed to meet different levels of satisfaction. Depending on the purpose of the plant samples and the nature of the problems, diagnostic levels vary from basic to more in-depth. Fig. 3.9 illustrates that a plant sample with leaf necrosis can be diagnosed at three levels. If the plant sample is submitted by a client who only wants to know if it is an abiotic or an infectious disease, the first two levels should serve the purpose very well. However, if there is a serious problem associated with a crop, the cause needs to be determined through a higher level of diagnosis. In this case, the specific pathogen or exact abiotic factor must be identified so preventive measures and treatments can be prescribed. As mentioned earlier, a full diagnosis is always needed for an unknown disease, but it is overkill for commonly known diseases. Similarly, identification of a pathogen to a race or pathovar level is needed for a grower who wants to know what resistant varieties should be used, but it may not be necessary for a client who just wants to know which fungicide can be applied. For example, a diagnosis of a high population level of soybean cyst nematode (*Heterodera glycines*) from a soybean crop may be sufficient for a grower who is choosing a nematicide treatment, but further diagnosis to a specific race is needed for a grower who prefers to use appropriate resistant varieties. The diagnosis of regulatory samples has much higher

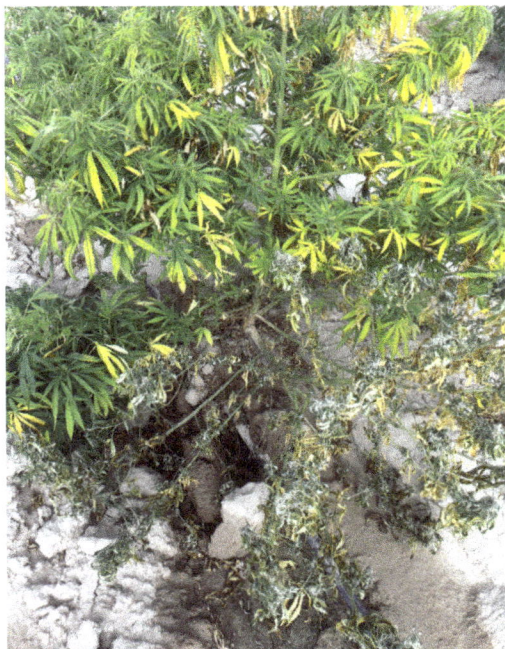

**Fig. 3.8.** A hemp crop infected by *Pythium* crown rot (*P. aphanidermatum*) showed obvious leaf yellowing but less noticeable crown rot in many plants. Note that a branch with severe crown rot was completely wilted while the rest of the plants were still alive, with yellow foliage. In this case, initial leaf or top branch samples submitted for lab diagnosis were irrelevant to the problem presented in the crop. A follow-up field visit found that the problem of leaf yellowing and plant wilting was due to a crown rot disease. Further field observations revealed that approximately 5–10% of the plants were affected by crown rot. Representative whole-plant samples were taken to the lab for further testing and the disease was confirmed to be *Pythium* crown rot.

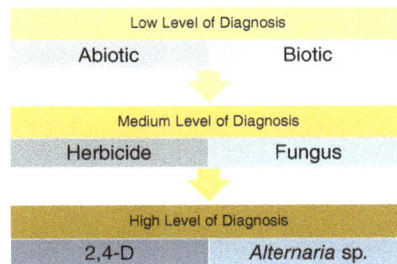

**Fig. 3.9.** Identifying the nature and cause of a disease to three satisfactory levels. Note that, in this illustration, a leaf sample exhibiting necrosis can be diagnosed to one of three levels in the scenario of an abiotic disorder or infectious disease, depending on the purpose of the sample.

standards than other samples in terms of regulatory consequences. Therefore, regulatory samples are frequently diagnosed at the level of DNA sequences or a specific biovar. For example, *Ralstonia solanacearum* race 3 biovar 2, a regulatory pathogen causing bacterial wilt, must be identified to be race 3 biovar 2 before a regulatory action is taken. For many routine plant or soil samples, a basic to medium level of diagnosis is sufficient and in most cases it is all the information a client wants to know. For example, a diagnosis of aphid infestation on a home-yard peach tree is acceptable to the homeowner without further identifying the aphid to a specific scientific name.

## Plant Diagnostic Flow

If one plant disease is defined as one plant species infected by one pathogen species, the number of infectious plant diseases is quite large. Shurtleff and Averre (1997a) calculated that the number of plant diseases would exceed 6.6 billion based on the 300,000 known vascular plants and 22,000 known plant pathogens. If including arthropod damage, abiotic disorders and genetic chimeras, the total number of plant problems is huge, but this calculation only reflects the maximum number of plant diseases that is possible if each and every plant species can be infected by all known pathogens. However, almost all species of plants are limited to being affected by only a number of infectious diseases and abiotic disorders, rarely exceeding 100 in total. For example, the tomato crop has approximately 65 infectious disease described and about 18 abiotic disorders identified (Jones *et al.*, 2014), suggesting the actual plant diseases are much less than the theoretical number. When a disease or problem does occur, there are many possible causes (Fig. 3.10) and narrowing a long list of possibilities to just a few is one of the greatest challenges for plant diagnosticians. Nevertheless, an overall process is established to make plant diagnostics more practical (Riley *et al.*, 2002).

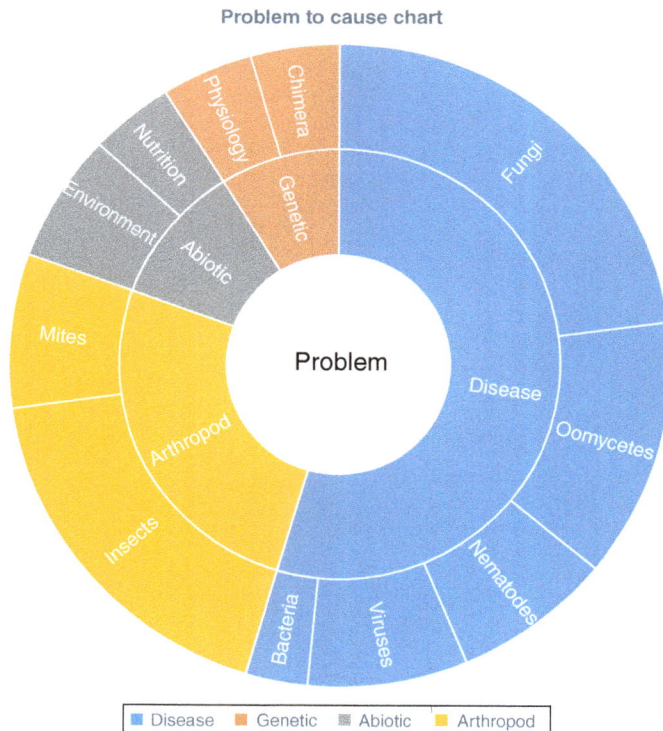

**Fig. 3.10.** Problem–cause chart of plant diagnostics. Note that diseases and arthropods are major causes of plant problems.

## Step 1: Identify plants

Each species of plant has only a limited number of diseases and insect pests because of the specificity established between the host plant and pathogens. For example, many fungal pathogens are highly host-specific, even though host jumps have occurred frequently (Borah *et al.*, 2018). Similar specificity exists in bacterial, viral and even nematode pathogens. By knowing the species of the plant, one can narrow an unknown number of diseases down to a short list based on diseases described from that specific plant species. Each species of plant has a unique Latin scientific name. For example, a hemp plant is named *Cannabis sativa*. However, often a plant species can have multiple common names, or a common name contains multiple species in the same genus (Brako *et al.*, 1995). For example, dogwood is a common name referring to a number of species in the *Cornus* genus such as *Cornus sanguinea* (common or blood-twig dogwood), *C. florida* (flowering dogwood), *C. nuttallii* (Pacific dogwood) and *C. kousa* (Kousa dogwood). Appropriately identifying the plant to a species is the first step of plant diagnosis.

Once the plant species is known, the diagnostician may need to obtain information on the variety or cultivar of a plant. Growers who submit a plant sample should provide variety information if available. Because some varieties are resistant to certain diseases, knowledge of a variety can rule out certain diseases to which it is resistant. On the other hand, some are very susceptible to certain diseases. For example, witches' broom disease is more prevalent in the hemp variety 'Cherry Blossom' (see Chapter 10) than in other varieties. With the information on varieties or cultivars, a trend of specific disease predominantly associated with specific varieties may be found and such trend can help future diagnosis and disease management.

There are many plant species and varieties and not everyone knows all the plants. Some ornamental plants have unique appearances that to some may look unhealthy. Therefore, it is very important to determine what a healthy plant in that species or cultivar normally looks like. Only when a normal plant is defined can a diagnosis of a sick plant be made. Some decorative plants have been developed to exhibit a uniform golden colour or certain colourful patterns on leaves for ornamental purpose. These 'symptoms' generally show up in all plants of the same cultivar, are normal and should not be confused with phytoplasma or viral infection. If these patterns are randomly distributed in a population of plants, or show up only in certain parts of the plant, a disease may be suspected.

## Step 2: Consider all possible agents

A common mistake often made by an inexperienced diagnostician is trying to find a pathogen from a symptomatic plant specimen without first considering all possible agents. As shown in Fig. 3.10, a plant problem can be caused by biotic (fungi, bacteria, viruses, nematodes and arthropods) and abiotic (genetics, nutrition and environmental conditions) factors. The approach of 'all things considered' should be used at this stage to determine if the case is an abiotic disorder or an infectious disease. Special attention should be paid to the current environmental conditions, such as dramatic weather changes, recent chemical sprays, fertilizer applications, irrigation changes, etc. For indoor cultivations, the history of fertilizer and pesticide use should be checked. Examining these factors helps to rule out some abiotic factors. Hemp growers and cannabis cultivators should also use this strategy to assess all possible abiotic factors and relay this information to the lab where samples are submitted. If abiotic factors are ruled out, possible infectious diseases and/or arthropod pests should be considered.

## Step 3: Consult the literature

There is a wealth of information regarding plant diseases on each plant species, available both online and offline. The most common resource is the disease and pest compendia published by the American Phytopathological Society (APS). Each compendium is a compilation of almost all known diseases affecting each crop or a group of related plant species. By reading the list of plant diseases, one can get an idea of what type of disease may fit the case. Another resource is websites that post plant pest and disease information. Please note that information on some sites is helpful but may not be peer-reviewed for accuracy. Professional web contents developed by academic institutions (an *.edu* domain), government agencies (a *.gov* domain), or non-profit organizations (an *.org* domain) are good online resources growers and plant diagnosticians can access instantly. One of the most valuable online resources for plant diagnosticians is the US Department of Agriculture (USDA) fungal database

(Farr and Rossman, 2019). This site can search most fungal pathogens reported from specific plant species worldwide and can help determine if a fungal disease is new on a plant species or in a geographical area. In addition, many books on plant diseases and pests can be referenced.

The purpose of consulting existing literatures is to learn what type of diseases and abiotic disorders have been described from the specific plant species one is dealing with and to get an idea of what type of problem the sample or the case may belong to. By comparing existing disease symptoms with some diseases described in the literature, one can narrow down to a very small list of possible diseases. For example, a hemp crop showing leaf mosaic and malformation without any necrotic or rotting symptoms on any part of the plant can be suspected for virus infection or chemical injuries. Many fungal and bacterial diseases can be ruled out based on their characteristics described in the literature. One common mistake when using literature information is trying to find images posted online or in the book that may look similar to the sample. Although it may work occasionally, comparing a sample to a picture without reading the full description of a disease can be misleading.

### Step 4: Define disease characteristics

Any disease or problem has its own unique signature that may include the timing of occurrence, distribution in the field, symptom progression, severity and its relationship to recent weather conditions, soil types and nutrients (Shurtleff and Averre, 1997a). In the initial diagnosis process, growers should observe the pattern and prevalence of a disease. If a large area of plants is affected, it is very likely to be an abiotic disorder, especially when it occurs within a few days. For example, a hemp crop damaged by cold weather exhibited a uniform discoloration and death of foliage on almost all plants within a week (Fig. 3.11). However, if a disease was scattered or localized in the field, there is a high chance that it was an infectious disease. For example, a hemp crop affected by phytoplasma witches' broom disease exhibited a scattered pattern of symptomatic plants (Fig. 3.12). In addition to the disease pattern in a field, the pattern of symptoms on a plant provides a clue to the cause. If a symptom occurs uniformly on certain parts of a plant, it is generally an abiotic issue. If the symptom is randomly distributed in a plant, a biotic disease is likely.

Symptom progress is another characteristic that can differentiate an abiotic disorder and a biotic disease. In general, a biotic disease progresses either slowly or rapidly, depending on the type of disease, and spreads to nearby plants. In the case of the hemp crop shown in Fig 3.12, phytoplasma infection spread gradually, causing more and more plants to became symptomatic during the season. Most soilborne diseases progress slowly and are often shown as a patch (see Fig. 3.4), while airborne and vectorborne diseases spread quickly. Remember, an infectious disease takes time to develop symptoms (see Fig. 2.5) and is generally observed as a trend of spreading from plant to plant. It rarely kills a plant overnight. In contrast, an abiotic disorder can occur suddenly and lacks symptom progression unless the problem is caused by nutritional issues. For example, hemp crop damage shown in Fig. 3.11 remained as it was and did not spread to other plants when the cold weather was over.

To better define disease characteristics, an experienced diagnostician always asks questions. The most frequent question would be: 'When was the problem noticed?' This question is to get an idea of whether the problem occurred suddenly or has occurred for a long time. If a problem occurs suddenly and very recently, abiotic factors need to be considered first before jumping into an infectious disease scenario. The follow-up questions are generally about cultural practices, chemical/pesticide usage, recent weather changes and any events associated with the crop. Once abiotic factors are ruled out, more in-depth questions may be asked. For example, is the problem staying as it is or progressing over time? How prevalent is the problem in the field? How old are the affected plants? What percentage of plants are affected? Is the crop grown from seeds, cuttings or commercial seedlings? Is the problem randomly distributed or clustered in patches? All these questions are helpful in determining the type of problem and possible causes.

Sometimes information obtained through phone conversations may not be enough. In this case, growers and diagnosticians may work together during a field visit to gather more information. In the field, cultural practices such as crop planting techniques, fertilization and pesticide application schedules and irrigation types and frequencies are reviewed to determine if any of these factors are related to the disease. Environmental conditions such as air temperatures, drought stress levels, soil type and conditions, weed conditions and presence

**Fig. 3.11.** Uniform foliage damage on a hemp crop caused by low temperatures. Note that all plants exhibited similar symptoms of freeze damage.

or absence of insect vectors are also assessed to determine if any environmental factors contribute to disease development. The varieties and cropping history are checked to see if certain varieties are susceptible to the disease or if disease pressure from previous crops causes the current crop to fail. Another purpose of field visits is to determine whether symptoms occur on a single or multiple plant species. If more than one plant species is affected with similar symptoms, the problem is likely to be an abiotic disorder. In many cases, a field visit can often pinpoint the cause right away, after examining affected plants throughout the field. In Fig. 3.13, a diagnostician dug out a wilted cucumber plant and observed both root-knots caused by a species of root-knot nematode (*Meloidogyne* sp.) and the

severe root rot apparently caused by a *Pythium* species. By examining multiple wilted plants in several locations of the field, the diagnostician gained a solid idea on what might have caused the problem and took representative samples for further lab analysis. Without a field visit, a diagnostician might have found the problem difficult to diagnose based on scanty information and sometimes unrepresentative samples, such as several wilted leaflets.

### Step 5: Study symptoms and signs

A symptom is the visible reaction of the host to an infection of pathogens or non-infectious factors. It defines an abnormality of a plant that is generally

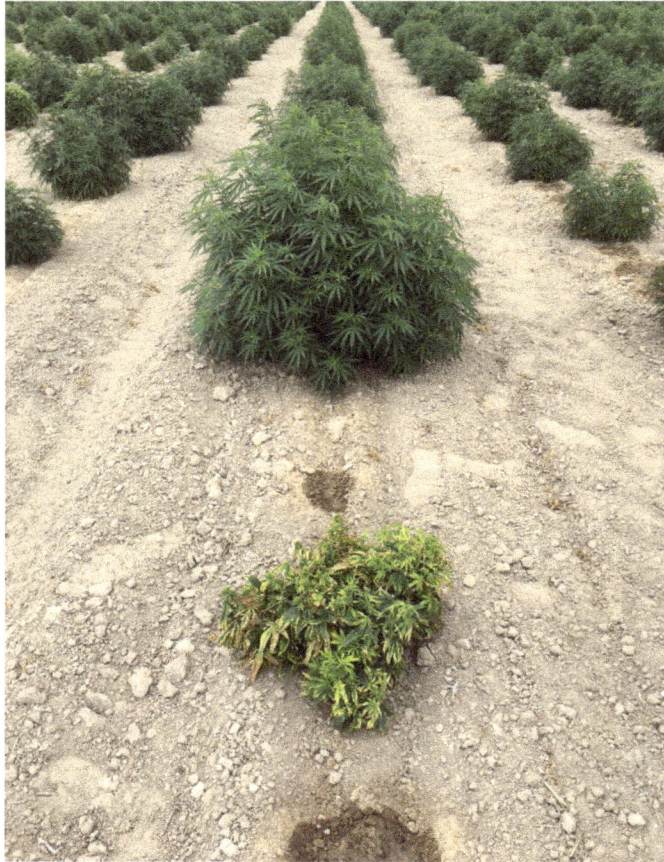

**Fig. 3.12.** A stunted plant randomly distributed in a hemp field. Note the only one symptomatic plant affected (in pictured area) by phytoplasma-caused witches' broom disease.

visually noticeable (Fig. 3.14). A 'sign' refers to the visual manifestation of the pathogen, such as fungal fruiting bodies and mycelium, bacteria ooze, cyst nematodes on the root, or parasitic plants (Fig. 3.15). Although 'sign' is used to describe pathogen appearance on a plant, the presence of insects or mites on plant tissue can also be a sign of infestation. Abiotic diseases do not have any pathogen or pest signs unless there is an infectious disease involved.

Not all infectious diseases show signs of causative organisms. Viral diseases do not have any visible signs, because viruses are submicroscopic and intracellular infectious parasites maintained only inside a living cell. Many fastidious bacteria, such as phytoplasmas, are obligate parasites and do not produce any identifiable signs on a plant. Nematodes do not have visible signs either, except cyst nematodes and some semi-endoparasites like citrus nematodes and reniform nematodes. Under a dissecting microscope, however, some nematodes, such as root-knot nematodes, stem nematodes and seed and leaf gall nematodes, can be directly observed without an extraction process. Many fungal pathogens produce characteristic structures on the surface of infected tissues. For example, rust fungi form raised pustules in which masses of coloured spores are produced. Pustules are visible because they are often located on leaf surfaces and rupture epidermal leaf tissue. Other fungal signs include pycnidia, mildews, sooty moulds and smut (soot-like

**Fig. 3.13.** Field examination of a wilted cucumber plant (*Cucumis sativus*) root system to detect potential causes of wilting. Note that roots had both root-knots and root rot and therefore root tissue and rhizosphere soil were taken for further lab analysis.

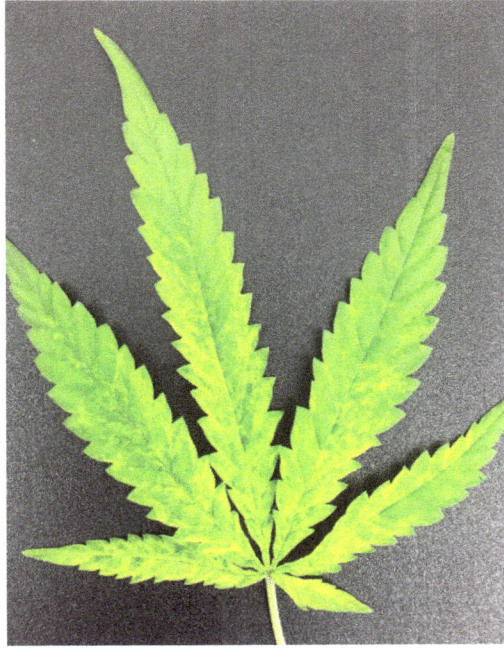

**Fig. 3.14.** A *Cannabis* leaflet showing the symptom of chlorosis. Note that only the symptom is visible without any sign of a pathogen or pest.

spore masses) (see Plate 2.2). Unlike fungal signs, bacterial signs are often presented as slimy and oozing appearance and/or a strong odour. A common method for observing bacterial signs is the bacterial streaming procedure (Fig. 3.16). By cutting a tiny piece of infected plant tissue and placing it in a drop of water, a flow of a large number of bacterial cells from the tissue to the water can be observed under a compound microscope.

Pathogen signs play important roles in plant disease diagnosis. Often, positive observations of a fungal or nematode sign can quickly determine the cause. Unlike symptoms, which require much more speculation on the potential causes, the presence of pathogen signs explicitly shows an association between the organism and the disease. Therefore, searching for pathogen signs from diseased plants is a vital part of visual and microscopic examinations.

Plant disease symptoms can be classified into five groups: (i) underdevelopment; (ii) overdevelopment; (iii) necrosis, rot, or death; (iv) alteration of normal appearance; and (v) wilting. Each group consists of different types of symptoms that may show up in certain parts of a plant. Almost all parts of plants can be infected by certain diseases. For hemp and cannabis plants, plant parts from the root, crown, stem, branches and leaves to flowers and seeds can be affected by pathogens, arthropods and abiotic agents. However, from the diagnostic perspective, symptomatic tissue may not contain causative agents. Therefore, identifying the source of a symptom within a plant system is more meaningful than just observing the symptom. For example, many foliage symptoms are a result of root damage. Obsessively studying a leaf sample without considering root diseases may result in a diagnosis frustration.

### Overdevelopment

Overdevelopment is characterized by the overgrowth of a part of a plant. This may include stem or crown galls, burls, or fasciation. The cause of overgrowth can be biotic, abiotic, or physiological.

**Fig. 3.15.** A hemp plant exhibiting basal stem rot caused by dual infections of *Fusarium oxysporum* and *F. solani*. Note that stem rot is the symptom of the disease and the visible white mycelia of *Fusarium* on the surface of the affected stem are the sign of the pathogens.

**Fig. 3.16.** Bacterial streaming observed from the stem tissue of a hemp plant infected by an *Enterobacter* species. Note that there were three points from which the stream originated (two on the top and one on the right).

For example, a portion of oak stem can overgrow into a giant gall that is five times larger in diameter than a normal stem (Fig. 3.17). In cannabis plants, the stem can grow up to several times larger in diameter than a normal stem under a certain light condition (Fig. 3.18). Leaf or branch proliferation (witches' broom) and hairy roots are other types of overdevelopment. They are often stimulated by certain types of pathogens. In hemp crops, a phytoplasma infection stimulates excessive leaf budding and proliferation (Fig. 3.19) (Schoener and Wang, 2019). The bacterium, *Agrobacterium rhizogenes*, can induce hairy roots in hemp plants under experimental conditions (Wahby *et al.*, 2013).

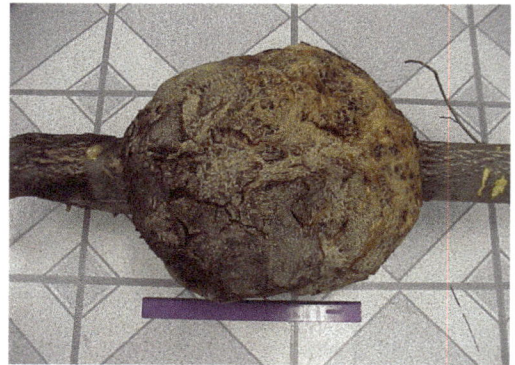

**Fig. 3.17.** A giant gall over 12 inches (30 cm) in diameter on an oak tree branch. Note that the gall is solid wood tissue that overgrew due to an unidentified factor.

**Fig. 3.18.** An enlarged stem of a cannabis plant caused by changing light conditions. Note that the tissue of enlarged stem is healthy.

### Underdevelopment

Underdevelopment refers to a plant less developed than normal. An underdeveloped plant usually exhibits symptoms of stunting, small leaves, shortened internodes and abnormally small root system. An underdeveloped plant may also have inadequate chlorophyll production and exhibit foliage yellowing or chlorosis. Underdevelopment is commonly caused by nutrient deficiency and environmental stress, but it can also be caused by various pathogens, including nematodes, viruses, phytoplasmas and certain bacterial/fungal pathogens. Since the symptom is systemic rather than localized in a certain part of a plant, an accurate diagnosis requires a comprehensive investigation of potential causes, both biotic and abiotic. In hemp crops, underdevelopment of a plant often shows up as stunting and leaf yellowing (Fig. 3.20) and the cause is generally associated with pathogens.

### Tissue necrosis

Necrosis is defined as cell death in plant tissue. This is generally caused by environmental stress, pathogen infection, chemical burn or physical impact. This type of cell death is considered to be traumatic due to external factors. In living multicellular organisms, there is another type of cell death called apoptosis. Apoptosis is a naturally occurring, highly regulated or programmed cell death by the organism. It is normal for an organism to kill certain cells in response to cell stress or signals from other cells.

However, excessive apoptosis can cause abnormal development while insufficient apoptosis causes cell proliferation, such as cancer. In plants, tissue necrosis is a consequence of collective cell death caused mostly by external stimuli or factors. The symptoms of necrosis may show up as leaf spot, blight, dieback, rot, canker, vascular streaking, or discoloration.

SPOTS AND LESIONS  Spots and lesions are visible signs of tissue necrosis in leaves, stem, fruits, or roots. Most leaf spots are brown, but they can be black, red, or other colours, depending on the plant species and the cause. Spots can be caused by some species of fungi, bacteria and viruses. For example, a hemp plant infected by the fungus *Alternaria* sp. exhibits extensive brown leaf spots (Fig. 3.21). Diagnosing spots may be difficult, because a lot of abiotic factors contribute to spotting. Abiotic spotting is generally non-specific and its diagnosis should focus on external factors rather than the tissue itself. For pathogen-caused spots, a pathogen sign may be visible from the tissue or can be isolated from the marginal tissue of a spot.

BLIGHT  Blight is the rapid death of a part of or the whole plant and it is used to describe some diseases such as leaf blight (Fig. 3.22), needle blight and fire blight. Unlike localized spots, blight occurs in a large area of a plant in a relatively short period of time. A blight disease can destroy an entire crop. For example, late blight of potato, caused by *Phytophthora infestans*, once destroyed many potato crops and led to the Great Irish Famine

**Fig. 3.19.** Excessive branching of young leaves from a single bud of a hemp plant showing witches' broom. Note that each individual leaf is underdeveloped and the number of leaves is excessive.

**Fig. 3.20.** An underdeveloped hemp plant in a field. Note that the plant exhibits severe stunting (short internodes), reduced leaf size and yellowing.

**Fig. 3.21.** A hemp plant leaflet exhibiting brown spots of varying shape and size. Note that some lesions are coalesced to form a big spot.

from 1845 to 1849. Another notorious blight disease is the chestnut blight that destroyed many mature American chestnuts in North America. Both abiotic and biotic factors cause blight, so diagnosing a blight condition is complex. However, most blight diseases are caused by fungi, bacteria, or oomycetes. Environmental stressors such as high salinity in soil and drought can cause needle blight in some conifer trees.

**DIEBACK** Dieback is the progressive death of shoots or branches starting from the top and moving down to the lower portion of a plant or tree. Initial dieback may begin with the death of the tip tissue of twigs, branches, or shoots. Dieback occurs on roots too. In this case, tissue death starts at the root tip and gradually moves toward the main roots. However, dieback is mostly used to

**Fig. 3.22.** Turfgrass leaf blight caused by *Ascochyta* sp. Note that whole grass plants were killed by the fungus.

describe branch dieback in woody plants. In woody plants, dieback is generally caused by stem canker diseases, borer insect damage, root problems and abiotic factors such as pesticide misuse. In annual crops, dieback can be caused by fungal, bacterial, or even nematode pathogens and sometimes by sudden harsh environmental conditions or herbicide damage (Fig. 3.23). Dieback is a term with the emphasis more on the progression of tissue death than a unique symptom, so its diagnosis requires a more comprehensive investigation. Often, the part of plant with dieback may not contain causative agents, because dieback is due to a condition in other parts of the plant.

**CANKER** Canker refers to a localized area where the plant tissue is necrotic or completely dead. Cankered tissue is usually discoloured, sometimes raised or sunken, and often dry or hard. It differs from rot by having a definite line of demarcation. Canker occurs mostly on stems, branches and twigs and is surrounded by living tissues. Both woody and herbaceous plants can be affected by canker diseases. A cankered area may expand into a large area or girdle a stem, causing wilt or death of the whole plant. Similarly, a tree branch girdled by a canker results in death of that branch. There are generally three types of canker diseases. The most common type is dry canker, where plant tissues are dead but firm (Fig. 3.24). The second type is called bleeding canker, where dark brown sap oozes out from the surface of a cankered area (Fig. 3.25). The third type is sooty canker, where black masses of fungal spores are present inside the bark or between the bark and sapwood (Fig. 3.26). Both bleeding canker and sooty canker occur on fruit, shade, or forest trees but are rare on herbaceous crops. Canker diseases are mostly caused by pathogens. However, in perennials or woody trees, environmental stresses play a predisposition role in the development of canker diseases, especially those caused by weak or opportunistic pathogens. In *Cannabis* plants, it is common to see stem canker diseases caused by fungal pathogens (see Chapter 8).

**ROT** Rot is the process of tissue deterioration, disintegration or decaying mostly via the actions of microorganisms such as bacteria, fungi and nematodes. The most common rots seen in plants are bud rot, fruit rot, crown rot, stem rot and root rot. Rot is generally caused by fungal or bacterial pathogens. Oomycetes such *Phytophthora* and

**Fig. 3.23.** Onion seedling dieback due to herbicide injury. Note that death of leaf tip tissue occurs first and then progresses down to the remaining parts of the leaf.

**Fig. 3.24.** A large cankered area on the stem of an azalea plant (*Rhododendron* sp.) caused by *Phytophthora citricola*. Note that there was an obvious line between the cankered and healthy areas. The cankered area was discoloured (brown) and firm.

*Pythium* are also common pathogens causing root and crown rot in many plant species. Sometimes, environmental conditions such as overwatering may cause root rot without the involvement of a primary pathogen. Once rot is initiated, disintegration and decaying may proceed rapidly to the entire organ of a plant. Unlike canker and lesion symptoms, rot moves to fresh tissue without a definite line between rotted and healthy tissue. For example, a hemp basal stem rot caused by a bacterial infection progressively moves up with only a transitional window from dark brown to light brown (Fig. 3.27). In this case, bacteria attack fresh tissue using

**Fig. 3.25.** A bleeding canker on a branch of a silver maple tree (*Acer saccharinum*) caused by *Phytophthora cactorum*. Note that two pieces of outer bark were removed to reveal a healthy (upper) and a cankered (lower) area underneath and that there was a clear margin between healthy tissue and the canker. In this case, canker occurred inside bark tissue with brown colour and the dark brown sap oozed on the surface. For a lab diagnosis, sample tissue should be taken from the marginal area of canker for isolation of a pathogenic organism.

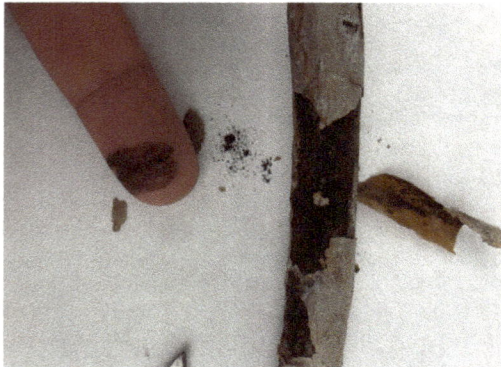

**Fig. 3.26.** Sooty canker on an ash tree (*Fraxinus* sp.) caused by the fungus *Nattrassia mangiferae*. Note that a mass of black fungal spores was produced underneath the cankered bark and would easily stain a finger with a light touch (left).

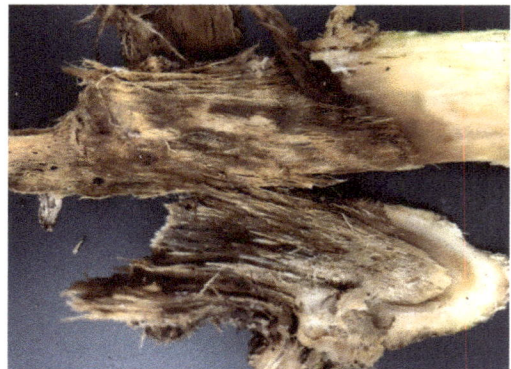

**Fig. 3.27.** Hemp basal stem rot caused by an *Enterobacter* species. Note that the tissue with a dark brown colour was completely rotted.

their secreted enzymes to destroy tissue integrity. Depending on the plant tissue, rot caused by bacteria can become soft, wet, slimy, or smelly. In contrast, rot caused by fungi generally lacks a strong odour and slimy appearance unless a secondary bacterial infection is involved. In the case of an oomycete infection, rot can become mouldy, as mycelia of *Phytophthora* or *Pythium* often grow out from the rotted tissue.

### Wilting

Wilting refers to the loss of rigidity of non-woody parts of plants, mostly exhibited on leaves, flowers, shoots and young seedlings. It is easily recognized in herbaceous crops and annual plants. In some woody plant species, however, wilting can appear as desiccated foliage with intact rigidity. For example, when the Canary Island date palm is infected by *Fusarium oxysporum* f. sp. *canariensis*, it does not exhibit a true wilting but rather a silver-coloured desiccation of fronds. Plant wilting can be caused by a lack of water, damage to the root system or stems and fungal and bacterial pathogens. When wilting is caused by water deficit, it can be a temporary symptom where the plant can recover if sufficient water is provided in time. This is called temporary wilting. In contrast, permanent wilting is a condition where a plant cannot recover and the damage to the plant is permanent (Fig. 3.28). Plant wilts caused by pathogens such as *Fusarium*, *Verticillium* and some bacteria are generally lethal and irreversible. These pathogens kill vascular tissue responsible for water transportation from the roots to foliage. Some nematodes also cause wilt disease, for example pine wilt caused by pine wood nematode (*Bursaphelenchus xylophilus*). Plants infected by root-knot nematodes may exhibit temporary wilting in the early afternoon during the summer but will recover the following morning when the temperature cools down. In summary, wilting can be caused simply by a lack of water but is also often caused by the destruction of the plant vascular system or root tissue due to pathogen invasions.

### Alteration of appearance

A plant infected by a pathogen or affected by an environmental agent may exhibit a different appearance from what is considered normal. In this case, the appearance of a plant or its parts is altered but there is no tissue necrosis or cell death. The common causes of alteration of plant appearance are virus infections and exposures to growth-regulator herbicides. When a plant is infected by a virus, it may exhibit mosaic (see Fig. 2.33), rings, ringspots, chlorosis, curling (see Fig. 2.3) and other abnormal appearances on the leaves and fruits. Similarly, when a plant is exposed to a post-emergence broadleaf herbicide, its leaves may become twisted, downward cupping, narrow and needle-like, or fan-like (Fig. 3.29). Diagnosing abnormal appearances of a plant requires an initial field assessment of possible virus/viroid infections and potential association with herbicide use, followed by a lab testing to confirm the suspected cause.

### Symptom complexity and disease complex

Sometimes a plant may exhibit multiple symptoms caused by one or more pathogens and/or abiotic factors. When diagnosing a disease complex, additional attention should be paid to which condition appears to be the primary issue, which one appears to be secondary, and whether both conditions work synergistically. Also, symptom expression is complex as it may be affected by either environmental conditions or the pathogen's virulence. For example, high temperatures may subdue leaf mosaic caused by viruses. More commonly, some symptoms observed in a plant are a result of a condition in another part of the plant. For instance, a local problem (e.g. root decay) can lead to foliage blight and even the death of the whole plant. Because all organs in a plant are interconnected, a localized infection can result in systemic symptoms. Being aware of and practising the 'one organ affects another' theory can help identify the origin of a problem.

### Step 6: Laboratory examination and analysis

After field investigations and symptom observations, representative samples should be further examined and tested to confirm the initial diagnosis. In a plant diagnostic lab, many procedures performed are classified as routine clinical observations (such as the visual and microscopic examination of fungal pathogens or insects). This type of observation is not a test by definition, but it is the most used diagnostic procedure in a plant clinic. Many plant problems are diagnosed quickly and correctly by microscopic examination only. However, in many cases, additional tests and procedures are needed to pinpoint

**Fig. 3.28.** A hemp plant exhibits wilting due to a water deficit. Note that some leaves have marginal scorch.

the exact cause. Table 3.1 lists a number of diagnostic tests commonly used in a plant clinic, covering routine microscopic observation, traditional culture methods and modern molecular diagnostic techniques. These methods can be classified into seven different approaches and each approach represents a

**Fig. 3.29.** An aspen tree injured by 2,4-D herbicide. Note that the appearance of leaves is altered into a fan-like shape.

different level of difficulty and complexity. Level 100 procedures are the most commonly performed by lab diagnosticians. They usually do not require sophisticated equipment. However, these procedures require that diagnosticians have basic knowledge in plant disease diagnostics. Level 200 procedures are routinely performed and require additional equipment such as biosafety cabinets, incubators and nematode extraction systems. Level 300 includes a group of serology-based tests for detecting specific pathogens. They are mostly used in virus, bacterial and oomycete detections. Level 400 encompasses a group of DNA-based methods used to detect a broad range of pathogens. They require a complete set of equipment and molecular biology skills. Procedures at Level 500 and beyond are generally applied to a full diagnosis, requiring both time and funding support. They are more research-oriented to detect an unknown transmissible agent in plants and demonstrate its pathogenicity. Although many diagnostic procedures and methods are available, not all of them are used in a single diagnostic case. Some cases may be diagnosed at Level 100, but some may require a combination of multiple higher levels of testing. The details of each diagnostic method are described in Chapter 5.

**Table 3.1.** Seven levels of methods and procedures commonly used in plant diagnostic labs.

| Procedure and method | Level code | Areas of diagnostics applied |
|---|---|---|
| Image observation | 100.01 | General diagnosis |
| Visual examination | 100.02 | General diagnosis |
| Microscopic examination | 100.03 | General diagnosis |
| Moist chamber pathogen induction | 100.04 | General diagnosis |
| All-purpose culture for fungi | 200.01 | Isolation of potential pathogens |
| Selective culture for fungi/oomycetes | 200.02 | Isolation of potential pathogens |
| All-purpose culture for bacteria | 200.03 | Isolation of potential pathogens |
| Selective culture for bacteria | 200.04 | Isolation of potential pathogens |
| Nematode extraction | 200.05 | Isolation of potential pathogens |
| Immunostrip test | 300.01 | Detection of potential pathogens |
| ELISA test | 300.02 | Detection of potential pathogens |
| Immunofluorescent labelling | 300.03 | Detection of potential pathogens |
| Biochemical analysis | 400.01 | Identification/detection of pathogens |
| Conventional PCR | 400.02 | Identification/detection of pathogens |
| Nested PCR | 400.03 | Identification/detection of pathogens |
| Real-time quantitative PCR | 400.04 | Identification/detection of pathogens |
| Reverse transcription PCR | 400.05 | Identification/detection of pathogens |
| Restriction polymorphism analysis | 400.06 | Identification/detection of pathogens |
| DNA barcoding – rDNA region | 500.01 | Universal/inclusive diagnosis |
| DNA barcoding – other genes | 500.02 | Universal/inclusive diagnosis |
| On-site investigation – field diagnosis only | 600.01 | Field diagnosis |
| On-site investigation plus lab testing | 600.02 | Comprehensive diagnosis |
| Bioassay – *in vitro* | 700.01 | Koch's postulates |
| Bioassay – greenhouse | 700.02 | Koch's postulates/virus detection |

Most plant diagnostic labs only perform testing or analysis to confirm if a pathogen or pest is a cause of a problem in a crop. For abiotic causes, a separate lab with different expertise should be sought to help diagnose abiotic causes. Soil and/or plant nutritional labs are most frequently used by growers for soil and nutrient testing. This type of lab can test soil pH level, perform complete soil analysis and determine plant nutrient contents. A chemistry lab has instruments to analyses various chemicals such as pesticide residues in plant tissues. Many abiotic problems can be visually diagnosed, but some complicated and high-consequence cases may require strict lab testing for confirmation.

### Step 7: Test pathogenicity

A pathogenicity study is to inoculate an organism into a group of plants of the same species where it is originally isolated from, and to determine whether the organism causes the same disease. This is a part of a full plant disease diagnosis under the rule of Koch's postulates, but many plant diseases can be diagnosed without performing a pathogenicity test. If an organism is identified from a known host, there is no need to perform pathogenicity studies. If the organism is isolated from a new host species or genus, or if the organism is a new species or genus, a complete study on its pathogenicity is needed to confirm its pathogen status. The pathogenicity test is a lengthy process and requires the same species of plants for testing. Under a routine diagnostic setting, it is not always achievable. More often, the plant materials needed (such as woody perennials and trees) may not be readily available for a pathogenicity test. For highly regulated plants such as hemp and cannabis, an inoculation test may require the regulatory approval to grow such testing plants. Also, some fastidious organisms such as phytoplasmas cannot be isolated and cultured, which makes it difficult to confirm the pathogenicity through Koch's postulates. In this case, if a phytoplasma is detected from a plant, it only indicates its association with an underlying disease condition.

### Step 8: Finalize diagnosis

All information obtained both in the field and laboratories from steps 1 to 7 needs to be complied with and further analysed, and the laboratory data need to be interpreted. With all the information in place, a conclusion may be made at this time as to whether the case is an abiotic or infectious disease. If it is an infectious disease, the causative organism should be identified. Similarly, if it is an abiotic disorder, a causative factor may be determined. The final diagnostic results and management recommendations should be provided in the report.

## The Power of Universal Diagnosis

### Limitations of specific detection

Specific detection of an organism is a part of plant disease diagnosis. Many labs perform specific tests to determine whether an organism of interest is present in a specimen. In some cases, a test can determine an organism to the subspecies levels such as races, strains, or pathovars. This high acuity of detection is often required for research projects or high-consequence regulatory samples. However, for cause-finding plant diagnostics, starting with a test for a specific organism may not be practical unless the organism has over a 90% chance of being the cause. In the past, researchers have developed methods for detecting an organism at the subspecies level. Although useful, these methods do not fit into diagnosing open-ended cases.

### Next-generation sequencing

Universal diagnosis has emerged in both medical and plant diagnostics. For example, whole-genome next-generation sequencing (WG-NGS) techniques are used to detect and identify non-targeted microbes without any *a priori* hypothesis and this is highly likely to become a standard procedure in microbiological diagnosis for human patients (Lecuit and Eloit, 2015). In plants, NGS technologies offer a generic approach (non-specific method) to virus discovery and identification that does not require any prior hypothesis on the targeted pathogens (Pecman *et al.*, 2017). Before NGS, specific methods such as serological or molecular tests were widely used in virus diagnostics, but they target only one or a few viral species and require prior knowledge of the pathogens being tested. Meanwhile, traditional non-specific approaches such as indicator test plants and electron microscopy only classify viruses to the genus level. These methods are not comparable to the NGS approach, where all viruses and viroids can be potentially detected and identified to a species or strain level. Universal detection with NGS has changed the game of virus/

viroid detection and many new viruses and viroids and new hosts have been reported using NGS. In hemp and cannabis crops, NGS will be a powerful technique in virus/viroid diagnosis, since there is not much prior knowledge available on the viruses and viroids associated with *Cannabis* plants at this time.

## DNA barcoding

DNA barcoding is another powerful technique for the molecular identification of organisms. Although this technique could be overshadowed in the future by the rapid development of next-generation sequencing (Taylor and Harris, 2012), its technical simplicity, diagnostic speed and independence in bioinformatics expertise make it a popular choice for species identification. The procedures of DNA barcoding include DNA extraction from an organism, PCR (polymerase chain reaction) to amplify a fragment of a target gene, DNA sequencing of the PCR product, comparison with DNA sequence data in GenBank (https://www.ncbi.nlm.nih.gov/) and identification of the organism based on sequence similarity data generated through the Basic Local Alignment Search Tool (BLAST) (https://blast.ncbi.nlm.nih.gov/Blast.cgi). GenBank is a publicly accessible online database for DNA/RNA/protein sequences of living organisms and is a great resource for finding all types of DNA sequences needed for organism identifications.

DNA barcoding is widely used in the identification of plants, fungi, oomycetes, bacteria (Lebonah *et al.*, 2014), nematodes, insects, mites, fish, birds and mammals. It can even be used to detect and identify a subset or genus of organisms from host plants. For instance, DNA barcoding can be used to detect whichever phytoplasma is present in a plant without knowing a particular phytoplasma in advance (Makarova *et al.*, 2012). DNA barcoding has several advantages over traditional identification techniques. First, no taxonomic expertise is required for identification. Anyone trained in molecular techniques can perform the task. Secondly, many organisms at the time of identification do not have complete morphology required for traditional morphology-based identification, but DNA barcoding can be performed without morphological data. Thirdly, DNA barcoding is a generic process that does not rely on knowing the organism first. For example, a fungal specimen can be identified using DNA barcoding without any prior knowledge of the fungus. Bacterial species and genera can also be identified

from isolated strains using DNA barcoding (Barghouthi, 2011). However, there are some limitations. For example, if an organism does not have DNA sequence data in GenBank, DNA barcoding will not be helpful. Sometimes, a pure organism is difficult to obtain from certain plant specimens so an organism-specific DNA may not be available. A major shortfall of DNA sequence-based identification is that some DNA sequence deposits may not be validated with their morphological data (Kang *et al.*, 2010). These shortfalls may lead to an incorrect interpretation of search results.

## Broad diagnostic target

There are many other universal diagnostic protocols developed for a broad range of pathogen detections. For example, the universal detection of all phytoplasmas using a pair of universal primers offers a significant advantage in diagnosing phytoplasma diseases (Gundersen and Lee, 1996). Similarly, phytophthora screening by either immunostrip, ELISA, or PCR (Bilodeau *et al.*, 2014) is an example of universal diagnosis where a given taxon of the organism can be targeted by single test. In some cases, universal diagnosis provides sufficient information for prescribing a treatment. For instance, a positive diagnosis of *Phytophthora* disease without knowing the specific species is sufficient to prescribe a chemical treatment. In human disease diagnosis, a broad diagnostic target for Zygomycota and Ascomycota is developed for detecting a broad range of pathogenic fungi, allowing for a quick but satisfactory diagnosis to justify anti-fungal drug treatment (Burnham-Marusich *et al.*, 2018).

## Sample Submission

Sampling is a crucial part of plant diagnosis. Incorrect samples often lead to an incorrect diagnosis. When a plant shows symptoms, a thorough examination and preliminary investigation should be conducted in the field to determine which parts of the plant are affected and what type of sample should be taken for a lab diagnosis. In some cases, this process may be complicated and growers may seek advices from a lab diagnostician. When taking samples, be sure to collect representative and problematic plant tissue and maintain the sample in excellent condition. Samples that are completely dead, dry, or irrelevant to the problem are not useful for a lab diagnosis.

## Timing of sampling and submission

Plant samples have a short shelf life, usually only staying fresh for a week in a refrigerator. Therefore, it is best to contact a diagnostic lab to determine when to collect and drop off the sample. The acceptable time frame from sampling to processing in the lab is less than a week. Sampling and processing within the same day is ideal. It is highly recommended to take samples on Mondays or Tuesdays so that samples can arrive at a lab on a workday. Avoid sending samples on Friday, as they may stay in the mail for too long. A fresh sample always has the most accurate symptoms, whereas a deteriorated sample may lose key characteristics of symptoms and ends up with an overgrowth of secondary microorganisms that mask the actual cause of the problem.

## What to take

Before sampling, check all plant parts for symptoms and make a preliminary diagnosis on which part of a plant should be sampled. If a source of infection cannot be identified, the whole plant with intact root system and rhizosphere soil should be taken and submitted. If the whole plant cannot be submitted, such as a large tree or a valuable plant, representative tissue samples from each affected area can be taken for a lab analysis. The sample size varies significantly from case to case, but in general, five to ten leaflets and/or three to five shoots or small branches representing mild to severe symptoms are sufficient for typical herbaceous plants. For seedling diseases, a couple of germination trays may be submitted. Lawn disease diagnosis requires a 6" (15 cm) diameter disc taken from the edge of the disease patch. For hemp and cannabis plants, the whole plant is preferred as many diseases are often associated with the root and crown even though symptoms mostly show up in the foliage. For nematode analysis, soil samples should be taken and submitted.

## Sample storage

All plant tissue samples, once cut off from a plant, should be immediately placed in a sealable plastic bag to prevent the sample from drying out. The sample should then be stored in an ice chest or cooler. Avoid exposing samples to direct sunlight during the summer. Do not put anything else inside the bag that may increase or decrease normal sample moisture. Avoid adding water to the sample bag, as it may cause bacterial or fungal growth and deteriorate the sample quality during the transition period. If a sample cannot be submitted immediately, store it in a cool environment (10–15°C) or in a 4°C refrigerator before shipping. Never freeze plant tissue samples unless this is instructed by a lab diagnostician, as it may prevent the positive isolation of certain pathogens. Do not freeze soil samples intended for nematode analysis. Extraction and identification of nematodes from soil relies on live nematodes and a temperature below zero is likely to kill all nematodes.

## Additional information

Additional information such as disease photos, cultural practice history, pesticide use history and any other information not reflected in the sample submission form is useful. This is especially important when a problematic tissue or portion of the plant cannot be sampled and submitted. When taking photos, try to take some close-up pictures of the problematic area of the plant and also long-range images of the entire plant or a portion of the field to show the disease pattern. Additional written documents detailing observations of disease development in the field is also helpful.

## Common mistakes

Common mistakes in sample submissions include submitting dead samples without any live tissue, completely rotten tissue, dried branches, insufficient material such as a single needle or leaf, unrepresentative samples, tissue that is irrelevant to the problem and sometimes totally healthy tissue. All these mistakes can be avoided by systemically investigating a plant health problem in the field and performing a preliminary diagnosis before taking a sample. Sometimes, good samples are taken but packaged wrong, causing the sample quality to deteriorate. Remember, a diagnosis based on a sample can only be as good as the sample.

## References

Barghouthi, S.A. (2011) A Universal method for the identification of bacteria based on general PCR primers. *Indian Journal of Microbiology* 51, 430–444. doi: 10.1007/s12088-011-0122-5

Bilodeau, G.J., Martin, F.N., Coffey, M.D. and Blomquist, C.L. (2014) Development of a multiplex assay for genus- and species-specific detection of *Phytophthora* based on differences in mitochondrial gene order. *Phytopathology* 104, 733–748. doi: 10.1094/PHYTO-09-13-0263-R

Bomberger, R.A., Duranovic, X. and Wang, S. (2016) First report of basal petiole rot and leaf blight caused by *Phytophthora nicotianae* on *Chamaerops humilis* in North America. *Plant Disease* 100, 212. doi: 10.1094/PDIS-05-15-0582-PDN

Borah, N., Albarouki, E. and Schirawski, J. (2018) Comparative methods for molecular determination of host-specificity factors in plant-pathogenic fungi. *International Journal of Molecular Sciences* 19, 863. doi: 10.3390/ijms19030863

Brako, L., Rossman, A.Y. and Farr, D.F. (1995) *Scientific and Common Names of 7,000 Vascular Plants in the United States*. APS Press, St Paul, Minnesota.

Burnham-Marusich, A.R., Hubbard, B., Kvam, A.J., Gates-Hollingsworth, M., Green, H.R. *et al.* (2018) Conservation of mannan synthesis in fungi of the Zygomycota and Ascomycota reveals a broad diagnostic target. *mSphere*, 3, e00094-18. doi: 10.1128/mSphere.00094-18

Farr, D.F. and Rossman, A.Y. (2019) Fungal Databases, US National Fungus Collections, ARS, USDA. Available at https://nt.ars-grin.gov/fungaldatabases/ (accessed 29 December 2019).

Gundersen, D.E. and Lee, I.-M. (1996) Ultrasensitive detection of phytoplasmas by nested-PCR assays using two universal primer pairs. *Phytopathologia Mediterranea* 35, 144–151.

Jones, J.B., Zitter T.A., Momol, T.M. and Miller, S.A. (2014) *Compendium of Tomato Diseases and Pests*, 2nd edn. APS Press, St Paul, Minnesota.

Kang, S., Mansfield, M.A., Park, B., Geiser, D.M., Ivors, K.L. *et al.* (2010) The promise and pitfalls of sequence-based identification of plant-pathogenic fungi and oomycetes. *Phytopathology* 100, 732–7377. doi: 10.1094/PHYTO-100-8-0732

Lebonah, D.E., Dileep, A., Chandrasekhar, K., Sreevani, S., Sreedevi, B. *et al.* (2014) DNA barcoding on bacteria: A review. *Advances in Biology* 2014, 541787. doi: 10.1155/2014/541787

Lecuit, M. and Eloit, M. (2015) The potential of whole genome NGS for infectious disease diagnosis. *Expert Review of Molecular Diagnostics* 15, 1517–1519. doi: 10.1586/14737159.2015.1111140

Makarova, O., Contaldo, N., Paltrinieri, S., Kawube, G., Bertaccini, A. *et al.* (2012) DNA barcoding for identification of 'Candidatus Phytoplasmas' using a fragment of the elongation factor Tu gene. *PLoS ONE* 7, e52092. doi: 10.1371/journal.pone.0052092

Pecman, A., Kutnjak, D., Gutiérrez-Aguirre, I., Adams, I., Fox, A. *et al.* (2017) Next generation sequencing for detection and discovery of plant viruses and viroids: comparison of two approaches. *Frontiers in Microbiology* 8, 1998. doi: 10.3389/fmicb.2017.01998

Riley, M.B., Williamson, M.R. and Maloy, O. (2002) *Plant Disease Diagnosis*. Plant Health Instructor, Teaching Notes, American Phytopathological Society, St Paul, Minnesota. doi: 10.1094/PHI-I-2002-1021-01

Schoener, J.L. and Wang, S. (2019) First detection of a phytoplasma associated with witches' broom of industrial hemp in the United States. *(Abstr.) Phytopathology* 109(S2), 110. doi: 10.1094/PHYTO-109-10-S2.1

Schoener, J.L., Wilhelm R., Rawson, R., Schmitz, P. and Wang, S. (2017) Five *Fusarium* species associated with root rot and sudden death of industrial hemp in Nevada. *(Abstr.) Phytopathology* 107(S5), 98. doi: 10.1094/PHYTO-107-12-S5.1

Shurtleff, M.C. and Averre, C.W. (1997a) *The Plant Disease Clinic and Field Diagnosis of Abiotic Diseases*. The American Phytopathological Society, St Paul, Minnesota.

Shurtleff, M.C. and Averre, C.W. (1997b) *Glossary of Plant-Pathological Terms*. APS Press, St Paul, Minnesota.

Taylor, H.R. and Harris, W.E. (2012) An emergent science on the brink of irrelevance: a review of the past 8 years of DNA barcoding. *Molecular Ecology Resources* 12, 377–388. doi: 10.1111/j.1755-0998.2012.03119.x

Wahby, I., Caba, J.M. and Ligero, F. (2013) *Agrobacterium* infection of hemp (*Cannabis sativa* L.): establishment of hairy root cultures. *Journal of Plant Interactions* 8, 312–320. doi: 10.1080/17429145.2012.746399

# 4 Building a Diagnostic Laboratory

## The Purpose of a Diagnostic Laboratory

A diagnostic laboratory is an essential part of the clinical plant pathology profession. Many diseases require a complete laboratory analysis to conclude a diagnosis. A lab-based plant diagnosis is reliable and legally defendable, but a diagnosis based solely on visual examination of symptoms is more like an opinion or guess. Many agricultural crop diseases significantly impact crop yields and commodity quality and they need to be diagnosed rapidly and accurately. A well-equipped diagnostic laboratory staffed with skilled diagnosticians is a great asset to growers and anyone who needs clinical plant diagnostic services.

Hemp and cannabis are high-value cash crops but are also highly regulated. Due to internal or external rules, cultivators may find it difficult to ship cannabis specimens out-of-state for disease diagnosis. Some laboratories do not even have a permission to accept cannabis samples. Such restrictions cause frustrations and delay in cannabis plant health diagnosis. However, for some large cultivators with million-dollar investments, a self-built lab may fit their own diagnostic needs well. One of the significant advantages of a self-owned lab is the ability to develop an in-house disease monitoring and testing programme. For example, a cannabis cultivation facility may use its in-house lab to screen mother plants or cuttings for *Fusarium* pathogens, to monitor insect or mite infestations and to diagnose a number of diseases associated with cannabis plants. The benefit of such a lab is the convenience and readiness for growers to quickly monitor, detect and analyse common diseases without relying on external resources.

## Levels of Plant Diagnostic Labs

A plant diagnostic lab can range from a single-staff lab to a well-staffed advanced lab. It depends on clients' needs and the number of samples to be diagnosed. In general, plant diagnostic labs can be classified into three levels: basic, standard and advanced. The basic lab usually has one or two personnel and mostly relies on microscopic examinations of specimens to diagnose a problem and identify an organism. This type of lab serves homeowner clients and arborists well, as they prefer to obtain a general diagnosis quickly. A standard diagnostic lab employs a small team of diagnosticians with different expertise and is well-equipped to perform diverse testing and analyses. This type of lab performs microscopic examinations, microorganism cultures, serology-based tests and routine polymerase chain reaction (PCR) analyses. These services are provided to a broad range of clients, including homeowners, farmers, extension agents, arborists, pesticide applicators and regulatory agencies. An advanced diagnostic lab is staffed with a team of diagnosticians with expertise in general plant pathology, mycology, nematology, virology, bacteriology and entomology. Each diagnostician is responsible for a specific type of diseases or pests. Labs at this level generally have a spacious layout, with separate rooms for each function, and are well-equipped to perform a broad range of diagnostic techniques, including DNA sequence-based diagnostics. Advanced diagnostic labs can perform a full diagnosis fulfilling Koch's postulates and often conduct clinical research to address important diseases. Standard and advanced labs may implement their own lab quality management systems and some may obtain diagnostic lab accreditation such as the STAR-D accreditation administrated by the National Plant Diagnostic Network (NPDN, created in 2002 by USDA National Institute of Food and Agriculture). Lab personnel may also perform annual proficiency tests for certain regulatory pathogens. Table 4.1 summarizes common methods and procedures used in the three levels of diagnostic labs (also see Table 3.1). Note that, as levels increase, diagnostic processes are

DOI: 10.1079/9781789246070.0004

**Table 4.1.** Three levels of plant diagnostic labs and their mostly used diagnostic methods.

| Procedures and methods | Level Codes | Basic | Standard | Advanced |
|---|---|---|---|---|
| Image observations | 100.01 | ✓ | ✓ | ✓ |
| Visual examinations | 100.02 | ✓ | ✓ | ✓ |
| Microscopic examinations | 100.03 | ✓ | ✓ | ✓ |
| Moist chamber pathogen induction | 100.04 | ✓ | ✓ | ✓ |
| All-purpose culture for fungi | 200.01 | ✓ | ✓ | ✓ |
| Selective culture for fungi | 200.02 | ✓ | ✓ | ✓ |
| All-purpose culture for bacteria | 200.03 | ✓ | ✓ | ✓ |
| Selective culture for bacteria | 200.04 | ✓ | ✓ | ✓ |
| Nematode extraction | 200.05 | ✓ | ✓ | ✓ |
| Immunostrip test | 300.01 | | ✓ | ✓ |
| ELISA test | 300.02 | | ✓ | ✓ |
| Immunofluorescent labeling | 300.03 | | ✓ | ✓ |
| Biochemical analysis | 400.01 | | ✓ | ✓ |
| Conventional PCR | 400.02 | | ✓ | ✓ |
| Nested PCR | 400.03 | | ✓ | ✓ |
| Real-time quantitative PCR | 400.04 | | | ✓ |
| Reverse transcription PCR | 400.05 | | | ✓ |
| Restriction polymorphism analysis | 400.06 | | | ✓ |
| DNA barcoding – rDNA region | 500.01 | | | ✓ |
| DNA barcoding – other genes | 500.02 | | | ✓ |
| Next-generation sequencing | 500.03 | | | ✓ |
| On-site investigation – field diagnosis | 600.01 | | | ✓ |
| On-site investigation plus lab testing | 600.02 | | | ✓ |
| Bioassay – *in vitro* | 700.01 | | | ✓ |
| Bioassay – Greenhouse | 700.02 | | | ✓ |

gradually shifting from observation-based diagnosis to more research-oriented diagnosis.

## Lab Layout and Functional Areas

A typical diagnostic lab has several functional areas. These include an area for sample receiving and visual observations, a cold room for sample and media storage, an area for large pieces of equipment, a preparation room, a room for microorganism isolation and culture, an area for microscopic examinations, a pre-PCR room and a major working area containing several working benches. These areas can be contained within a single large room, but it is best to have each functional area in a separate room.

### Sample receiving and observation table

A stainless-steel table (8'L × 4'W × 3'H, i.e. 2.4 m × 1.2 m × 0.9 m) can serve as a working area to assess received samples and further dissect tissues from the original specimens (Fig. 4.1). The table should be placed in an area close to a sink equipped with tap water and deionized water, as plant samples with root systems often need to be washed before observing or plating. To avoid sample traffic inside a lab, it is ideal to have the table near the cold room where samples are stored. A regular lab bench is not large enough for this purpose, especially when a large hemp plant or a tree branch is brought in for diagnosis. The table surface should be sterilized with 70% alcohol solution after each sample is processed. On and near the table, commonly used tools and forms should be available. These include sample processing forms, a date stamp, knives, scissors, axe, chopping board, stapler, Ziploc (zippered self-sealing plastic) bags, razor blades, spray bottles, Kimwipes, parafilm, regular or quarter-divided Petri dishes, pens, sticky notes and other stationery.

### Walk-in cold storage room

A cold room is an essential for a plant diagnostic lab. Without it, multiple refrigerators may be needed

**Fig. 4.1.** A stainless-steel table serving a sample receiving and observation area. Note that a series of shelves are provided for better organization. This table can also serve as a dissection table, especially when a large plant needs to be cut to check for vascular discoloration or a tree sample needs to be examined for borer insects. A large table also makes sample photography easy (usually with a set of different-coloured boards used as backgrounds).

for sample storage, but some large specimens do not fit. A cold room is also useful for performing certain lab procedures that require a 4°C environment. A standard cold room is equipped with 120V/240 outlets to run small equipment, and with a sink and multiple layers of shelves. In Fig. 4.2, the cold room provides ample space for storing samples of various sizes, culture media, seeds and other items requiring a low temperature. Under normal operating conditions, a walk-in cold room controls the temperature and humidity at the level set by the user, but it requires regular monitoring and maintenance.

### Equipment area

Many essential pieces of equipment are noisy and take up space. An area designated for such equipment

makes a diagnostic lab more organized and keeps major working areas quiet. Common stand-alone equipment includes refrigerators, −20°C and −80°C freezers, various incubators, ovens, growth chambers, germinators, ice machine and autoclaves. Some instruments make noises and also generate heat while operating and should be hosted in a well-vented room. Certain equipment should be placed in areas of the lab that best fit their functions. For example, tabletop microcentrifuges, the ELISA plate washer and reader, the Nanodrop (spectrophotometer) and water bath should be placed in appropriate functional areas of the main working room. The gel documentation system and ultraviolet (UV) illuminator box are used for DNA visualization in agarose gel and thus should be in a designated post-PCR area.

### Preparation room

A plant diagnostic lab should have a designated room for preparing media, making solutions and buffers and storing chemicals. The room should be equipped with cabinets and working benches. Some key equipment may be hosted in this room for convenience, such as ice-machine, stand-alone autoclave, nanopore water filter system, microwave, juice press, fume hood, freezer, stirrer and pH meter.

### Microorganism isolation, culture and examination room

Plating plant tissue on a culture medium to isolate pathogens from a sample is a routine procedure in plant diagnostics. A separate area should be designated for this procedure. If a room is designated, cabinets and standard bench tops should be provided for supply storage and working space. A biosafety cabinet should be installed in this room, along with various incubators suitable for fungal and bacterial culture. Culture plates, after plating and culturing, are often observed under a microscope to identify the organisms. For convenience, a set of dissecting and compound microscopes should be hosted in this area to avoid unnecessary traffic (Fig. 4.3). Most labs have a computer and digital camera system connected to microscopes that facilitate picture taking when observing a specimen.

### Electrophoresis and gel extraction area

If conventional PCR tests are routinely run in a lab, a working area should be designated for DNA gel

**Fig. 4.2.** An example of a walk-in cold room serving as a storage space for plant samples and temperature-sensitive laboratory materials. Note that multiple shelves offer better organization of samples in different diagnostic categories such as disease diagnosis, insect identification, weed identification and pesticide analysis.

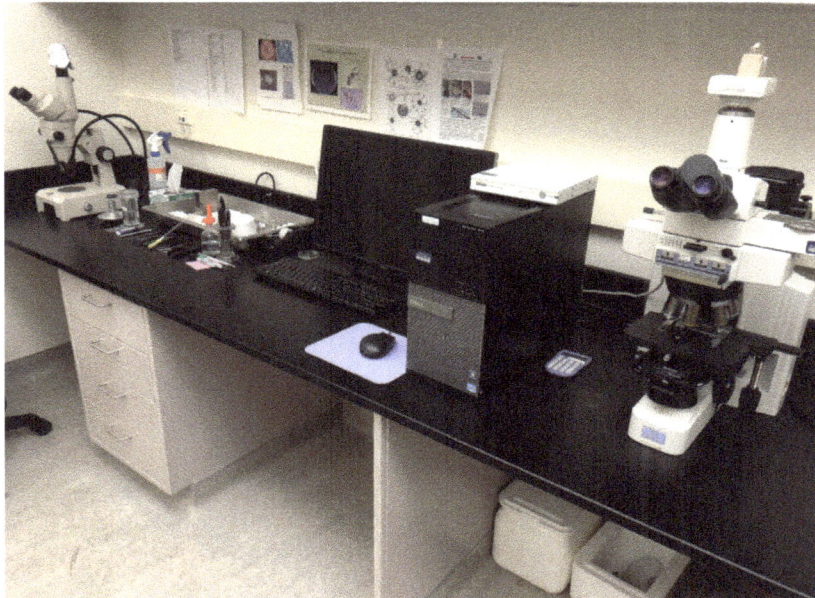

**Fig. 4.3.** An example of a setting for microscopic examination. Note that both dissecting microscope and compound microscope can be connected to a computer for image capture. Necessary supplies are in place to assist microscopic examinations.

electrophoresis and gel cutting. Since the running tris-acetate-EDTA (TAE) buffer contains ethidium bromide (abbreviated as EtBr or EB), a strong mutagen, working surfaces should be clearly marked as potential EtBr-contaminated areas for caution. If any spills occur, clean the area with activated charcoal followed by a regular detergent. Gel cutting should be performed in the same area to keep the rest of lab free from EtBr contamination. When cutting DNA bands from a gel, always do so in a designated area followed by cleaning. All used agarose gel should be disposed of properly following the institution's guide. All used EB-containing TAE buffer should be decontaminated before disposal (see Chapter 5).

Ethidium bromide is an intercalating agent and binds to DNA via intercalation between the base pairs (Sigmon and Larcom, 1996). When the gel stained with EtBr is under a UV light source, the DNA bands can be visualized as orange fluorescence. In some labs, EtBr is added into the running buffer directly during electrophoresis so that the pattern of DNA bands can be seen under a UV light immediately after electrophoresis. In others, an EtBr solution is used to stain the gel after electrophoresis. EtBr is widely used for nucleic acid stains in many molecular biology labs, but it is notoriously toxic and mutagenic. Because of it, a number of alternative products are available for use in DNA visualization. However, when handling DNA-binding chemicals, care should be always taken to avoid direct contact with skin or eyes. An emergency shower/eye washer should be available for use inside the lab in case any spill or contact occurs.

The electrophoresis area should have enough bench space to run multiple gels at the same time and an ample amount of drawers to store all accessories associated with electrophoresis, such as combs, plates, levels, trays and tools (Fig. 4.4). A UV box should be nearby and UV-protective eyewear should be available. A container storing TAE buffer and a container for storing used TAE buffer should be in place to avoid unnecessary transportation, especially when the buffer contains EtBr. TAE buffer can be reused for non-critical sample runs, for example running a DNA sample to check the size or pattern.

In plant diagnostic labs, electrophoresis of PCR products is a routine procedure. Therefore, the working area and pipettes used for sample loading are exposed to PCR products. PCR product contamination, also called amplicon carryover contamination (Aslanzadeh, 2004), is a serious problem in labs and sometimes hard to eliminate. When a lab

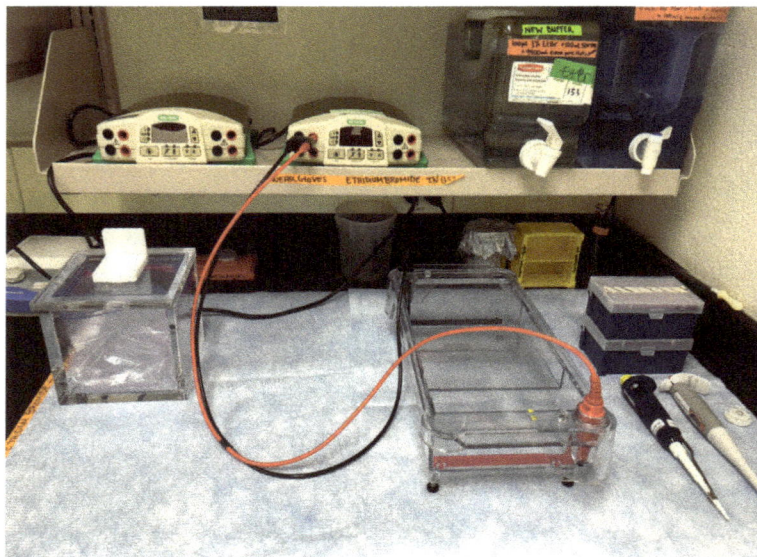

**Fig. 4.4.** An electrophoresis area with power supply and electrophoresis cells. Note that a designated pipette set should remain in this area and be used for gel loading only, to prevent post-PCR product contamination in the lab.

Chapter 4

repeatedly performs PCR tests using the same primer pair to detect an organism, amplification due to amplicon carryover contamination may occur in healthy or negative control samples. Therefore, a designated area for electrophoresis and gel extraction with scheduled cleaning to remove DNA and PCR products is one of the key considerations in planning a diagnostic lab layout. If possible, this post-PCR area should be completely separated with an area designated for performing pre-amplification procedures (pre-PCR area).

### Pre- and post-PCR areas

A molecular diagnostic lab routinely uses the PCR technique to amplify specific regions of microbial DNA or RNA to detect infectious agents. A typical PCR generates approximately one billion copies of the targeted sequence and the smallest aerosol contains about 1 million copies of the product (Persing, 1991). If uncontrolled, PCR products may accumulate quickly in the lab environment and will contaminate laboratory reagents, equipment and ventilation systems. To prevent amplicon carryover contamination, a lab should have both mechanical and chemical barriers. The most critical barrier is the strict separation of the pre-PCR area where samples and reagents are prepared and the post-PCR area where amplification products are analysed. These two areas can be located in two distant areas of the lab and, if possible, in two separate rooms. Between these two areas, all traffic (when performing PCR) should be unidirectional from the pre-PCR area to the post-PCR area and associated equipment, devices, laboratory coats, gloves, pipette sets, tips and reagents should be designated for use in each area. All wastes must be handled in place before disposal. Although physical separation offers an effective barrier between pre-PCR and post-PCR areas, diagnosticians should be constantly aware of where the template DNAs (the sample DNAs) and amplified DNAs are, and which items these DNAs are associated with during each step of PCR analysis.

Creating a chemical barrier is another step to minimize amplicon carryover contamination. PCR working stations, regular lab benches, racks, tuber holders and openers, trays and pipettes should be cleaned with 10% bleach solution (active ingredient: sodium hypochlorite) followed by 70% ethanol to remove bleach. Small items such as tuber racks and openers should be soaked in 10% bleach solution overnight after each use and then washed and rinsed with water. Bleach solution causes oxidative damage to DNA, preventing DNA from amplifying.

UV light irradiation was the first method used to eliminate amplicon carryover contamination (Sarkar and Sommer, 1990). UV light induces thymidine dimers and other covalent modifications in DNA, thus inhibiting the DNA from being used as a template for further amplification. UV light can be used to destroy any contaminating nucleic acids by irradiating a PCR reaction tube containing all the amplification reagents (buffer, primers, $MgCl_2$, dNTPs, enzyme) except sample DNA for 5–20 min (Aslanzadeh, 2004). However, this procedure is not commonly used in plant diagnostic labs. Rather, UV light irradiation is often used to sterilize areas such as the PCR workstation. When UV light is used for this purpose, be cautious of UV hazards and avoid exposure.

### Main working area

The main working area has multiple lab island benches with cabinetry. Depending on the number of lab staff, each diagnostician should have a lab bench as their own workstation to perform daily bench work. In each bench, a set of equipment and supplies are in place (Fig. 4.5), typically including a benchtop microcentrifuge, a vortex, a mini centrifuge for quick spins, a set of pipettes (1–2 µl, 2–20 µl, 20–200 µl, 200–1000 µl) with a set of pipette tips, a tip disposal box, a container for waste liquid, microcentrifuge tube racks, a tube opener, marker pens, a calculator and a timer. The shelves above the bench offer space for storing buffers and solutions, testing kits, homogenizer tubers, sterilized tube racks and pipette tips, and other supplies needed for the workstation. As mentioned earlier, the lab bench is not appropriate for handling raw plant materials such as a whole hemp plant or tree branch. The bench is solely dedicated to downstream analyses such as molecular testing. A large lab may have two to four working islands that provide four to eight benches for use by up to eight diagnosticians at the same time. As always, a trash can should be provided at each bench for quick disposal of waste. The main working area is the most frequently used space in a lab and usually handles a variety of activities. A clean and well-organized setting can substantially improve working efficiency and prevent contamination.

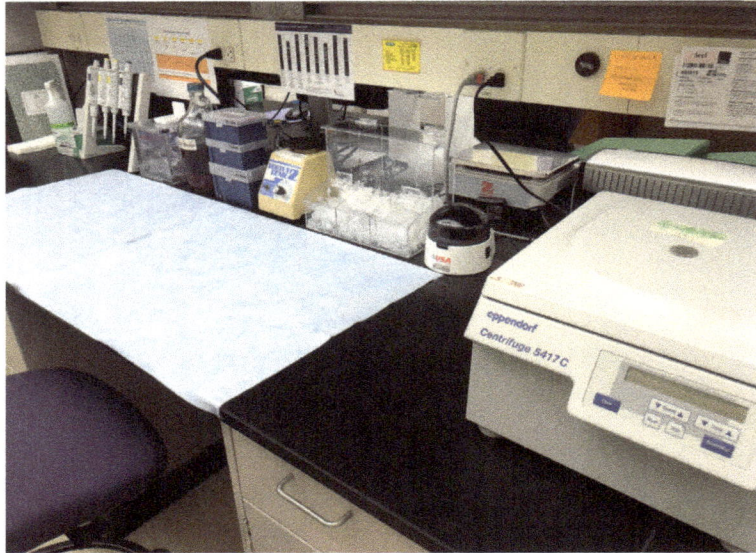

**Fig. 4.5.** An example of a lab bench serving as a diagnostician's workstation to perform a variety of diagnostic testing. Note that drawers are also available for storage of essential supplies used in this bench.

### Nematode extraction room

Extracting nematodes from soil and plant tissue is the first step to diagnosing nematode problems in plants. For labs handling a large amount of soil or odorous raw plant materials, a separate room may be designated for nematode extraction only. In some cases, a nematode extraction area is designated in a fully equipped nematology lab near the regular plant diagnostic lab. An ideal nematode extraction room should be equipped with a walk-in cold room for sample storage, a set of large stainless-steel sinks for wet-sieving nematodes from the soil, a set of mist chambers for extracting nematodes from plant tissue and a centrifuge with a rotor holding 28 × 50 ml centrifuge tubes (see Nematode Extraction System in this chapter). To help with sorting a high volume of samples, one or two large stainless-steel tables as described in Fig. 4.1 should be placed close to the wet-sieving sink. The tables will be most useful if placed in the centre of the extraction room. Multiple sets of sieves with three different diameters (7.62 cm, 15.24 cm and 20.32 cm) are required for wet sieving. Both a dissecting microscope (for cyst observation) and a compound microscope as well as associated supplies (such as slides, cover slips, dissecting tools, water bottle and glass dishes) should be in place to accommodate nematode identification after extraction. Glassware such as beakers of various sizes (though mostly 50 ml beakers), funnels, squeezable water bottles, trays and buckets should be available. To further clean nematode extracts, sugar flotation is often used. Therefore, granular sugar and a container for the sugar solution should be supplied. A microwave, stirrer and a large beaker (4000 ml) are often used to make the sugar solution. The layout of the nematode extraction room should reflect the typical workflow: take samples from the cold room, sort them on the table, proceed to wet-sieving or mist-chamber, centrifuge and then examine under a microscope.

### Diagnostic library and offices

A diagnostic library usually has a collection of literature on plant pathology, entomology, agronomy and molecular biology. The literature is a handy daily resource for diagnosticians when diagnosing a disease or identifying a pathogen or pest. All diagnostic literature should be placed in a designated area for easy access. Many diagnosticians use online resources to support diagnosis. A collection of folders containing important diagnostic literature or bookmarking all relevant webpages allows for quick and easy access to the information. A lab diagnostician should have an office that is usually outside or adjacent to the lab.

## Lab Equipment and Systems

There is a list of key instruments that are required for a plant diagnostic lab. However, most instruments are very expensive. The necessity of each piece of equipment depends on the diagnostic task and level of the lab. Advanced labs are generally equipped fully with all needed instruments. Certain tests, such as enzyme-linked immunosorbent assay (ELISA), require a set of instruments including a juice press machine, microplate washer and microplate photometer. For most tests, all critical instruments must be in place before the tests can be performed. Sometimes, a set of apparatus is built into a system for a specific diagnosis. For example, a misting system used in nematode extraction consists of multiple chambers with a controlled sprinkler system, a drain and racks with an array of Baermann funnel apparatus.

### Eyewash and safety shower station

The eyewash station is an essential device in a plant diagnostic laboratory. It should be located in an easily accessible place. Eyewash stations are used to flush the eyes if hazardous chemicals are splashed into the face. The stream of water from the spray should be directed into the eye for a period of 15 min before seeking medical attention (UOW WHS Unit, 2020). Safety showers are designed to wash an individual's head and body to remove hazardous chemicals that may have splashed onto the skin. They may be also used to wash away contaminants from contaminated clothing.

### Autoclaves

Autoclaves use elevated temperature and pressure to kill microorganisms associated with solids, liquids, or hollows, and autoclaving has become an essential process for decontamination and sterilization in the lab. A plant diagnostic lab must have an autoclave to sterilize media, solution, tools, tips and waste. There are two common types of autoclaves: stand-alone and tabletop. Some autoclaves are manually controlled and some are electronically controlled. Small labs can be equipped with a tabletop autoclave that will satisfy daily sterilization needs. Stand-alone autoclaves are used to sterilize large items or a high volume of medium.

### Ice maker

For molecular diagnosis, a flake ice maker is needed to provide ice. Ice is used to cool down tubes and reagents and keep them at 4°C or below. Many procedures in molecular diagnosis require reagents to stay in ice to keep them cold. The purpose of cold temperature is to maintain the stability of reagents and prevent them from degrading at room temperature. The common temperature-sensitive reagents are enzymes, primers, probes, DNA, RNA, antibodies, competent cells and certain buffers such as ligation buffers. Ice is also used to hold frozen reagents or cells so they can gradually melt. For example, when performing a PCR reaction, all frozen buffers, $MgCl_2$, primers, DNA templates and nuclease-free water are placed on the top of ice to thaw gradually before a quick spin down. In cloning, ice is used to thaw competent cells slowly before transformation and suddenly chill the cells after transformation. Many lab procedures require ice and an ice bucket filled with flake ice is a common item seen on a lab bench.

### Biosafety cabinet

Almost all plant diagnostic labs have a biosafety cabinet for isolating and transferring microorganisms and handling biohazard materials. A biological safety cabinet is a ventilated enclosure designed to protect the user, the product and the environment from aerosols arising from handling hazardous biological agents. It is an essential piece of equipment to ensure biosafety. Because a plant diagnostic lab handles a broad range of microorganisms associated with plants and environments – some of which may be potentially hazardous to humans – it should meet one of the recommended biosafety levels set for microbiological and biomedical laboratories. There are four biosafety levels recommended for infectious agents (NRC, 1989). A level one laboratory is a basic facility that can handle standard microbiological practices and most works are performed on an open bench but adhere to standard laboratory practices. A level two lab has limited access, requires laboratory coats and protective gloves, performs decontamination of all infectious wastes and posts biohazard warning signs in the facility, in addition to meeting level one requirements. It requires partial containment equipment (such as class I or II biological safety cabinets) to conduct laboratory procedures that may have high aerosol potential. Both level one and level two are basic requirements and a plant diagnostic laboratory should achieve level two biosafety. Levels three and four have much stricter requirements, such as maximum containment equipment or facility with controlled access, and have additional practices and techniques, for example changing between street clothing and laboratory clothing and showering on exit. These higher levels are often required for work with dangerous and exotic agents that may cause serious illness to individuals. Biosafety cabinets need annual certification to verify that they can continue protecting the user, the product and the environment.

### Fume hood

Every lab should be equipped with a fume hood to protect users from potential exposure to harmful vaporous chemicals. A fume hood is a ventilated enclosure designed to contain gases, vapours, fumes and dusts so that there is a safe barrier between the user and the materials being handled. When chemicals or dusty samples are handled inside the fume hood, vapours, fumes and other airborne contaminants are directed by an exhaust fan to the outside via the laboratory exhaust system. A fume hood may also contribute slightly to laboratory ventilation, as indoor air flows through the hood to the outside. In addition to protection from harmful vapours, the clear sliding window of a fume hood, also called a sash, protects the user from chemical splashes, fires and minor explosions. Most acids and harmful chemicals such as formaldehyde should be handled inside a fume hood. A typical fume hood also has a designated storage cabinet for certain flammable and erosive chemicals.

### Freezers

Two freezers, at −20°C and −80°C respectively, are needed in a plant diagnostic lab. The −20°C freezer is used to store temperature-sensitive reagents or ingredients required to be cold during use. Commonly stored items include homogenized tissue samples, DNA or RNA samples, lyophilized or resuspended DNA primers or probes, PCR ingredients, nuclease-free water, antibiotic powder or aliquoted solutions, reagents used for subcloning, self-made antibody or serum, aliquoted buffers or medium solutions and

bottled ethanol. The −20°C freezer is often used to temporarily preserve working products related to a molecular diagnosis before continuing the next day or week. Because the freezer is for keeping the samples/products at −20°C constantly, nearly all −20°C lab freezers are featured with manual defrosting rather than automatic defrosting. That said, when the frost becomes a quarter to half an inch (6–12 mm) thick the freezer needs to be defrosted manually. To ensure the temperature inside the freezer is constantly maintained, a thermometer may be placed inside to monitor the temperature on a daily or weekly basis. Or, it can be equipped with a real-time temperature-monitoring system offered by some companies, where the history of temperature fluctuations is recorded and users are immediately notified when the temperature is out of range.

A −80°C freezer offers a much lower temperature for long-term storage of some critical samples or products to protect stability or viability. Most items safe in −20°C can be stored in −80°C. Suitable items to be stored at −80°C include virus-infected plant tissue (ground or intact), bacterial cells in glycerol solution, competent cells, DNA primers and probes, any DNA or RNA samples or products, certain enzymes and many molecular biological reagents. Due to the ultra-low temperature, users should wear protective gloves and minimize exposure to the temperature. Similar to the −20°C freezer, manual defrosting is required periodically.

### Refrigerators

Multiple refrigerators at 4°C with or without a frozen section are needed even if the lab is equipped with a cold storage room. They provide back-up spaces in case the cold room is out of order. Many laboratory items are safe to store in refrigerators, such as samples, bulk buffer solutions, certain chemical reagents, working products, ELISA test kits, immunostrip, bacterial plates, liquid media, agar plates, DNA extraction columns and ribonuclease (RNase). Some tests require certain procedures performed at 4°C. Which temperature (4°C, −20°C, or −80°C) should be used is based on the product storage requirements, storage duration, product stability, or testing protocols.

### Microscopes

At least one dissecting microscope and one compound microscope are needed to perform routine microscopic diagnosis. Microscopes are used to observe tiny creatures associated with plant specimen that naked eyes cannot see. These two microscopes should be placed together for smooth transitioning between them and both should be linked to a computer with an imaging software to take pictures.

## Balances, magnetic stirrers and pH meters

A diagnostic laboratory needs a set of balances for weighing mass. There are three types of balances, each measuring a specific range of amount. The most commonly used (and least expensive) weighs from 0.1 g (readability) up to 400 or 1000 g (capacity). It can be used to weigh medium powder, chemicals and other materials. The second type of balance is called an analytical balance, which measures mass in milligrams. Certain types of ingredients, such as antibiotics, are usually measured as milligrams. Plant tissue samples or agarose gel cut for DNA extraction are also weighed in milligrams before processing. A typical analytical balance has a readability of 0.0001 g and its weighing capacity is usually between 50 g and 200 g. A lab may also require a basic scale to measure mass in kilograms. These three types of balances should be sufficient to cover most laboratory weighing needs.

Magnetic stirrers are frequently used in a lab to mix ingredients in the water or solution. A stir bar is immersed in the liquid and spin at a speed adjustable through the stirrer control button. Magnetic stirrers offer a quiet, efficient and quick mix or homogenization of a solution. It is best to have a set of different-sized stirrers to hold small, medium or large beakers for a broad range of solution-making. Because some ingredients may be difficult to dissolve in water or a buffer solution at room temperature, a magnetic stirrer with a heating option is often used to increase the solution temperature and speed up dissolving.

A pH meter is always needed when making buffer solutions that require a certain pH range. A set of reference buffers at three pH levels (e.g. 4.00, 7.00 and 10.00) is needed to calibrate the pH meter before each use. In a typical preparation room, balances, magnetic stirrers and pH meters are often placed on the same bench, as solution-making involves a coherent process of weighing, mixing and finally measuring a pH value.

## Centrifuges

Each benchtop should be equipped with one microcentrifuge. Most microcentrifuges hold twenty-four 1.5 ml tubes and can reach a speed up to 20,000 rpm, sufficient to meet routine centrifuging needs in the lab. A much smaller centrifuge is also equipped at each bench to quickly spin down newly thawed or vortexed liquid in the tube. Both types of centrifuges should be placed at each workstation as 'personal' centrifuges. A lab may be equipped with a refrigerated centrifuge to handle temperature-sensitive samples. A refrigerated centrifuge is more expensive than a regular one and it can be equipped with multiple changeable rotors that can hold 1.5 ml, 15 ml, 50 ml or even 100 ml tubes for various centrifuging needs. A nematode extraction lab should have a centrifuge with a rotor holding four buckets, each bucket holding seven 50 ml tubes. This type of centrifuge can spin up to 28 samples each run (Fig. 4.6).

**Fig. 4.6.** An example of a centrifuge that contains a rotor holding four buckets and twenty-eight 50 ml tubes. Note that seven tubes are placed in a rack and then placed in the bucket. Each tube is numbered so that it corresponds to a specific sample.

## Incubators and ovens

Various incubators must be equipped for a plant diagnostic lab to culture microorganisms at certain temperatures. Although most incubators have similar temperature ranges, each incubator should be set at a temperature specific to a certain type of organism. For example, fungal cultures are usually incubated at 18–22°C and optimal growth of bacterial cultures requires a higher temperature up to 37°C. An incubator with an orbital shaking platform inside is commonly used for culturing bacteria or recovering competent cells after heat shock.

Ovens of various sizes are used to hold products or media in higher temperatures. They can be used to dry out specimens or to pasteurize or sterilize pots and containers. A common use of an oven is to place autoclaved medium with a solidifying agent, such as agar, at 57°C to prevent it from solidifying before pouring onto plates. In a molecular biology lab, ovens are frequently used to maintain reagents at above room temperature.

## Growth chambers and greenhouse

A growth chamber is a unit that has a controlled temperature and lighting. It is used to grow a small number of young plants for inoculation tests. As Koch's postulates are not routinely performed in many plant diagnostic labs, a growth chamber can be used for other purposes. A greenhouse facility has many advantages over a growth chamber. It is spacious and can be used to grow large test plants or perform large-scale experiments. Greenhouses are routinely used to grow indicator plants for virus testing and to perform winter grow-out tests for certifying seed potatoes. Certain diseases and nematodes can be maintained on susceptible hosts in the greenhouse for further studies. Although greenhouse facilities are often used for research, a small greenhouse adds a unique function to plant diagnostics, especially when a detected pathogen requires further research on its pathogenicity and host range.

## ELISA testing system

ELISA stands for enzyme-linked immunosorbent assay. The complete system requires a list of equipment and supplies in place (Table 4.2). To detect viruses, bacteria, or oomycetes, infected tissue must be prepared to generate a liquid form that contains

**Table 4.2.** A list of equipment and supplies required for an ELISA testing system.

| Equipment | Supplies and reagents |
| --- | --- |
| Juice press or tissue grinder | Pipette tips |
| Automatic microplate washer | Microplates and plate cover |
| Microplate photometer or ELISA plate reader | V-shape solution basins |
| Refrigerator | Buffers |
| 8- or 12-channel pipettes | Antibodies or reagent sets |
| A set of single channel pipettes | Positive and negative controls |

infected agents before loading it into an antibody-coated microplate. One vital piece of equipment is the juice press or tissue grinder. A juice press grinds plant leaf tissue in seconds and can be cleaned between samples by pressing the washing button. A juice press works very well for handling a large quantity (e.g. over 100 to thousands) of fresh plant tissue samples in a short period of time. It is much better than mortars and pestles that are usually used to grind a small batch of samples. Certain hard tissues, such as seeds, may require specialized grinders. The preparational grinding of plant material for ELISA tests should follow the instructions for the commercial kit used for testing.

Other associated equipment includes the microplate washer and microplate photometer. Although many commercial kits offer instructions that do not require these two pieces of equipment (instead using hand washing and visual assessment), having both will accelerate the test and significantly save time, especially when conducting multiple tests with several hundred samples. Once the test is done, a microplate photometer will read optical density (OD) under a certain filter wavelength to indicate the degree of colour change in the test. The OD number is quantitative data that can help determine if a sample with a weak reaction is positive or negative. Visual determination based on colour change works well for a very strong reaction, but it is difficult when the reaction is marginal. In addition, an 8-channel or 12-channel pipette is needed to perform an ELISA test. An 8-channel pipette can increase efficiency eight times compared with a single-channel pipette, and a 12-channel increases efficiency 12-fold. To use a multi-channel pipette, a solution should be placed in a V-shaped plastic basin.

## Nematode extraction system

There are two nematode extraction systems commonly used, depending on the type and purpose of the sample. One is the wet-sieving and sugar flotation extraction system. This method requires a custom-designed stainless-steel sink with a bowl size of 15 in deep × 2.5 ft long × 2.5 ft wide (0.4 m × 0.75 m × 0.75 m) (Fig. 4.7), a 2 ft (0.6 m) draining board on each side of bowl and a splash board 6.5 ft long and 1 ft high (2 m × 0.3 m). There are two cabinets mounted under the sink to provide storage space. The system also requires a set of different-sized stainless-steel sieves and a centrifuge with a rotor holding 50 ml tubes × 28 (Fig. 4.6). A general set of sieves for nematode wet-sieving extraction includes 20-, 60-, 325- and 500-mesh with 8 in (20 cm) diameter made with stainless-steel. A similar set of small sieves with 3 in or 6 in (7.5 cm or 15 cm) diameters (easy gripping during sieving) are also used to collect nematodes after centrifuging. The 20-mesh sieve is used to collect large plant materials and debris trapped on the screen, 60-mesh is used to collect cysts, 325-mesh for most vermiform nematodes and 500-mesh for almost all vermiform nematodes. The system requires additional equipment, such as a microwave to make hot water and a stirrer to make sugar solution for centrifugation.

The other system is called the mist chamber nematode extraction system. In this system, an array of funnels and tubes similar to a Baermann funnel are set up inside a chamber (Fig. 4.8). A Baermann funnel is a simple device used to extract nematodes from a soil sample or plant tissue. Assembling this device only requires a rack to hold a funnel sealed at the lower end with a rubber tube and a clip (see Fig. 5.15). On the top of the funnel, a stainless-steel wire cloth (mesh screen) is placed followed by one or two layers of Kimwipes or other paper tissue that can last 24–72 h in water. When soil or plant tissues are placed on top of the paper layers and immersed in water, the nematodes move down due to gravity, pass through the paper layer and settle at the bottom of the rubber tube. This device can be used routinely for a small number of samples. When there are hundreds of samples or a large quantity of plant materials to be extracted, a series of mist chambers each hosting up to 24 samples can speed up the extraction process significantly. In the mist chamber, each funnel

**Fig. 4.7.** A stainless-steel sink used for extracting nematodes from soil. Note that there is a set of small, medium and large sieves for wet sieving different types of nematodes. A bucket, a tray of twenty-eight 100 ml beakers and a wash bottle should be available for wet-sieving procedures. The sink has ample surface on both sides to perform wet sieving and prepare centrifugation.

**Fig. 4.8.** An example of mist chamber used for nematode extraction from plant tissue. Note that each funnel is assembled from top to bottom: extension tube, paper layer, metal screen (for support), funnel and long glass tube. The extension tube is used only when a sample quantity is large and cannot be held on top of the funnel.

is assembled slightly differently from a typical Baermann funnel device: the bottom portion of the funnel is inserted into a long glass tube and placed into a rack. The rack is then placed inside a chamber with multiple mist nozzles on the top. When the nozzles start spraying mist onto the sample, nematodes move down with the water to the bottom of the glass tube. Because misting spray is programmed to work for only 90 s every 10 min, the accumulated water in the tube eventually flows out so slowly that the nematodes at the bottom are intact. After about 12–72 h, nematodes can be collected from the tube by pouring the water into a 325-mesh sieve and washing the nematodes into a beaker. Each mist chamber has a drain to remove surplus water during the process.

Both the Baermann funnel method and the mist extraction system are efficient at extracting all life stages of nematodes from soil and plant tissue, except sedentary life stages of nematodes such as root-knot and cyst nematode females, but may not be efficient for some sluggish nematodes.

To set up a mist chamber nematode extraction system, a plumber should be hired to connect the water source and install additional irrigation pipes, mist nozzles, interval timer, water solenoid valve, water line strainer or filter, and a switch to turn the system on or off. A water heater may be required to maintain the mist temperature between 20°C and 24°C, as a cold mist temperature may affect nematode mobility and therefore have a negative impact on the efficiency of the extraction system. Additional wiring, pipes and fittings may be needed to assemble the system based on local code requirements. Some special supplies used in this system should be obtained from available venders. Table 4.3 lists the purpose and specification of all components and supplies needed to build the mist extraction system. The components and specification are based on instructions written by Adam C. Weiner, a senior plant nematologist at the California Department of Food and Agriculture, with slight modifications.

### Molecular diagnosis instruments

Molecular diagnosis uses techniques and procedures that require specific equipment and supplies. A typical molecular diagnosis process starts from DNA/RNA extraction to polymerase chain reaction (PCR) followed by PCR product (amplicon) purification and DNA sequencing. For an amplicon over 1 kb, cloning the product into a vector may be

**Table 4.3.** Components and supplies needed for the mist chamber nematode extraction system.

| Items | Specifications | Purpose |
|---|---|---|
| Acrylic chamber enclosure | 122 cm W × 82 cm D × 92 cm H, 0.635 cm thick, clear, front open for easily removing racks, funnels or samples, but installed with a plastic curtain to contain sprays inside | A confined space for racks in which funnels and tubes are placed, to hold overhead pipe and nozzles, to contain mist sprays within the unit and to drain the excessive water |
| Water heater (optional) | A small (38 l) water heater can support one to three mist units | To maintain mist temperature at 21–24°C for highest efficiency of nematode recovery |
| Water solenoid valve | One valve is sufficient for multiple mist units | To turn water on/off based on the timer |
| Interval timer | 90 s on every 10 min cycle | To regulate mist frequency |
| Mist nozzles | 15 l per hour at 40 psi | Mist spray to samples |
| Water line filter or strainer | Any | To filter out sands, rust and other particles that may clog the nozzles |
| Electrical switch | Any | To shut down the system when not in use |
| Wiring, pipes and fittings | Any | To assemble the system |
| Supporting rack | 61 cm L × 31 cm H × 26 cm W. | To support six funnels |
| Funnels | 130 mm top diameter, stem 140 mm long, inside bowl angle 60° | To hold samples |
| Glass tubes | 25 mm OD × 200 mm L | To collect water and nematodes |
| Wire screen or stainless-steel mesh | Stainless steel, eight mesh per inch, 0.7 mm wire diameter, cut to fit funnel top size. | To support samples |
| Funnel tissue | Kimwipes or other cleaning tissues, non-toxic to nematodes | To block soil or degraded plant tissue from getting into the tube but allow nematode to penetrate and move into the tube |
| Funnel extension | 127 mm OD × 102 mm L, 6.4 mm thick | To increase holding capacity of samples |

needed before sequencing (to obtain 5'- and 3'- end of amplicon sequences as well as the middle portion of the sequence through primer walking strategy) (see Chapter 5 and Fig. 5.26). The typical instruments required to perform these procedures are listed in Table 4.4. Some of these instruments are also used in general laboratory procedures and have been described in an earlier section of this chapter.

## Lab Supplies

Although there are many types of supplies used in different labs, there is a core set of supplies used by most diagnostic labs. For examples, Petri dishes and culture media are used in all plant diagnostic labs to culture microorganisms from plant tissue. Depending on the level of a diagnostic lab, the types of lab supplies may vary significantly. For example, a molecular diagnostic lab is loaded heavily and frequently with pipette tips, microcentrifuge tubes, DNA extraction kits, buffers and more, while a microscopy-based diagnostic lab is mainly

loaded with slides and tools used for sample observations. The attainment of a specific supply, either reusable or consumable, is based on its use in a diagnostic procedure and/or a testing protocol, and the specifications of that supply should meet the needs of testing procedures. Table 4.5 lists some items that are usually used in plant diagnostic labs. The list can be used as a guideline for purchasing routine lab supplies but note that some items listed may represent a group of products with different specifications and purposes of use. A complete and lab-specific supply list can be generated by recording all items (including vendor names and category numbers) used in each and every testing or procedure and this list can be used for tracking their usage and subsequent reordering.

## Lab Techniques

After a diagnostic lab is equipped with adequate instruments and necessary supplies, it still requires various techniques to be strategically employed for

**Table 4.4.** Major instruments used in molecular diagnosis.

| Instrument | Purpose |
| --- | --- |
| Tissue homogenizer | To grind and homogenize tissue rapidly and thoroughly for DNA extraction |
| Tabletop centrifuge | To spin down or separate contents in a tube |
| Vortex mixer | To mix liquid in a tube quickly by an oscillating circular motion |
| Thermocycler | To perform polymerase chain reaction |
| PCR workstation | To offer a sterile and clean environment for pre-PCR procedures to prevent contamination |
| Water bath | To incubate contents in tubes at designated temperature |
| Benchtop thermomixer | To incubate contents in tubes at designated temperature with shaking |
| Electrophoresis system | To run DNA/RNA samples on a gel |
| UV transilluminator | To visualize DNA/RNA on a gel |
| Gel documentation system | To visualize and take image of DNA/RNA bands on a gel |
| Ultrapure water purification system | To provide highly purified water suitable for common molecular biology lab applications |
| Ice machine | To provide ice for common molecular biology lab applications |
| Autoclave | To sterilize tips, tools, media and other lab items |
| Spectrometer | To measure DNA/RNA concentrations |
| Ultra-low freezer | To store temperature-sensitive products or reagents |
| pH meter | To measure and adjust the pH level of buffers |
| Pipette | To handle liquid amount from 1 µl to 1000 µl |
| Balances | To weigh small mass in grams or milligrams with readability up to 0.0001 g |
| Incubator with orbital shaking platform | To culture bacterial or competent cells |

daily diagnostics. To make an analogy: the equipment and supplies are analogous to computer hardware and the techniques used to run a diagnostic lab are like software used to run a computer. The use of various diagnostic techniques is largely dependent on the skills of diagnosticians and, to some extent, determines the capacity and capability of the lab. This section highlights common approaches used in plant diagnostics. Details of diagnostic methods are presented in Chapter 5.

### Visual assessment

In plant diagnostic labs, there are several levels of technical approaches used in disease diagnosis and pathogen identification. The first and most commonly used one is the visual assessment. This is an essential skill to visually characterize a problem during the initial diagnosis. Visual assessment is a basic diagnostic technique and is practised by clinical plant pathologists. Many plant samples or cases of crop problems require this technique to pinpoint the nature of the disease before prescribing relevant tests to confirm the cause. Without this skill, a sample may be unnecessarily tested for something that is not related to the problem. Visual assessment also includes image observations and field investigations.

### Morphological approach

Morphologically identifying encountered pathogens or pests is a traditional but fundamental technique in a plant clinic. It makes diagnostic tasks much easier if no additional testing is needed. Morphological identification requires a certain level of taxonomy expertise, relevant literature, reference specimens and a good microscope that has tools to measure the organism accurately. Most plant clinics employ visual assessment and morphological identification to diagnose common plant disease and pest problems. As mentioned in Chapter 3, many of these cases only require a diagnosis level to the genus or species at most, which is achievable through the morphological approach.

### Microbiological approach

The majority of plant diseases are caused by microorganisms such as fungi, bacteria and viruses, and microbiological methods are routinely used in diagnosis. Microbiological approaches include the induction of microorganism growth *in vitro* or *in situ*, baiting, trapping, general culturing, selective or semi-selective isolation, single spore isolation, inoculation tests and characterization of plant-associated microorganisms. The analysis of a

**Table 4.5.** A list of supplies commonly used in a plant diagnostic lab.

| Reusable items | Consumable items |
|---|---|
| Alcohol burners | Aluminium foil |
| Beakers | Antibiotics |
| Bottles | Autoclave indicator tapes |
| Calculators | Bench mat |
| Chisels | Biohazard bags |
| Counters | Centrifuge tubes (15 ml and 50 ml) |
| Dissecting knives and needles | Cheesecloth |
| Flasks | Chemicals |
| Floating racks | Cloning reagents |
| Forceps | Culture media |
| Freezer boxes | Culture tubes |
| Freezer racks | DNA extraction kits |
| Funnels | DNA primers |
| Garden shears | DNA probes |
| Glass tubes | ELISA and immunostrip kits |
| Graduated cylinder | Ethanol |
| Haemocytometers | Filter papers |
| Ice buckets | Gel extraction kits |
| Inoculating needles or loops | Glass beads |
| Jugs | Gloves |
| Knives | Labels |
| Lab coats | Lens paper |
| Lab stool | Marker pens |
| Label printers | Microcentrifuge tubes (1.5 ml) |
| Lighters | Microplates |
| Liquid containers | Microscopic slides/covers |
| Loppers | Nuclease-free water and TE buffer |
| Microcentrifuge tube openers | Parafilm |
| Microtube cutters | PCR plates and tubes |
| Mortar and pestles | PCR purification kits |
| Pipettes and stands | PCR reagents |
| Pruners | Petri dishes |
| Racks | Pipette tips |
| Scissors | Razor blades |
| Sieves | Reference materials |
| Soil samplers | Sample bags |
| Spraying bottles | Tapes |
| Stirring bars | Tissue homogenization tubes/ matrix tubes |
| Thermometers | Vials |
| Timers | Weighing dishes |
| Utility tray | Weighing paper |
| UV-protection eyewear | Wipes/paper towels |

range of microorganisms associated with diseased plants is a common approach to diagnose a disease complex and identify the role of each organism in disease development and suppression.

## Serology-based approach

Serology-based detection techniques are widely employed in plant diagnostics. ELISA, lateral flow devices, or simply an immunostrip are routinely used in disease diagnosis. The serology approach also includes immunofluorescent labelling, a technique using primary antibodies followed by a fluorescent dye-conjugated secondary antibody to detect a specific pathogen in plant tissue or biological vectors under a fluorescence microscope (Wang and Gergerich, 1998). The fluorescence microscope can be used to detect a pathogen in plant tissue without antibodies. For example, the 4′,6-diamidino-2-phenylindole (DAPI), a molecule that binds AT-rich DNA, is used to detect phytoplasmas in plant tissue (Andrade and Arismendi, 2013). If the phloem tissue contains phytoplasma DNA, DAPI will bind to it and emit a blue fluorescence visible under a fluorescence microscope. Another fluorescence microscopy technique is called fluorescence *in situ* hybridization (FISH). It was developed by biomedical researchers to detect and localize a specific DNA sequence in chromosomes for biological research and clinical diagnosis. In microbiology, FISH is used to simultaneously visualize, identify, localize and enumerate microbial cells (Moter and Göbel, 2000). Although the FISH technique requires competency in DNA sequence analysis and labelling of designed DNA probes, it has been used (though not commonly) in plant pathogen detection (Fang and Ramasamy, 2015).

## Molecular approach

Molecular detection techniques, especially those detecting a broad range of species, are powerful in plant disease diagnosis. These techniques can be easily set up in the lab and routinely used to test any group of pathogens rapidly. Prior to the molecular techniques widely adopted, a biochemical approach to analyse microbial metabolite production, fermentation, or other phenotypic characteristics was used in the identification of pathogens, especially for culturable bacteria. Unlike biochemical methods, the molecular approach is based on genotypic characteristics such as DNA/RNA sequences mostly in conservative regions or housekeeping genes. The methods commonly used in a lab include conventional PCR, real-time PCR, reverse-transcript PCR and isothermal amplification. Conventional PCR uses two primers to amplify a short and targeted DNA fragment from a DNA sample. Positive amplification is visualized on an agarose gel under a UV light after electrophoresis. Real-time PCR, also known as quantitative PCR (qPCR), can monitor the amplification of a targeted DNA molecule during the PCR without gel electrophoresis. Because qPCR uses sequence-specific DNA probes labelled with a fluorescent reporter dye and generates a cycle threshold value (Ct), it can be used to quantitate DNA copies in a sample. Reverse transcription PCR (RT-PCR) is a molecular technique starting with reverse transcription of RNA into DNA followed by the amplification of a specific DNA target. It is mainly used to analyse gene expression profiles in an organism or to detect RNA viruses. Unlike PCR, which requires an alternating temperature to amplify a DNA fragment, isothermal amplification is usually carried out at a constant temperature.

### Bioinformatics approach

Many public or private genomic centres offer quick DNA sequencing services. A PCR product either directly purified or subcloned into a vector can be submitted for sequencing at the cost of a few dollars, depending on the fee schedule of a service provider. Unless there is a huge number of sequencing samples to be handled daily in a particular lab, there is no need to acquire a DNA sequencing machine and perform the work that otherwise can be delegated to a centralized service provider. DNA sequencing data, presented as hundreds or thousands of nucleotides in four bases (A, G, C, T), do not offer direct answers if they are not analysed by bioinformatics tools. Computer work using various sequence analysis software or online tools is an essential part of the bioinformatics process used in DNA-based diagnosis. For specific research projects, *in silico* data mining and analysis may be needed before *in vivo* or *in vitro* biological experiments are conducted. Next-generation sequencing (NGS) is a powerful technique in both discovering and detecting microorganism populations in a plant, organ, or organism, and it has the potential to replace conventional methods, such as morphology- or biochemical-based identifications and locus-specific DNA detections, with a genomic-based pathogen characterization. However, NGS in the clinical setting has some disadvantages, such as the requirement of computing infrastructure and expertise in bioinformatics to analyse and interpret the sequence

data (Behjati and Tarpey, 2013). Though the actual sequencing cost may be affordable to certain plant diagnostic labs, the volume of data to be managed and extracted for clinically or diagnostically important information may cost much more than the benefit the information could bring.

## Laboratory Quality Management System

A diagnostic lab in a clinical setting must have a system to manage the practice and its quality. The system is often called a quality management system (QMS). Without a QMS, a lab procedure could go wrong, clients' samples could be messed up and a sample might be unprocessed without knowing. As suggested earlier, the equipment and supplies are analogous to computer hardware and the techniques are analogous to software. Using this analogy, the QMS is analogous to antivirus software. Antivirus software is a program developed to protect computers from malware such as computer viruses, worms, spyware and many other malicious programs, and it functions to monitor and remove viruses from the computer. A QMS has a similar function in a diagnostic lab, ensuring that a lab is run smoothly, efficiently and as error-free as possible. A QMS also helps correct wrongdoings and prevent recurrence.

A lab implementing a QMS should have a quality manger (QM) who will be primarily responsible for running the QMS in a lab. A typical lab QMS includes a lab quality manual, quality procedures, working instructions, standard operation procedures and forms. The quality manual should outline and list the structure of the documentation used in the quality system. It may contain various sections addressing the laboratory mission, administrative requirements, management requirements and technical requirements. Management requirements may include document and record control, customer feedback, control of non-conformances and internal audits. Technical requirements govern the personnel, lab infrastructure, equipment maintenance, test methods, sample handling and reporting. A QMS can be tailored to fit the needs of a specific diagnostic lab.

## Diagnostic Lab Accreditation

With a documented quality system and technical competence, a lab can seek formal accreditation from an accreditation body or organization such as the ISO (International Organization for Standardization). A formal accreditation offers significant benefits to the laboratory. First, some industries require tests to be done in an accredited laboratory. Secondly, many laboratories seek accreditation to increase their service share in a market. Thirdly, certain companies accredited under certain standards such as ISO may require the lab to supply test results also ISO/IEC 17025 accredited. In general, an accreditation provides formal recognition for the competence of the lab.

A plant diagnostic laboratory is heavily involved in biological testing and may use ISO/IEC 17025 requirements (ISO, 2020). The ISO/IEC 17025:2017 specifies the general requirements for the lab to carry out tests and/or calibrations with technical competency. It is applicable to laboratories performing tests and/or calibrations. To seek an accreditation under the ISO/IEC 17025 standards, the laboratory must implement a laboratory management system and demonstrate technical competence.

However, many plant diagnostic laboratories function as clinics where samples can be diagnosed without a defined test performed. For example, a disease may be quickly diagnosed by a diagnostician based on their clinical expertise and the local trend of the disease. In this case, the scope of the work is opinion-based rather than a strict test (see Chapter 3). Because of this, a plant diagnostic lab is not just a testing lab, but also a clinic. The ISO/IEC 17025 requirements may not be fully applicable to certain functions of a diagnostic clinic. In the USA, the National Plant Diagnostic Network (NPDN) offers a STAR-D accreditation to qualified plant diagnostic laboratories. STAR-D stands for a system for timely, accurate and reliable diagnosis. It is tailored for a plant diagnostic laboratory. The STAR-D standards require the laboratory to meet administrative, management and technical requirements for a formal accreditation.

## References

Andrade, N.M. and Arismendi, N.L. (2013) DAPI staining and fluorescence microscopy techniques for phytoplasmas. In: Dickinson, M. and Hodgetts, J. (eds) *Phytoplasma. Methods and Protocols.* Vol. 938 in series *Methods in Molecular Biology.* Humana Press, Totowa, New Jersey. doi: 10.1007/978-1-62703-089-2_10

Aslanzadeh, J. (2004) Preventing PCR amplification carryover contamination in a clinical laboratory. *Annals of Clinical & Laboratory Science* 34, 389–396.

Behjati, S. and Tarpey, P. S. (2013) What is next generation sequencing? *Archives of Disease in Childhood. Education and Practice* 98, 236–238. doi: 10.1136/archdischild-2013-304340

Fang, Y. and Ramasamy, R.P. (2015) Current and prospective methods for plant disease detection. *Biosensors,* 5, 537–561. doi: 10.3390/bios5030537

ISO (2020) General requirements for the competence of testing and calibration laboratories. ISO/IEC 17025:2017. Available at: https://www.iso.org/obp/ui/#iso:std:iso-iec:17025:ed-3:v1:en (accessed 19 January 2020).

Moter, A. and Göbel, U.B. (2000) Fluorescence in situ hybridization (FISH) for direct visualization of microorganisms. *Journal of Microbiological Methods* 41, 85–112. doi: 10.1016/s0167-7012(00)00152-4

NRC (1989) Appendix A, Biosafety in Microbiological and Biomedical Laboratories. In: *Biosafety in the Laboratory: Prudent Practices for the Handling and Disposal of Infectious Materials.* National Research Council (US) Committee on Hazardous Biological Substances in the Laboratory. National Academies Press, Washington, DC. Available at: https://www.ncbi.nlm.nih.gov/books/NBK218631/ (accessed 10 January 2020).

Persing D.H. (1991) Polymerase chain reaction: trenches to benches. *Journal of Clinical Microbiology* 29, 1281–1285.

Sarkar, G. and Sommer, S.S. (1990) Shedding light on PCR contamination. *Nature* 343, 27. doi: 10.1038/343027a0

Sigmon, J. and Larcom, L.L. (1996) The effect of ethidium bromide on mobility of DNA fragments in agarose gel electrophoresis. *Electrophoresis* 17, 1524–1527. doi: 10.1002/elps.1150171003

UOW WHS Unit (2020) Emergency eyewash station and safety shower guidelines. Available at: https://staff.uow.edu.au/content/groups/public/@web/@ohs/documents/doc/uow148621.pdf (accessed 10 January 2020).

Wang, S. and Gergerich, R.C. (1998) Immunofluorescent localization of tobacco ringspot nepovirus in the vector nematode *Xiphinema americanum. Phytopathology* 88, 885–889. doi: 10.1094/PHYTO.1998.88.9.885

# 5 Diagnostic Procedures and Protocols

In plant diagnostics, there are procedures to be followed to diagnose a problem. Some procedures are not necessarily considered analytical tests. For example, visual examination is a procedure that can sometimes diagnose certain diseases or insect problems, but it is rarely considered a test. Similarly, dissecting a piece of root tissue to observe females of root-knot nematodes inside is considered a procedure rather than a test. However, the majority of diagnostic procedures are standard tests, especially in the area of serological and molecular diagnosis. These tests are developed by researchers and optimized for clinical use, and they generally come with standard protocols and positive or negative controls. In most cases, a final diagnosis requires a series of clinical procedures and standard testing. Some cases may require a completed diagnostic scheme that include various levels of tests (Fig. 5.1). A diagnostic scheme is often proposed and used for regulatory or complicated plant diseases and it is the highest level of diagnosis.

## Visual Examination

Visual examination is one of the most critical steps in all plant diagnostics, especially when a plant sample is submitted for an open-ended diagnosis instead of for a specific test. A visual assessment has specific objectives, key components and vital contributions to the ultimate diagnosis of a given problem.

## Objectives

A visual examination includes a complete check of all parts of plants or tissue samples to define the nature of a problem. In this procedure, diagnosticians should look for any abnormal symptoms expressed on leaves, flowers, stems, crown or roots, and the potential pathogen sign as well. If done correctly and carefully, diagnosticians should be able to figure out the relationship of various symptoms exhibited on different parts of a plant and the potential cause. This process will determine: (i) which part of the plant is the primary source of the problem; (ii) which part of tissue will be further sampled for testing; (iii) what tests or procedures are to be performed next; and (iv) which pathogen(s) or agents should be tested for.

## Components

In medical pathology, surgical specimens are always carefully examined, identified and described before being processed. The procedure is called gross examination or grossing (Bell *et al.*, 2008). Gross examination provides important diagnostic information, and anatomical pathologists use it (along with results of histological examination of processed tissues) to diagnose diseases. Similar to grossing, a visual examination and description of a plant sample is a critical step for plant pathologists to find disease information. A visual examination usually includes the following components.

1. Identify the sample received or collected with a unique identifiable number.
2. Describe the sample, including quality, quantity, type of specimen, and species.
3. Describe additional samples submitted from the same case.
4. Record normal and abnormal features or symptoms.
5. Record any sign of pathogens observed.
6. Record any arthropod pests observed.
7. Evaluate client's comments or observations on the disease or problem.
8. Evaluate additional information such as pictures or third-party test results submitted by the client.
9. Dissect plant sample, when necessary, to reveal internal symptoms.

DOI: 10.1079/9781789246070.0005

Sample collected

Microscopic examination → Bacterial streaming

Tissue rot            Negative

Plating — PDA plus streptomycin

PARP           Negative

Positive ——————— Morphological ID

Immunostrip test for *Phytophthora*

Positive*         *Pythium*

DNA barcoding

*Pythium aphanidermatum*

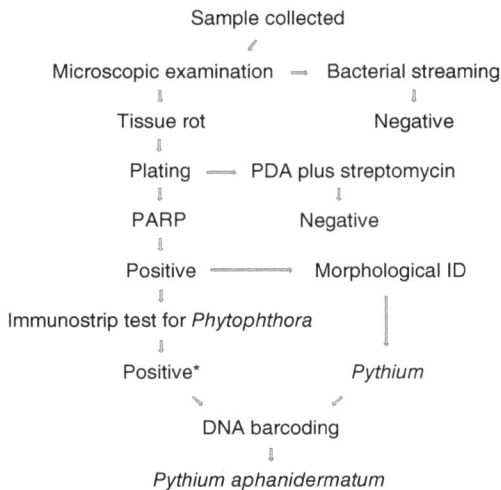

**Fig. 5.1.** An example of a diagnostic scheme proposed for hemp crown rot caused by *Pythium aphanidermatum*. In this case, the immunostrip test for *Phytophthora* cross-reacts with *P. aphanidermatum*.

10. Determine the location from which specific tissue should be taken for plating or testing.
11. Formulate a hypothesis on the potential cause of the problem.
12. Establish a diagnostic scheme.

### Procedures

To perform a visual examination, place a plant sample or whole plant on a clean observation table (see Fig. 4.1), closely examine every part of the plant or sample for abnormalities, starting from the top portion of the plant including leaves, buds, flowers, or fruits, to every branch, stem, crown and roots. If a symptom is observed and warrants an internal observation, cutting or dissecting the plant sample is often used to reveal internal symptoms. A knife, scissors, hand pruner, axe and a cutting board should be available when performing internal tissue examinations. These tools are frequently used to examine many types of symptoms both internally and externally by eye. For example, if a trunk sample of a woody plant exhibits numerous holes, an axe can be used to remove bark tissue to see if there are any insect larvae and galleries between the bark and sapwood. If a hemp plant shows a stem lesion or wilting, a knife is used to cut stem tissue to see if there is any stem canker or vascular tissue discoloration. Table 5.1 lists common plant symptoms observed during the initial examination and the procedures that should be taken next to further diagnose the problem.

### Systemic examinations

In the early stages of a plant disease, the initial symptoms may be subtle and difficult for diagnosing. Sometimes, the problem resembles several other diseases, making the diagnosis complicated. In this situation, it is necessary to execute detective skills and examine the whole plant to find the initial symptom that may offer a clue to the root cause. Here, a real case is presented to illustrate how a thorough visual examination led to the final diagnosis of hemp collar rot caused by *Plectosphaerella cucumerina* with the crop only exhibiting a mild leaf yellow symptom (see Chapter 8).

Several whole hemp plants were submitted by a grower who was concerned by leaf yellowing observed in the field. In the lab, plants were placed on the floor and examined from top to bottom (root). Overall, the plants had almost normal foliage with mild yellowing on newly grown leaves (Fig. 5.2). Leaves on the lower branches had mild necrosis along the leaf edge (Fig. 5.3). There was no visible insect or mite infestation, which was confirmed after microscopic examination. The rest of the foliage was perfectly healthy. The stems were normal without any visible symptoms of discoloration, lesions, or canker. Based on this examination, one may speculate that the yellowing of newly grown leaves may be caused by nutrient issues and that the mild leaf-edge necrosis may be related to drought or other environmental stress. However, the symptoms exhibited on both new and old leaves were likely caused by the same factor. In our clinical experience, if foliage has symptoms speculated for nutrient or drought issues, a lower stem or root system should be examined to search for or rule out any conditions that may affect water and nutrient movement in the plant.

The next step was to examine the stem, crown or collar, and root system. The upper stem and branch stem were healthy internally and externally. The lower stem surface had no discoloration and the crown area appeared to be healthy (Fig. 5.4). The root system was visually strong and healthy (Fig. 5.5).

**Table 5.1.** Visual symptoms commonly observed, their potential causes and follow-up diagnostic procedures.

| Organs | Visual symptoms | Possible causes or agents | Next diagnostic procedures for biotic agents |
|---|---|---|---|
| Leaf and flower | Blight | Leaf, stem, or root damage caused by fungi, bacteria or abiotic factors | Microscopic examination of roots, stem and leaves, plating diseased tissue |
| | Blotch | Abiotic factors or fungi | Microscopic examination of blotched leaves |
| | Chlorosis | Nutrient deficiency or virus | Testing for certain viruses |
| | Cluster, witches' broom | Insects or phytoplasma infection | Microscopic examination and test for phytoplasma |
| | Discoloration or bleaching | Sunburn or other abiotic factor | Microscopic examination of leaves |
| | Drop | Root rot or stem damage | Examination of the root system and stem |
| | Edema (oedema) | Excessive water | Microscopic examination of raised leaf tissue to rule out other conditions |
| | Fuzzy appearance | Fungal or oomycete infection | Microscopic examination of fuzzy areas |
| | Gall | Insects, mites, or nematodes | Microscopic examination of galls |
| | Holes | Insects, fungi, or abiotic factors | Microscopic examination |
| | Malformation | Herbicide injury or virus infection | Testing for viruses or herbicide |
| | Mosaic or mottling | Virus infection | Testing for viruses |
| | Ragged or damaged | Abiotic factors | Microscopic examination of damaged areas for pest signs |
| | Scorch | Drought, root rot, or *Xylella fastidiosa* | Examining root or test for *Xylella fastidiosa* |
| | Silver appearance | Insect or mite damage, abiotic factors | Microscopic examination of damaged areas for pest signs |
| | Sooty mould | Aphids and fungi | Microscopic examination of damaged areas for pest signs |
| | Spot/necrosis | Fungi, bacteria or abiotic factors | Plating diseased tissue for fungi, bacteria, or oomycetes |
| | Sticky or oily | Insects, bacteria, or sprays | Microscopic examination for signs of pathogens, pests, or substances |
| | Underdevelopment | Nutrient deficiency, viruses, phytoplasma, etc. | Testing for suspected viruses or phytoplasma based on symptoms |
| | Vein clearing | Herbicide, nutrient deficiency, viruses | Testing to rule out viruses |
| | Wilt | Root rot, vascular damage, drought | Examining roots and vascular tissue for discoloration, plating diseased tissue |
| Stem, branch and shoot | Bleeding | Phytophthora or bacterial infection, borer insects | Selective culture for oomycetes |
| | Blight | Root rot, stem injury, foliar diseases | Examining root, stem and leaves to determine which part causes blight |
| | Canker, large area | Fungi, oomycetes, bacteria | Plating diseased tissue for fungi, bacteria, or oomycetes |
| | Cracking, broken | Insects, abiotic factors | Microscopic examination of damaged areas for pest signs |
| | Dieback | Fungi, bacteria, abiotic factors | Examining root, stem and leaves to determine which part causes dieback |
| | Discoloration | Abiotic factors | Microscopic examination to rule out pathogens/pests |
| | Fuzzy appearance | Fungi, oomycetes, insects, mites | Microscopic examination for pest or pathogen signs |
| | Galls | Insects, mites, nematodes | Microscopic examination for pest or pathogen signs |

*Continued*

Table 5.1. Contiuned.

| Organs | Visual symptoms | Possible causes or agents | Next diagnostic procedures for biotic agents |
|---|---|---|---|
| | Holes | Insects, fungi, abiotic factors | Microscopic examination for pest or pathogen signs |
| | Lesions, isolated | Fungi, oomycetes, bacteria, viruses | Plating diseases tissue for fungi, bacteria, or oomycetes |
| | Mouldy | Fungi, oomycetes | Plating diseased tissue for fungi or oomycetes |
| | Overgrowth, proliferation | Phytoplasmas, abiotic factors | Testing for phytoplasma |
| | Rot | Fungi, oomycetes, bacteria, abiotic factors | Plating diseased tissue for fungi, bacteria, or oomycetes |
| | Short internode | Stem nematodes, phytoplasmas | Dissecting a stem to exam for nematodes under a microscope |
| | Swelling | Nematodes, insects, abiotic or physiological factors | Dissecting swelling tissue to examine for nematodes or insects |
| | Wilting | Root rot, vascular damage, drought | Examining roots and vascular tissue for discoloration |
| Crown | Discoloration | Fungi, oomycetes, bacteria | Plating diseased tissue for fungi, bacteria, or oomycetes |
| | Dry rot | Fungi, oomycetes | Plating diseased tissue for fungi or oomycetes |
| | Gall, tumour, overgrowth | Bacteria | Isolating *Agrobacterium* from gall tissue |
| | Mouldy, fuzzy appearance | Fungi, oomycetes | Plating diseased tissue for fungi or oomycetes |
| | Soft rot | Bacteria | Examining bacterial streaming, isolating bacteria |
| Root | Brown or black colour | Oomycetes, fungi, overwatering | Plating diseased tissue for fungi or oomycetes |
| | Death | Abiotic factors | Examining the entire plant to determine the cause |
| | Gall, knot | Nematodes, bacteria | Dissecting swelling tissue to examine for nematodes |
| | Lack of fresh root | Abiotic factors | Examining the entire plant to determine the cause |
| | Rot | Bacteria, oomycetes, fungi, overwatering | Plating diseased tissue for fungi, bacteria, or oomycetes |
| | Smelling | Bacteria, overwatering | Isolating bacteria |
| | Underdeveloped | Viruses, phytoplasmas, nutrient deficiency | Examining the entire plant to determine the cause |
| Seed | Deformation | Nematodes, fungi | Dissecting seeds to observe abnormal texture and nematode signs |
| | Low germination rate | Fungi, oomycetes, bacteria | Plating seeds for fungi, bacteria, or oomycetes |
| Seedling | Damping off | Fungi, oomycetes | Plating diseased tissue for fungi or oomycetes |
| | Mouldy | Fungi, oomycetes | Plating diseased tissue for fungi or oomycetes |

Although both roots and stem appeared to be healthy, a further examination into the internal tissue of the root and stem needed to be performed, because tissue necrosis or infection sometimes starts from the inside tissue and may not be visible from the outside in the early stages.

After washing the soil off from the roots and collar area, a mild browning appearance (Fig. 5.6) was

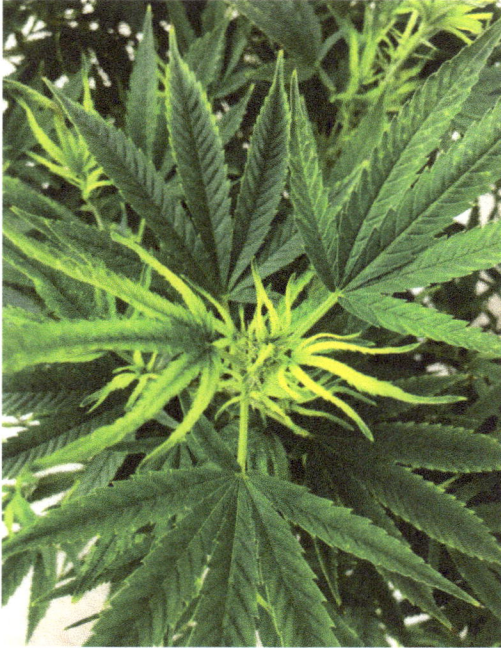

**Fig. 5.2.** A hemp plant exhibits yellowing and underdevelopment of newly grown leaves.

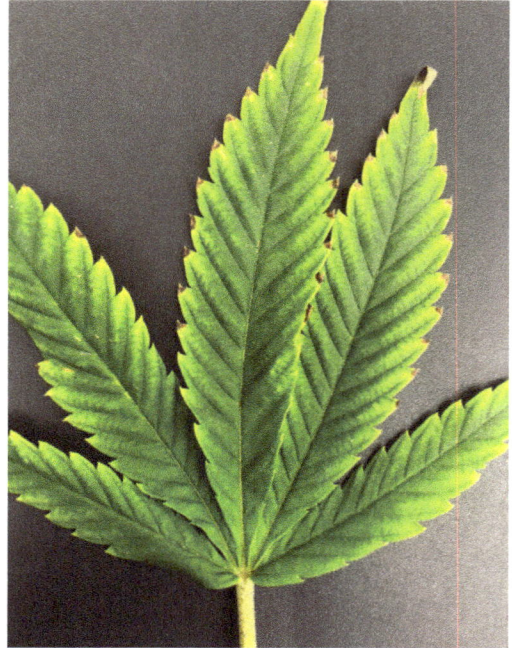

**Fig. 5.3.** Mild leaf-edge necrosis of a hemp leaf at a lower branch.

noticed and warranted a dissection to see what was happening inside the collar tissue. Cutting longitudinally into the vascular tissue revealed a deep and long necrosis starting from the junction area of root and stem and extending into the stem (Fig. 5.7). This internal damage – collar rot – could have been the cause of the problem since the collar rot affected water and nutrient movement in the plant.

Visual observation at this step found evidence of damage inside the collar area of the plant and this was an important finding, as it gave a clear clue on what might have caused the mild foliage symptoms. The association between foliage symptoms and internal collar rot were further demonstrated through visually examining a number of symptomatic and healthy hemp plants. From a diagnostic standpoint, presence of collar rot may quickly rule out nutrient deficiency or water deficiency, especially when all symptomatic plants have internal collar rot while healthy plants do not.

The final step of visual examination is to check for the sign of pathogens. A fungus-infected tissue sometimes exhibits mouldy growth of mycelium and/or spores that can be visually observed, but when fungal growth occurs inside the necrotic tissue, it may not be evident visually. In this hemp collar rot case, fungal structures (mycelia or spores) were only visible under a microscope (Fig. 5.8). Based on visual and microscopic examinations and fungal spores observed inside collar tissue, the sample was proceeded to fungal isolation and DNA-based identification (see Chapter 8).

## Microscopic Examination

### Objectives

Microscopic examination is the most-used procedure to diagnose a plant disease or insect problem. Similar to visual examination, a microscopic examination involves searching for any abnormality that we cannot see with the naked eye. There are two microscopes used in a clinic: the dissecting microscope and the compound microscope. A dissecting microscope is used to observe insects, mites, collective growth of fungi, bacterial fluid, nematodes and plant tissue necrosis in greater detail. Meanwhile, a compound microscope is used to observe fungal

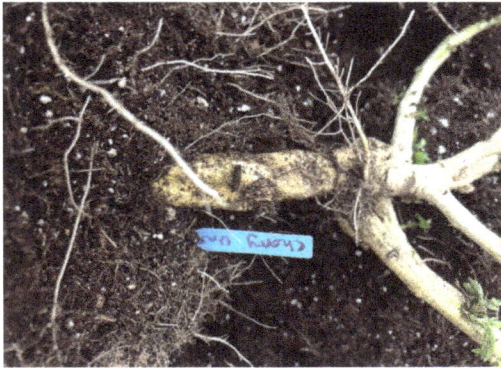

**Fig. 5.4.** A healthy appearance of the lower stem of a hemp plant showing mild leaf yellowing and leaf scorch.

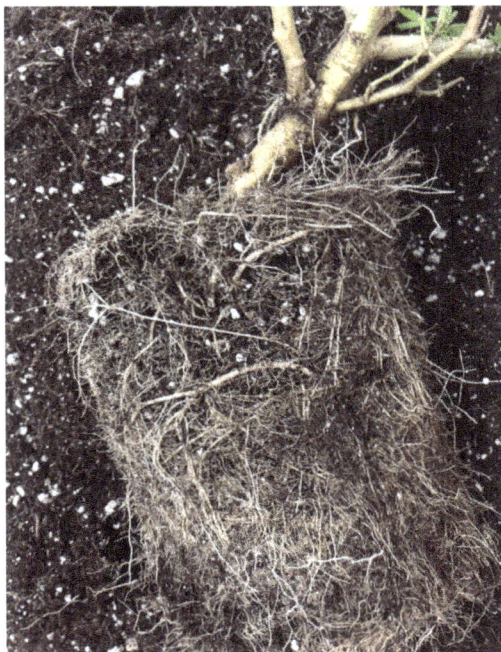

**Fig. 5.6.** Mild colour change of the collar area of a hemp plant. Note that it is normal to see a difference in colour between the stem (above ground) and the collar area (mostly under the ground).

**Fig. 5.5.** A strong root system of a hemp plant showing mild leaf yellowing and scorch.

**Fig. 5.7.** A narrow and extended internal tissue necrosis was observed inside the collar of a hemp plant.

observed under the microscope. For common diseases or pests, microscopic examination can often conclude a diagnosis without further testing.

## Components

There are several components of microscopic examination, all of which contribute to the final diagnostic judgment.

hyphae and spores, bacterial cells, nematode anatomical structures and small insects/mites. The objective of microscopic examination is to determine if any pest or pathogen is associated with an investigated plant part or tissue, and if there is one, what it might be, based on morphological characteristics

1. Observe and record the abnormality of examined tissue at higher magnifications and compare it with normal plant tissue.
2. Determine the type of plant tissue or cells affected by the disease or pest.

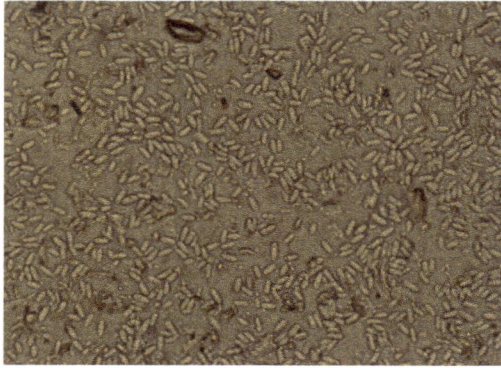

**Fig. 5.8.** Massive fungal spores observed under a microscope from the necrotic tissue of a hemp plant exhibiting collar rot.

**Fig. 5.9.** An aphid on the lower side of a hemp leaf observed under a reflected-light microscope. Note that the 3D shape of the leaf surface and aphid are clearly seen.

3. Determine if any pathogen or pest is present and/or associated with problematic plant tissue.
4. Check if bacterial streaming occurs for plant samples suspected for bacterial infection.
5. Dissect plant tissue to observe nematodes.
6. Observe pathogen growth or pest infestation level.
7. Observe the morphology of a pathogen or pest and measure its size or shape to obtain its morphometric data.

### Procedures

Observing plant specimens under a microscope is straightforward and the use of a microscope should follow the instructions of the manufacturer.

### *Direct observations*

A dissecting microscope should be equipped with two light sources: reflected light and transmitted light. Reflected light is used to observe plant specimens, insects, collective fungal growth, or mites. The image seen through the eyepieces is formed by light reflected from the surface of the specimen; therefore, the detail of the organism's shape and surface can be visualized (Fig. 5.9). Transmitted light is typically used to examine transparent or semi-transparent specimens, such as vermiform nematodes. However, to observe matured cysts of cyst nematodes, reflected light is often used to see the colour, texture and shape of the surface. When a cyst is crushed to release the eggs inside, transmitted light is needed as eggs are semi-transparent.

The choice of reflected or transmitted light is based on whether the specimen is transparent or not. When the correct light source is used, a specimen can be observed clearly with greater detail at appropriate magnifications. Also, the angle of reflected or transmitted light should be adjusted for better observations.

A compound microscope is typically equipped with a transmitted-light lamp house. To observe a specimen under a compound microscope, a slide must be made. For example, to examine a fungal culture under a compound microscope, place a small piece of the colony with or without medium on a glass slide, add a drop of water, cover with a slip, gently press the slip to flatten the specimen and load the slide to the microscope to observe the mycelia or spores. At higher magnifications, specimens such as fungal spores can be observed clearly (Fig. 5.10). To observe a live or moving organism such as a nematode, simply pick several nematodes into a drop of water on a slide, cover with a slip and examine under the microscope. If the nematode is too active to observe, place the slide above the flame of an alcohol burner for a few seconds to gently kill the nematode. Unlike a dissecting microscope, a compound microscope is designed to be more adjustable to observe a variety of specimens. Commonly equipped adjustable parts include a set of objective lenses ranging from scanning power (4×), lower power (10×), high-power (20× or 40×) to the oil immersion lens (100×). Compound microscopes also have adjustable eyepieces (adjusted to

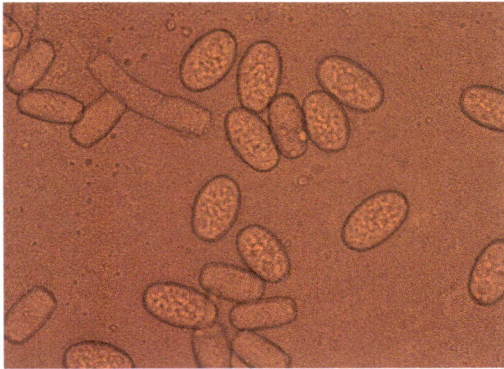

**Fig. 5.10.** Spores of a powdery mildew fungus from a hemp plant observed under a compound microscope. Note that under transmitted light, an object is typically viewed as two-dimensional.

offset any vision differences between a user's two eyes), coarse and fine adjustable knobs, condenser and field diaphragm. A clear image of a specimen can be observed when the specimen is well focused under the right amount of light with a good contrast. Specimens that are too thick or that block light will not generate a clear image of the structure.

### Bacterial streaming

Bacterial streaming is simple and inexpensive, and many bacterial diseases are often initially diagnosed by using this procedure before proceeding to other testing methods such as isolation or ELISA tests. Please note, this procedure is used as screening for a disease potentially caused by a bacterial pathogen; and the presence or absence of bacterial streaming is only used to preliminarily indicate or rule out a bacterial disease. Positive bacterial streaming does not provide information on which specific bacterium causes the disease, unless it is a known disease on a local crop.

When a plant specimen exhibits a water-soaked lesion, soft rot, necrotic spot or oozing, a small piece of plant tissue can be checked for bacterial streaming. To observe bacterial streaming, use a sterile razor blade to cut a small lesion from the symptomatic tissue, place it on a glass slide, slice it into multiple pieces, add a drop of water and a cover slip and observe under a compound microscope. If bacteria are present, a cloud of bacteria can usually be seen 'streaming' out from the infected

tissue. Bacterial streaming may be observed first at a lower magnification to see multiple points of streaming (see Fig. 3.16). For a large lesion, cut a piece at the marginal area that contains both necrotic lesion and healthy tissue. For a bacterial infection on the stem, the bacteria may stream out from specific areas of the stem tissue such as the vascular tissue. Bacterial streaming is easily observed with phase contrast or by adjusting the condenser and diaphragm of a compound microscope (Wright, 2020). This is especially important when observing samples with less obvious streaming. To demonstrate an association between a specific symptom and potential bacterial infection, always check several lesions of multiple samples. Keep in mind that there will be some cellular debris or plant sap that may ooze out from plant tissue, resembling bacterial streaming. Further clinical examinations of the disease and associated samples will determine the nature of the streaming.

Below is a typical protocol used for examination of bacterial streaming.

**1.** Cut a small piece of plant tissue that has water-soaked necrotic spots. Place the tissue on a glass microscope slide, slice it several times, add a drop of water and cover with a glass cover slip.
**2.** Under a compound microscope, locate the edge of the plant tissue. Adjust field of view and magnification as necessary. Adjust the condenser and diaphragm of the microscope to create a greater contrast.
**3.** If bacteria are present, a cloudy 'stream' of bacteria can be seen oozing or streaming from the affected plant tissue.
**4.** Repeat this process with multiple tissue pieces to demonstrate that all spotted tissue has bacterial streaming.
**5.** To check stem tissue for bacterial streaming, fill a beaker with distilled water and place it against a dark background, use a flamed and then cooled scalpel to cut the stem, place the cut end of the stem into the water and look for bacteria oozing out of the stem.

### Blotting

'Blot' used here does not refer to the blot used in molecular biology methods such as Southern, northern and western blots. Instead, it is a simple method to transfer fungal spores and mycelia from a plant surface to a glass slide so they can be observed under a compound microscope. Simply take a glass slide and add a drop of water at the centre of the slide. Cut a clear piece of tape 2 inches (5 cm) long from a tape dispenser, hold the tape and place it on the area of the plant specimen where mould or fungal growth is obvious or suspected. Make sure the sticky side is attached to the plant surface. Gently press the tape to ensure it is fully attached and then immediately peel the tape off. Fungal spores or mycelia, if any, are now attached to the sticky side of the tape. Place the tape (sticky side face down) on the slide and gently press the tape to ensure water is distributed between the tape and the slide. Observe the slide under a compound microscope. This simple procedure is helpful to quickly identify powdery mildew fungi, grey mould (*Botrytis* spp.), downy mildew fungi, *Alternaria*, *Rhizopus*, *Pythium*, *Phytophthora* and other fungi or oomycetes that grow out from infected plant tissue.

Blotting microorganisms from a plant surface onto a microscopic slide for further observation under a compound microscope can be done on a routine basis during the growing season, especially for cannabis production in the greenhouse. It can be used for early detection of powdery mildew and other airborne diseases. When a hemp crop is infected with powdery mildew, the initial symptom may not be visible, but a small number of hyphae and spores are already present on the leaf. By periodically blotting asymptomatic leaves and observing under a microscope, one can detect the disease at an early stage. Blotting also provides useful information on the composition of different fungi on leaf surfaces (Fig. 5.11), as well as mites and small insects.

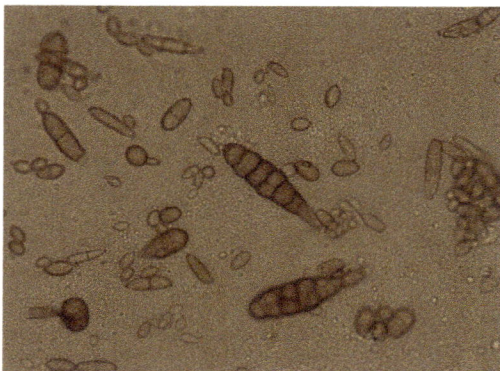

**Fig. 5.11.** Multiple types of fungal spores blotted on the tape from the infected hemp leaf. Note that there are multi-cell *Alternaria* spores in the blotting.

## Moist Chamber Induction of Pathogens

In plant diagnostic clinics, moist chamber induction is commonly used to induce growth of a pathogen from a plant specimen. It is an inexpensive procedure that can be done in a non-lab setting. If used correctly, it is a reliable method to detect a specific pathogen from the plants. For example, a moist chamber method was adopted to isolate and identify *Verticillium albo-atrum*, *V. dahliae* and *V. nigrescens* from herbaceous and woody hosts (McKeen and Thorpe, 1971). This method can be used to induce development of various fungi from a broad range of substrata. The principle of this method is to create a moist environment that allows inconspicuous fungi to grow out from the plant specimen. Once a fungus grows out from plant tissue, the characteristics of mycelia, spores or fruiting bodies can be observed under a microscope for further identification. This method is useful when a lab or a cannabis production facility is unable to culture plant fungal pathogens from plant tissue, and it can be used to detect many common fungi and oomycetes such as *Rhizopus*, *Alternaria*, *Botrytis*, *Fusarium*, *Pythium* and *Phytophthora*. However, it has limitations. First, moist chamber induction may take longer than a regular plating method. In a well-equipped laboratory, it is much faster and more effective to perform direct isolation of plant pathogens from pieces of plant tissue on a nutrient medium. Secondly, some saprophytic organisms may overgrow during the incubation, which may mislead the diagnosis. Thirdly, in certain types of plant specimens or diseases, moist chamber induction will not work. The use of this method entirely depends on the diagnostician's preference and availability of alternative methods in the lab.

The procedures to perform moist chamber induction vary case by case, depending on the type and size of a sample. If a small tomato plant exhibits leaf lesions and blight symptoms and a *Phytophthora* pathogen is suspected, the plant can be placed in a large clear plastic bag into which a small amount of water is added. Seal the plastic bag so that a relatively high humidity is maintained. In a couple of days, if the symptom is caused by a *Phytophthora* species, affected leaves may exhibit fuzzy mycelium growth and sporangia. For small leaves, they can be placed on a standard 100 mm Petri dish with a filter paper wetted with approximately 5–10 ml sterilized water. Seal the Petri dish with a piece of parafilm to maintain the moisture. Place the Petri dish at room temperature under light and incubate for a couple of days. For large plant specimens, a customized container can be used to accommodate the specimen and maintain a designed moisture level. In a non-lab setting, many containers used in everyday life can serve the purpose, such as plastic sandwich boxes, bottles with large openings, storage boxes, and plastic bags with or without zippers. The key to performing moist chamber induction is having a closable container or bag, adding the appropriate amount of water with absorbing materials such as filter paper or paper towels, and incubating at room temperature under light.

## Isolation and Extraction

Isolation is one of the most important procedures in clinical diagnosis and plant pathology research (Schuck *et al.*, 2014). Most plant specimens, after initial visual or microscopic examinations, may be plated on one or more types of culture media to isolate pathogenic microorganisms. Most plant pathogens can be isolated on culture media, but many obligate and fastidious pathogens cannot. The media commonly used in a plant diagnostic lab can be classified into three groups: (i) a general-purpose medium to isolate a broad range of organisms; (ii) a semi-selective medium to isolate a certain group of organisms (Kini *et al.*, 2019); and (iii) a highly selective medium to isolate specific organisms. The use of each type of medium is based on the organisms and purpose of diagnosis.

For organisms that cannot be isolated on a culture medium, they can be isolated through an extraction process. For example, viruses, obligate pathogens of plants, can be purified from plant tissue. Similarly, plant-parasitic nematodes can be extracted from plant tissue through either a moist-chamber extraction system or Baermann funnel method. For some fastidious bacteria such as phytoplasmas and *Xylella fastidiosa*, neither culture nor organism extraction can be practically achieved in routine diagnostics. In this case, the DNA of such an organism can be extracted for molecular diagnoses.

### Fungal and oomycete isolation

#### Procedures

To isolate a fungal pathogen or oomycete from a symptomatic plant, the symptomatic tissue must be collected from the plant specimen using a single-use razor blade. Place a small portion of the specimen that contains lesions or decayed tissue on a piece of clean parafilm, use a razor blade to cut ten pieces of tissue (approximately 4 mm × 4 mm) from the marginal area of diseased tissue (typically at the junction of damaged and healthy tissue) and place the tissue pieces in the first quarter of a 100 mm × 15 mm quad-plate (divided Petri dish). For most fungal pathogens, they infect and kill plant tissue and then move into the fresh plant tissue; therefore, the marginal area of a symptomatic tissue is an ideal location to isolate the primary pathogen. The number of pieces needed for plating depends on the medium to be used and the diagnostic purpose. Once the tissue pieces are collected from the specimen, they are ready to be plated using the following procedure.

1. Turn on a glass bead sterilizer inside the biosafety cabinet (BSC). It may take about 30 min to reach 250°C. Note that the bead sterilizer is a separate piece of equipment from the BSC.
2. Turn on the UV light to sterilize the interior surface of the BSC for 15–30 min.
3. Turn off UV light, turn on regular light and blower, and spray/wipe down biosafety cabinet counter top with 70% ethanol.
4. Inside the biosafety cabinet, add 1% sodium hypochlorite (NaClO) to the first quarter of the Petri dish to immerse all sample pieces.
5. Fill the remaining three-quarters with autoclaved water.
6. With sterilized forceps, submerge the tissue pieces in the NaClO solution for 30–60 s. The forceps should be sterilized by dipping in 95% ethanol and placing in the bead sterilizer for 15 s before and after each use.
7. Rinse the tissue pieces by transferring into the next quarter of the Petri dish filled with sterilized water. Allow tissue pieces to sit in water for 30–60 s by using forceps to submerge pieces intermittently.
8. Repeat step 7 until all the tissue pieces have been rinsed and transferred to the fourth quarter; remove them from the Petri dish and place them on a piece of autoclaved filter paper to absorb extra water. The tissue pieces are surface sterilized and ready to be plated.
9. Place the pieces on the surface of pre-made culture medium solidified in the Petri dish, such as potato dextrose agar (PDA) and corn meal agar amended with pimaricin, ampicillin, rifampicin and pentachloronitrobenzene (PARP) plate. Up to five pieces can be plated on PDA and up to ten pieces on the PARP plate.

**10.** Carefully, so as to not dislodge the tissue pieces from the medium surface, wrap the plates with parafilm and place the plates upside down in a 22°C incubator.

**11.** Check for growth in 2–5 days and transfer each representative colony into a new culture plate to obtain a pure culture for further identification.

### Medium selection

There are many types of culture media formulated differently to isolate specific fungal pathogens from plant tissue or soil (Shurtleff and Averre, 1997). From the diagnostic perspective, a small set of commonly used media is sufficient to handle the vast majority of samples suspected for fungal or oomycete infections. PDA is the most used medium to isolate a broad range of fungal and oomycete species and it is considered a general-purpose medium. Bacteria can grow on the PDA plate if it is not supplemented with acid or antibiotics. However, it is not recommended to use PDA to isolate bacteria from plant tissue. In general practice, PDA medium amended with streptomycin (100 µg/ml) is used to isolate fungi and certain oomycetes without growth of bacteria. If a sample is suspected for a fungal pathogen, the PDA amended with streptomycin should be used. Certain oomycetes are sensitive to streptomycin and their growth may be inhibited on the streptomycin-amended medium. To isolate species of oomycetes, the PARP medium is popularly used. The PARP medium is a cornmeal agar amended with PCNB (pentachloronitrobenzene), ampicillin sodium salt, rifampicin and pimaricin. Note that rifampicin is toxic and its waste should be disposed of properly. Any sample suspected for oomycete infections may be plated on PARP to isolate *Phytophthora* or *Pythium* species that cause many diseases on crops, including hemp and cannabis. Colonies grown on PARP plates can be transferred to cornmeal agar (CMA), PDA without being amended with streptomycin, and V8 juice agar medium for further studies. Although PDA and PARP are the main media used to isolate fungi and oomycetes, there are other media that are needed for specific diagnostic purposes (Table 5.2).

### Amending media with antibiotics

A stock solution should be made, if possible, at a concentration that is easy to add into the medium without doing maths. For example, in Table 5.3, all antibiotic solutions except tetracycline can be added in the format $x$-microlitres of antibiotic solution into $x$-millilitres. The stock solution should be made with sterilized deionized water and then filtered through a 0.2 µm pore size membrane filter (to be sterile). Some antibiotics are light-sensitive, so the tube should be wrapped in aluminium foil or the solution should be stored in an amber tube. The following is a general protocol for making a sterile 100 mg/ml ampicillin solution.

**Table 5.2.** Commonly used culture media in a plant diagnostic lab.

| Medium name | Abbreviation | Purpose |
|---|---|---|
| Potato dextrose agar | PDA | General culture of fungi and bacteria |
| Potato dextrose agar amended with streptomycin (100 µg/ml) | PDA[+strep] | General culture of fungi |
| V8 juice agar | V8 | Growth of oomycetes or sporulation of fungi |
| Cornmeal agar | CMA | Maintenance of oomycetes |
| 2% Water agar with streptomycin (100 µg/ml) | WA[+strep] | Isolation of a single spore to obtain single spore isolates |
| PARP medium | PARP | Selective isolation of oomycetes |
| Peptone PCNB agar | PPA | Selective isolation of *Fusarium* |
| Nutrient agar or broth | NA | General culture of non-fastidious bacteria |
| Luria–Bertani agar | LB agar | Culture and maintenance of *E. coli* used in molecular biology |
| LB agar amended with ampicillin | LB[+amp] | Selection of transformed competent *E. coli* clones |
| Luria–Bertani medium | LB broth | General culture of competent cells[a] |
| Super broth medium | Super broth | General culture of competent cells |
| SOC medium | SOC | Recovering competent cells after heat shock |

[a]Competent cells are used in cloning of a PCR product in DNA sequence-based identification.

**Table 5.3.** Amending culture media with an antibiotic at a desirable final concentration in medium.

| Stock solution | Add amount | Concertation in 250 ml of medium |
|---|---|---|
| 100 mg/ml ampicillin | 250 µl | 100 µg/ml |
| 50 mg/ml kanamycin | 250 µl | 50 µg/ml |
| 10 mg/ml tetracycline | 75 µl | 3 µg/ml (light sensitive) |
| 100 mg/ml streptomycin | 250 µl | 100 µg/ml |

1. Add 20 ml of molecular-grade water into a 50 ml Falcon centrifuge tube.
2. Add 2 g of ampicillin sodium salt (choose water-soluble product) to water.
3. Vortex to mix.
4. Transfer solution into a syringe fitted with a 0.2 µm pore-size membrane filter.
5. Push syringe and filter solution into a sterilized tube.
6. Aliquot solution into 1.5 ml microcentrifuge tubes and store at –20 °C.

### Assessment of colony growth

In 2–5 days after plating, the primary fungal pathogen or oomycete may grow out from most, if not all, tissue pieces. When only a single species or the same type of colony grows out from over 70% of tissue pieces plated, this suggests a strong association of the given organism with the symptomatic tissue. However, if a plate only shows a low percentage of fungal growth with some random colonies of different fungal species, this indicates that those fungi are not responsible pathogens. In this case, the diagnostician may rethink the diagnostic hypothesis and approach. In hemp disease diagnosis, it is not uncommon to see two to multiple types of fungal colonies growing out from tissue pieces taken from a single diseased plant (Fig. 5. 12). Some may grow faster than others, with slow-growing pathogens showing up in an extended time frame. In some cases, multiple types of media may be needed to isolate different groups of pathogens and the frequency of each type of colony in each type of medium may be counted to assess their correlations to the disease. Isolation of multiple organisms from a single plant or crop may suggest a disease complex in which more than one organism contributes to the disease.

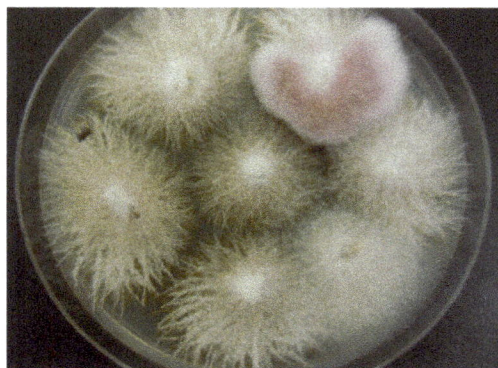

**Fig. 5.12.** Three species of *Fusarium* (*F. redolens*, *F. tricinctum* and *F. equiseti*) were isolated from seven tissue pieces obtained from a single root exhibiting root rot. Note that both *F. redolens* and *F. tricinctum* (pink colour) were growing out from the same pieces of tissue (at 1 o'clock position). The colony at 5 o'clock position was *F. equiseti*. All other colonies were *F. redolens*.

### Medium formulas and procedures

Commonly used media such as PDA, CMA, LB agar, LB broth and SOC can be purchased from commercial vendors. Most are in the form of powder. To make them, simply dissolve the medium powder into the deionized water following the manufacturer's instructions. Some media are in liquid form and ready to use. Below is a general protocol to make medium plates using 2% water agar as an example.

1. Add 10 g of agar to 500 ml deionized water in a large beaker and mix thoroughly.
2. Microwave until agar is dissolved.
3. Pour solution into two 500 ml flasks, each with 250 ml of medium.
4. Cover top of flask with two layers of aluminium foil, attach a piece of autoclave indicator tape, label each flask with the name of medium and autoclave using the programmed liquid cycle.
5. After autoclaving, allow the medium to cool down in the biosafety cabinet to 55°C (when bare hands can comfortably hold the flask).
6. Slowly pour approximately 20–25 ml of medium into each 100 mm × 15 mm Petri dish and let the medium solidify at room temperature inside the biosafety cabinet.
7. Label plates and store them upside down in a clean container at 4°C.

Selective media or media incorporated with antibiotics (antifungal and antibacterial) require an extra step of adding a specific amount of stock antibiotic solution into the autoclaved medium at approximately 55°C before pouring into the Petri dish. Below is a protocol to make PARP medium plates for isolation of *Phytophthora* and other oomycetes (500 ml volume).

1. Add 8.5 g cornmeal agar to 500 ml deionized water in a large beaker and mix thoroughly.
2. Microwave until ingredients are dissolved.
3. Pour solution into two 500 ml flasks, each with 250 ml of medium.
4. Cover the tops of the flasks with two layers of aluminium foil, place a piece of autoclave indicator tape on, label each flask with PARP.
5. Loosely wrap a 100 ml beaker in aluminium foil for use in preparation of the antibiotics mix.
6. Autoclave the flasks containing medium and the 100 ml beaker using the programmed liquid cycle.
7. After autoclaving, allow the medium to cool down in the biosafety cabinet to 55°C.
8. Weigh the following antibiotics and antifungal ingredients and add them to the autoclaved 100 ml beaker:
   - 0.125 g ampicillin sodium salt (water-soluble)
   - 0.05 g PCNB (pentachloronitrobenzene)
   - 0.005 g rifampicin
   - 200 µl of 2.5% aqueous pimaricin solution.
9. Inside the biosafety cabinet, pipette 4 ml of sterilized room-temperature water into the beaker containing the four antibiotic and antifungal ingredients (step 8) and mix thoroughly.
10. When the medium cools down to 55°C, add 2 ml of the antibiotics mix into each flask of 250 ml medium and mix by lightly swirling to prevent bubbles.
11. Slowly pour approximately 20–25 ml of medium into each 100 mm × 15 mm Petri dish and let the medium solidify at room temperature inside the biosafety cabinet.
12. Label plates and store them upside down in a clean container at 4°C.
13. PARP plates are good for one month at 4°C. Plates are considered expired after 30 days or when the medium is thin, cracked or dried.

Certain antibiotic solutions can be made in advance, aliquoted, and stored at –20°C for easy use. Below are the procedures to make stock streptomycin sulfate or other antibiotic solutions at a concentration of 100 mg/ml, and how to add them into the medium solution to achieve a concentration of 100 µg/ml.

To make streptomycin sulfate stock solution (100 mg/ml):

1. Weigh 2.0 g streptomycin sulfate and transfer it into a clean 100 ml glass beaker.
2. Add 20 ml nanopure water into the beaker and stir for 30 min or until the ingredient is dissolved.
3. Collect the solution in a 20 ml sterile syringe, place a sterile Millipore filter (filter unit 0.22 µm) on the top of the syringe and push the syringe to filter the solution into a sterile 50 ml tube.
4. Aliquot filter-sterilized solution into sterilized 1.5 ml microcentrifuge tubes and store at –20°C.

To amend a medium with 100 mg/ml stock streptomycin sulfate or ampicillin solution:

1. Thaw, vortex and then spin down the solution.
2. Add equal amount (in µl) of stock solution to equal amount (in ml) of medium. For example, if the medium amount is 250 ml, add 250 µl of antibiotic solution to it to achieve a final concentration of 100 µg/ml.
3. Shake medium slightly to distribute the antibiotic solution evenly.

PPA plates are used to selectively isolate *Fusarium* species from the soil and plant tissue (Leslie and Summerell, 2006). PPA inhibits growth of most fungi and bacteria but allows the growth of *Fusarium* species. Below is the protocol to make PPA plates.

1. Add the ingredients to a large beaker and mix thoroughly.
   | | |
   |---|---|
   | Peptone | 7.5 g |
   | $KH_2PO_4$ | 0.5 g |
   | $MgSO_4 \bullet 7H_2O$ | 0.25 g |
   | PCNB (pentachloronitrobenzene) | 0.375 g |
   | Agar | 10.0 g |
   | Deionized water | to 500 ml |
2. Pour solution into two 500 ml flasks, each with 250 ml of medium.
3. Cover the tops of flasks with two layers of aluminium foil, place a piece of autoclave indicator tape on, label each flask as PPA, and autoclave using the programmed liquid cycle.
4. After autoclaving, allow the medium to cool down in the biosafety cabinet to 55°C.
5. Add 2.5 ml of streptomycin sulfate stock solution (100 mg/ml) and 3 ml of neomycin sulfate stock solution (10 mg/ml) into each flask. Gently swirl each flask to mix without creating bubbles.

6. Slowly pour approximately 20–25 ml of medium into each 100 mm x 15 mm Petri dish and let the medium solidify at room temperature inside the biosafety cabinet.

7. Label plates and store them upside down in a clean container at 4°C.

## Bacterial isolation

### Serial dilution method

Isolation of bacteria from infected plant tissue can be done in different ways depending on the organism suspected and the preference of diagnosticians. The traditional way is to macerate a small piece of tissue in a tube of sterile water, make a series of 1:10 dilutions three times in sterile tubes, place 0.5 ml of each dilution into a separate Petri dish, add cooled agar medium, gently mix to evenly distribute bacterial cells in the agar medium and let the medium solidify (Agrios, 1997). After a few days of incubating (plates upside down) at 28–30°C, single colonies appear on the plates. If a bacterial pathogen is present in the plant tissue, the plate with 1:10 dilution may show crowded colonies, depending on the number of bacteria in the tissue. However, the plate with 1:1000 dilution may have far fewer colonies, which allows easy picking of any representative colonies from the plate for further subculture or purification. The initial tissue pieces must be surface sterilized using similar procedures described in the section on fungal isolation, above.

### Streaking plate method

Some labs use an inoculating loop to streak a loop of bacterial suspension on nutrient agar or other culture media. Select pieces of infected plant tissue and briefly grind them in a small amount of sterilized water (e.g. 1 ml) using a mortar and pestle. Dip an inoculating loop into the ground solution and then streak one-quarter area of the plate. The first streak contains a high number of bacteria cells. Sterilize the loop or use a new disposable loop, cross over the initial streak once and streak another quarter of the plate. Repeat the procedure two more times until the full plate is streaked in four quarters (Fig. 5.13). After a few days of incubating (plates upside down) at 28–30°C, single colonies may appear on the fourth streak, and the first and second streaks will exhibit bacterial smears.

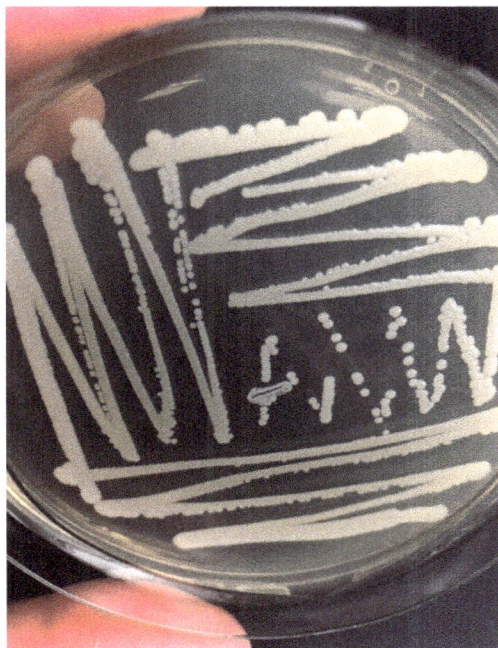

**Fig. 5.13.** A bacterium species isolated from a hemp plant exhibiting extensive crown rot. Note that single colonies appear in the fourth streak area.

### Glass beads spread

Instead of pouring cooled culture medium into a diluted bacterial preparation as described in the serial dilution method, a small amount of diluted bacterial preparation can be transferred to a culture plate by a pipette. Add three sterilized 5 mm solid-glass beads into each plate and gently swirl and shake the plate for 30 s to allow the beads to roll over the entire plate. Dump the beads into a biohazard container, incubate the plate at the desired temperature and examine the individual colonies on the plate within 24–48 hours (see Fig. 2.21). To obtain distant single colonies on the plate, different volumes (e.g. 10 µl, 30 µl and 60 µl) from 1:100 or 1:1000 dilution preparations can be added into three plates, respectively. Often, the least amount of volume (10 µl) gives well-separated colonies for subsequent pick-up.

### Single colony pick-up and culture

A single colony can be picked up using an inoculating loop and used to streak bacterial cells on a new culture plate for further culture. More frequently, a single colony is further cultured in a liquid medium to obtain a larger volume of bacterial cells for downstream diagnostic procedures. In a biosafety cabinet, use a 2–20 μl pipette loaded with a sterilized tip. Point the tip to a colony, gently scratch the medium surface to pick the colony and then dip it into a 3 ml LB broth or nutrient broth in a sterile 17 × 100 mm culture tube. Close the tube with cap, place it in a rack inside the incubator and incubate with shaking (200–250 rpm) at the desired temperature overnight. Bacterial cells can be collected by centrifuging at 8000 rpm for 5 min.

## Virus isolation and purification

Nowadays, the diagnosis of viral diseases is mostly performed using ELISA kits or immunostrips to test a specific virus or a few grouped viruses. As discussed in Chapter 3, an ELISA kit is used to test a specific suspected virus. For example, there is little information on hemp or cannabis viruses and some labs or cultivators have tried many different ELISA and immunostrip kits to test a handful of viruses without any positive results. For an unknown viral disease, a universal diagnostic approach should be used, which includes isolation of the virus, propagation on an indicator plant, purification of virus particles and observation under an electron microscope. These diagnostic methods are traditional but still valuable tools for diagnosing and characterizing a new virus or disease.

### Isolation

There is a list of indicator plants that are susceptible to a broad range of plant viruses (Table 5.4). These plants should be grown in a greenhouse to produce healthy and virus-free seeds for routine bioassay tests. To isolate a virus from a symptomatic plant specimen onto indicator plants, young indicator plants should be grown in a greenhouse at the appropriate age. Too old or too small indicator plants are not optimal for the successful inoculation of a given virus. In general, cucumber plants at the cotyledon stage are the best for virus inoculation, and tobacco plants at 2–4-leaf stage with cross-plant size of 10 cm are ideal. However, *Chenopodium* plants with up to 10–12 leaves can still be inoculated. After successful inoculation, some indicator plants produce systemic symptoms such as mosaic or mottle (see Fig. 2.32), while some

produce necrotic lesions (Fig. 5.14). Both systemic symptoms and localized lesions indicate the presence of a transmissible virus in a plant specimen, and characteristic symptoms on a given set of indicator plants offer important diagnostic information on what type of virus is infecting the plant.

**Table 5.4.** Indicator plants commonly used to isolate and maintain viruses.

| Common name | Scientific name |
| --- | --- |
| Benthi tobacco | *Nicotiana benthamiana* |
| Chenopodium (wild) | *Chenopodium amaranticolor* |
| Common bean | *Phaseolus vulgaris* |
| Cowpea | *Vigna unguiculata* |
| Cucumber | *Cucumis sativus* |
| Pea | *Pisum sativum* |
| Pepper | *Capsicum annuum* |
| Petunia | *Petunia hybrida* |
| Quinoa | *Chenopodium quinoa* |
| Tobacco | *Nicotiana tabacum* |
| Tomato | *Solanum lycopersicum* |

**Fig. 5.14.** Localized lesions on a *Chenopodium* plant leaf after it was mechanically inoculated with *Tomato ringspot virus*.

Different individuals may follow slightly different procedures when inoculating indicator plants with a virus, but they all follow a commonly accepted protocol (Browning, 2009). Below is the general procedure used to mechanically inoculate a suspected virus from infected plant tissue to a series of herbaceous indicator plants.

1. Place young indicator plants on a bench, wear a protective face mask and sprinkle a thin layer of carborundum powder (silicon carbide) on the leaves to be inoculated.
2. Place one or several symptomatic leaves, localized lesions, or dried and frozen ($-80°C$) infected plant tissue in a sterile mortar, add 1 or 2 ml of 0.05 M phosphate buffer (pH 7.0) or sterilized water and grind the tissue with a sterile pestle.
3. Wearing gloves, dip a finger into the plant sap prepared in step 2, apply it to a leaf to be inoculated, place another hand under the leaf to provide support and gently rub the leaf 4–6 times. Instead of a finger, a sterile cotton swab may be used.
4. After inoculating all the indicator plants, briefly rinse inoculated leaves with sterilized water using a squeezable plastic wash bottle. To increase virus infection, place inoculated plants in the dark for 2 h and then move them to a bench in the greenhouse. Alternatively, inoculate the plants in late afternoon.
5. Observe any symptoms after 2–3 days. Localized lesions may show up in a few days, but systemic symptoms may take up to 3 weeks to develop, depending on the virus and the titre (the concentration of viruses in a sample).

### Propagation and purification

If one or more indicator plants show typical viral symptoms after inoculation, this suggests that the donor plant or plant sample contains an infectious virus. A single necrotic lesion, if it occurs on an indicator plant, may be ground and used to inoculate susceptible plants to maintain the virus in the greenhouse for further studies. To extract and purify virus particles from plant tissue, the virus must be inoculated into a propagation host that is easy to grow and supports a high concentration of viruses in the tissue. For many viruses, species of *Chenopodium*, *Cucumis*, *Nicotiana*, *Petunia*, *Phaseolus* and *Vigna* are used as propagation hosts for virus purification. Purification procedures vary from virus to virus and are affected by the type of

host plant. In general, the process involves tissue homogenization, low-speed centrifugation to remove the macromolecular host material, virus precipitation with PEG (polyethylene glycol) and further purification by density gradient centrifugation. Certain purification procedures may damage virus particles. Peyret (2015) described a gentle protocol to extract virus-like particles (VLPs) from plant tissue and suggested that it can be used to purify an unknown virus. Depending on the research purpose, a virus preparation can be further purified and concentrated by removing other contaminants. However, there is no absolutely pure virus preparation. In most plant diagnostic labs, it is unrealistic to try to obtain a highly purified virus preparation, due to equipment limitations. Fortunately, a partially purified virus preparation can be used for certain diagnostic procedures, such as electron microscopic examination. The following protocol is for the purification of *Turnip mosaic virus* (TuMV) (Choi *et al.*, 1977), which can be used as a starting point for purifying some unknown viruses from plants, especially for rod-shaped viruses.

1. Freeze infected leaves at $-20°C$.
2. Homogenize frozen leaf tissue in 0.5 M phosphate buffer (pH 7.5) containing 0.01 M Na-EDTA and 0.1% thioglycolic acid.
3. Filter through two layers of cheesecloth and collect crude sap.
4. Centrifuge crude sap at 3200 $\times g$ for 10 min and collect supernatant.
5. Add 1% Triton X-100, 4% PEG and 0.1 M NaCl (all final concentrations) to the supernatant and stir at room temperature for 2–3 h.
6. Centrifuge at 8500 $\times g$ for 15 min and save pellet.
7. Resuspend pellet in 0.5 M phosphate buffer (pH 7.5) containing 0.01 M $MgCl_2$.
8. Centrifuge at 8500 $\times g$ for 10 min and collect supernatant.
9. Centrifuge supernatant at 65,000 $\times g$ for 90 min and save pellet.
10. Resuspend pellet in 0.5 M phosphate buffer (pH 7.5) containing 0.01 M $MgCl_2$.
11. Centrifuge at 8500 $\times g$ for 10 min and collect supernatant.
12. Load supernatant into a 10–40% sucrose density-gradient column.
13. Centrifuge at 61,000 $\times g$ for 2 h.
14. Collect the virus from the specific zone of the gradient and dilute with 0.01 M phosphate buffer (pH 7.0).

**15.** Centrifuge at 69,000 ×*g* for 90 min and save pellet.

**16.** Diffuse pellet in 0.01 M phosphate buffer (pH 7.0).

**17.** Centrifuge at 8500 ×*g* for 10 min and collect supernatant as purified virus.

### Observation under electron microscope

A diluted (1:10) virus preparation or plant sap from infected tissue can be negatively stained to observe the virus particles under an electron microscope. Add one drop of diluted virus preparation or fresh plant sap on a piece of parafilm, float several carbon-coated grids on the surface of the drop with the coated side touching the solution, and let grids sit for 1–2 min. Use self-locking forceps to remove the grid from the drop and touch it with a piece of filter paper to absorb extra liquid. Add a drop of 2% phosphotungstic acid, float the grids, let it sit for 30–60 s and then remove extra liquid using filter paper. Place the grids in a glass Petri dish with a filter paper in the bottom and incubate at 37°C in oven for 1 h. The grids can now be loaded onto an electron microscope for observation. In the observation field, there may be hundreds of uniform particles observed from the purified virus preparation (see Figs 2.27 and 2.28), but for fresh sap preparation, virus particles may be sporadically distributed in a background of other electron-dense materials. With helical rod-shaped virus particles, the central hollow is often revealed by the stain. As most spherical plant viruses are about 30 nm in diameter and rod-shaped or filamentous viruses can be up to 2000 nm in length, a magnification of 10,000–33,000× is needed to view individual virus particles.

### Nematode extraction

There are three methods commonly used in a nematology lab to extract live nematodes from soil or plant tissue: (i) Baermann funnel; (ii) mist chamber system; and (iii) wet sieving plus sugar flotation. For root-knot nematodes, root tissue can be dissected and observed directly under a microscope.

### Baermann funnel

This method can be used on a small scale to extract active nematodes from plant tissue or soil samples. To assemble this device, place a clamped rubber tube below the funnel and then place the funnel onto a rack. Place a small piece of metal mesh or window screen inside the mouth of the funnel and then place two layers of Kimwipes or other tissue paper on top of the screen to fully cover the mouth area. Now the device is ready for use (Fig. 5.15). For multiple samples, several funnels can be placed in an array of holes made on a wooden board. To extract nematodes, add soil or chopped plant tissue on top of the paper tissue and then add water to the funnel until samples are immersed. Fold the extra paper tissue to the inside of the funnel to avoid water dripping. Wait overnight or 48 h to allow nematodes to move down to the bottom of the funnel tube (due to gravity) and collect the first 5–10 ml of water from the bottom of the tube by slowly releasing the clamp. Examine nematodes under a microscope.

### Mist chamber

When nematodes need to be extracted from a large number of plant tissue samples, a mist chamber system can be used (see Chapter 4). This system is similar to the Baermann funnel method but with

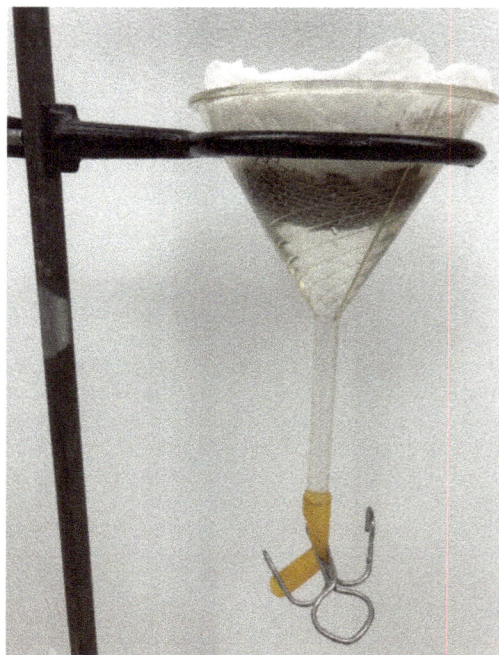

**Fig. 5.15.** A Baermann funnel set-up to extract nematodes from soil or plant tissue.

slight modifications. The following protocol can be used to set up a mist chamber extraction.

1. Place a glass tube into each slot of a holding rack and insert a funnel tube into the glass tube.
2. Place a stainless-steel mesh on top of the funnel mouth.
3. Place two to three layers of Kimwipes or tissue paper on top of the mesh.
4. Add 4" (10 cm) tall PVC pipe ring (as funnel extension) on top of the Kimwipes.
5. Add soil or chopped plant tissue into the PVC ring.
6. Place the entire rack (with 12 or more funnel set-ups) into the mist chamber.
7. Turn on the water and timer. Samples are sprayed for 90 s every 10 min.
8. After 48–72 h, collect tubes and pour all liquid and nematodes into a No. 325 sieve, gently collect nematodes from the sieve into a beaker using a wash bottle.
9. Observe nematodes under a microscope.

### Wet sieving

Most soil samples are processed by wet sieving followed by sugar flotation. Because all contents after wet sieving must be cleaned by centrifugation, samples can be processed in a group that matches the number of tubes in the centrifuge. For example, the centrifuge holding 28 tubes (see Fig. 4.6) can handle 28 samples in one load. A tray of 28 beakers arranged in 4 × 7 should be in place. Below are the procedures used for extracting vermiform nematodes from 28 soil samples. The procedures are performed at a large and specially designed stainless-steel sink with a durable rubber tube connected to the tap (see Fig. 4.7). For cyst nematode extraction, replace sieve No. 325 with No. 60.

1. Label the bottom left-hand corner of a fast-food tray with an 'L' and place 28 beakers (100 ml) in four rows of seven on the tray.
2. Prepare 28 or more pieces of paper in the size 2" × 2" (5 cm × 5 cm).
3. Pour 200 ml of soil into a 5-gallon (20 l) bucket and fill it with cool tap water until two-thirds full.
4. While filling water, use a hand to break soil particles until soil is completely dissolved, then agitate with a strong stream of water.
5. Let the bucket sit for 60–120 s.
6. Place a No. 20 sieve on the top of a No. 325 sieve.

7. Slowly pour the water into the nested 20/325 sieves and leave precipitated soil in the bucket.
8. Rinse No. 20 sieve gently with tap water and remove it from the No. 325 sieve.
9. Slightly tilt the No. 325 sieve and apply a gentle water stream to collect all contents into one area close to the edge, then use a wash bottle to collect all the contents into a 100 ml beaker.
10. Place the beaker back onto the tray, write the sample number on the 2" × 2" (5 cm × 5 cm) square of paper and put it underneath the beaker.
11. Dump the soil into the waste bin and clean the bucket, sieves and sink surfaces before processing the next sample.
12. To facilitate centrifugation with numerically labeled centrifuge tubes, the 28 samples can be arranged with the first top row (left to right) being samples 1–7, second row samples 8–14, third row samples 15–21 and fourth row samples 22–28.
13. Proceed to sugar centrifugation or store at 4–8°C.

### Sugar flotation

Sugar solution has a high enough density to float nematodes while most soil and debris may be precipitated during low-speed centrifugation, which allows further separation of nematodes from soil. There are two types of sugar solutions: light and heavy. Light solutions are for flotation of vermiform nematodes and heavy solutions are for cysts. To make a light sugar solution, add 500 ml granulated white sugar into 700 ml hot water and stir to dissolve. To make a heavy sugar solution, add 750 ml sugar into 600 ml hot water, then stir to dissolve.

There are two steps of centrifugation to obtain a clean nematode suspension that can be observed under a stereo (i.e. dissecting) microscope. The first step is to precipitate nematodes and soil into the bottom of the tube so that water can be removed. The second step is to float nematodes in the sugar solution while soil is pelleted on the bottom. The following is a general protocol for handling 28 samples for centrifugation.

1. Position the tray with 'L' always in the bottom left corner and align four tube holders in order, each with seven numerically labelled centrifuge tubes.
2. With a tray of 28 samples arranged in four rows of seven samples, transfer the top first row of samples to the first set of tubers labelled 1 to 7 by gently swirling the contents of the beaker and pouring it

into the corresponding tube. Similarly, transfer the second, third and fourth row into the corresponding sets of tubes.

3. Use a wash bottle to add water to the tubes so that each has about the same volume.

4. Centrifuge for 5 min at 1440 rpm.

5. Carefully pour the supernatant of each tube into a sink and save the pellet.

6. Add approximately 45 ml sugar solution to each tube and resuspend the pellet.

7. Centrifuge for 1 min at 1440 rpm.

8. Pour supernatant into nested 3" (7.5 cm) diameter 20/325 sieves.

9. Remove the No. 20 sieve after rinsing it with water and collect contents from No. 325 sieve.

10. Repeat with the rest of the samples until the tray's 28 beakers are filled with clean nematode suspensions.

11. Observe nematodes under a microscope or store at 4°C until use.

### Preserving nematodes

Nematode specimens can be preserved in 2% formaldehyde solution and stored at room temperature for at least a year without significant effect on the integrity of the structures.

1. Let a nematode suspension in a beaker sit for 30 min, then remove water slowly from top of the solution using a glass pipette until there are only 10–20 ml remaining.

2. Swirl the beaker gently and pour the contents into a small glass vial.

3. Allow the nematodes to settle to the bottom of the vial.

4. Use a pipette to remove the excess water gently from the vial until less than half of the vial is filled with water and nematodes.

5. Under a fume hood, add to the nematode suspension in the vial an equal amount of 4% formaldehyde to bring the final concentration to 2%.

6. Gently mix to kill and fix nematodes and then close the vial with a cap.

### Observing nematodes under a compound microscope

For instant observation of live nematodes extracted from soil or plant tissue, a few nematodes can be picked using a toothpick sharpened with a razor blade.

The tip of the toothpick should be about 3 times the diameter of the nematode to be picked. Place the nematodes in a drop of water at the centre of a glass slide, cover with a cover slip and then observe under a compound microscope. The following is a general guideline to transfer nematodes to a slide.

1. Concentrate nematodes in 10–20 ml of water by removing extra water after letting it sit for 30 min.

2. Swirl the beaker gently and pour the contents into a nematode counting dish.

3. Place the dish under a stereo microscope with transmitting light on.

4. Use left hand to adjust focus to see nematodes at the bottom of dish while the right hand holds a sharpened toothpick and points to the nematode to be picked.

5. Slightly touch the nematode and push it up into the water and at the same time adjust the focus to follow the nematode. Continue to do so until the nematode reaches the surface of the water.

6. Place the toothpick tip underneath the nematode (at the middle portion) and quickly pick it out of the water.

7. Gently touch the toothpick tip into a drop of water at the centre of a glass slide, slightly moving the tip to allow the nematode to be released into the water.

8. Continue steps 4–7 until desirable amount or types of nematodes are picked. Place a cover slip and place the slide under the compound microscope. Use the lower magnification to locate nematodes and then switch to higher magnification.

9. If nematodes are too active and hamper observation, place slide over the flame of an alcohol burner for a few seconds to gently kill the nematodes with heat.

### Making permanent slides

Nematodes can be permanently preserved in a slide for teaching or reference purposes. The following is a protocol to make permanent nematode slides.

1. Pick about ten nematodes of the same species from a fixed nematode suspension into a 1" (2.5 cm) diameter glass dish with 2% glycerol. Place the dish at room temperature to allow water evaporation.

2. After 1 week, place the dish into a desiccator to further remove water from the dish (1–3 days).

3. Add one drop of 100% glycerol in the centre of a square cover glass (22 mm × 22 mm), place three pieces of fibre glass (slightly thicker than nematodes)

in the drop (arranged as a triangle) and pick five to ten glycerol-saturated nematodes from the dish into the drop.

**4.** Cover the drop with a smaller round cover slip and seal with nail polish.

**5.** Insert the slide into an aluminum slide slot first and then add two pieces of smooth white cardboard paper (25 mm × 25 mm × 1.5 mm) from both ends to secure the glass slide in the middle. Gently press the curved edge to hold the cardboard paper in place (Fig. 5.16).

**6.** Write specimen information on the left-hand piece of paper and indicate the relative location of each nematode in the slide on the right-hand piece of paper.

**7.** Store the slide in a slide box at room temperature.

## Serology-based Tests

### Principle

Serology is a term used in medical sciences that defines the study of blood serum, the clear fluid that separates out when blood coagulates. When a person has been exposed to a particular microorganism, plasma B cells secrete large quantities of proteins called antibodies in response to specific intruding substances. Those substances are typically proteins, peptides (a short amino acid chain) and polysaccharides that are typically present on the surface of viruses and bacteria. Such molecules are called antigens, as they trigger an immune response. The antibodies are then transported to the site of the antigen to neutralize or destroy it. This is a unique mechanism in animal immune systems for defending against infections caused by pathogens such as viruses and bacteria. An antibody is a Y-shaped protein that recognizes a unique antigen and binds to it. The specificity between antigens and antibodies is analogous to the key–lock relationship, which means that each type of antigen triggers production of a specific antibody that only binds to that antigen. Based on this biochemical mechanism, many types of clinical tests have been established to detect infectious agents in both human patients and plants, and these tests are generally classified as serology based. For example, a serological test can determine whether a person has been exposed to a particular microorganism by detecting the presence

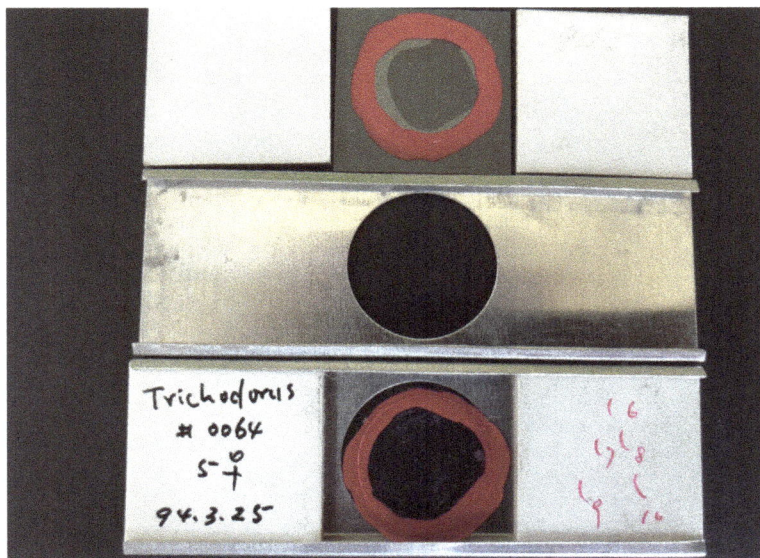

**Fig. 5.16.** An example of nematode specimens permanently mounted on a slide. Note that a completed slide (bottom) requires two pieces of white cardboard paper, one square cover glass, one round cover glass (top) and an aluminium slide with a curling edge and hole in the centre (middle). Additional materials include 100% glycerol and regular nail polish to seal the edge of the glass slide.

of the antigen and/or antibodies against the antigen. In plant diagnostics, antibodies specific to plant viruses can be produced in laboratory animals such as rabbits by inoculating the animals with the virus. Polyclonal antibodies are obtained directly from serum (bleeds), but the production of monoclonal antibodies requires more steps, such as creating monoclonal hybridoma cell lines. Both polyclonal and monoclonal antibodies can be used to detect specific pathogens in plants. Figure 5.17 illustrates the principles of two serology-based diagnostic methods: ELISA and immunofluorescent microscopy. With the same principle, other methods have also been developed and used in diagnostics; these include the immunostrip test and immunogold labelling (Gulati *et al.*, 2019).

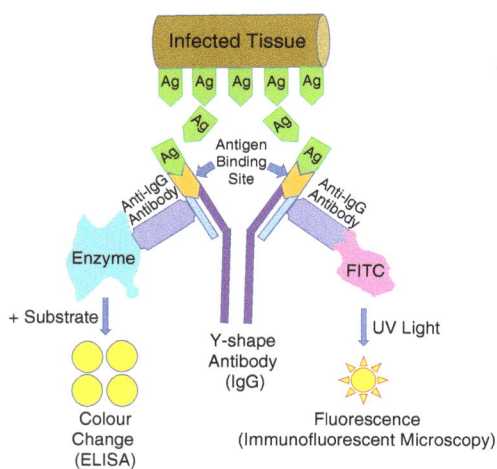

**Fig. 5.17.** The principle of serology-based methods for plant pathogen detection. The Y-shaped immunoglobulin G (IgG) contains two identical heavy chains and two identical light chains. It has two antigen (Ag)-binding sites. In a sample containing antigens, the IgG binds specifically to the antigen. To visually detect the antigen, a conjugated secondary antibody against the IgG (anti-IgG antibody) is added to bind the primary antibody (IgG). If the secondary antibody is conjugated with a fluorescent dye such as fluorescein isothiocyanate (FITC), a sample containing antigens emits fluorescence under a confocal or fluorescent microscope (right-hand portion of the figure). If the secondary antibody is conjugated with an enzyme such as alkaline phosphatase, wells of the ELISA plate containing antigens turn the substrate to colour (left-hand portion of the figure).

## Immunostrip test

Some commercial companies offer immunostrip products to test for viruses, bacteria and oomycetes. Immunostrips may be designed to test for pathogens at species or genus level. There are pros and cons to the single target and multi-target immunostrips, so the choice depends on the diagnostic hypothesis and test objectives. Flowers, leaves, stems, roots, seeds, or pure culture organisms can be tested using immunostrips. The test procedure is simple and is detailed in manufacturer's instructions. The following is a typical protocol for performing an immunostrip test.

**1.** Cut one square inch (2.5 × 2.5 cm) of infected plant tissue and grind it with the provided buffer.
**2.** Insert the immunostrip into the ground tissue solution. The solution moves from the bottom to the top of the strip.
**3.** Wait 5–10 min and check if there are any reaction lines showing up in the strip. There are two lines designed in the strip. The lower line is a test line, the upper line is a positive control line. Two lines suggest the positive control works (therefore the test is valid) and indicate a positive result. A single line (only positive control line) suggests a negative result.

## Immunofluorescent microscopy

Some microscopes are equipped with a UV light source and a set of filters to observe fluorescence from specimens. These are called fluorescent microscopes and are widely used in fluorescence-based detection of pathogen or diseases in both medical and plant diagnostics (Janse and Kokoskova, 2009; Sanderson *et al.*, 2014). Unlike immunostrips and ELISA tests, immunofluorescent microscopy can detect and locate a plant pathogen in plant tissues or vectors. It is an *in situ* detection method (see Fig. 2.31). The following protocol is used to test the presence of *Tobacco ringspot virus* particles in viruliferous *Xiphenema americanum* (Wang and Gergerich, 1998) and it can be used as basic guidelines to develop a specific *in situ* detection protocol for other pathogens in plant tissues.

**1.** Place the nematodes on a clean glass slide in a small amount of 2% formaldehyde and cut the nematodes into pieces with a razor blade.
**2.** Transfer and suspend nematodes in 500 µl of blocking buffer for 15 min at 4°C (0.14 M NaCl,

0.01 M phosphate buffer, 3% bovine serum albumin and 0.2% Triton X-100, pH 7.2).

3. Centrifuge to pellet nematodes and remove blocking solution.

4. Add 200 µl of 1:50 diluted primary antibody (in blocking buffer) against *Tobacco ringspot virus* into the microcentrifuge tube containing the nematode fragments and incubate on an orbital shaker at 28°C for 18 h.

5. Wash the nematode fragments with blocking buffer four times for 10 min each at room temperature.

6. Add 200 µl of a 1:50 dilution of fluorescein isothiocyanate (FITC)-conjugated goat anti-rabbit immunoglobulin G (in blocking buffer) and incubate on an orbital shaker for 20 h at 28°C.

7. Wash nematode fragments with blocking buffer four times for 10 min each at room temperature.

8. Dry nematodes in a vacuum dryer for 15 min.

9. Mount nematodes on a glass slide in 10 µl of 50% glycerol in phosphate-buffered saline.

10. Examine the specimens with an epifluorescent microscope.

### Confocal laser scanning microscopy

Confocal microscopy is widely used in biomedical research to visualize and locate expression of targeted proteins in single cells or tissue. It is also a powerful tool to study plant–pathogen interactions using fluorescently tagged molecules in both fixed and living cells and intact tissues (Hardham, 2012). For example, a *Cauliflower mosaic virus* (CaMV) fused with a green fluorescent protein (GFP11) was used to track the spread of the virus infection from the inoculated leaf to other leaves by whole-plant imaging. Infected cells can be displayed in real time by confocal laser scanning microscopy, which allows tracking of the CaMV protein *in vivo* in the context of an authentic infection (Dáder *et al.*, 2019). Similarly, fluorescent derivatives of the bacterial pathogen *P. syringae* were used to study *in planta* dynamic interactions between pathogenic, avirulent and non-pathogenic strains using confocal microscopy (Rufián *et al.*, 2018). However, genetic tagging of pathogen proteins with a fluorescent protein is a research tool rather than a diagnostic method. To use confocal microscopy in plant diagnostics, immunofluorescent procedures can be established to observe and locate a specific protein or pathogen in plant or animal tissue (Fig. 5.18). In this case, tissue pieces must be fixed, embedded and

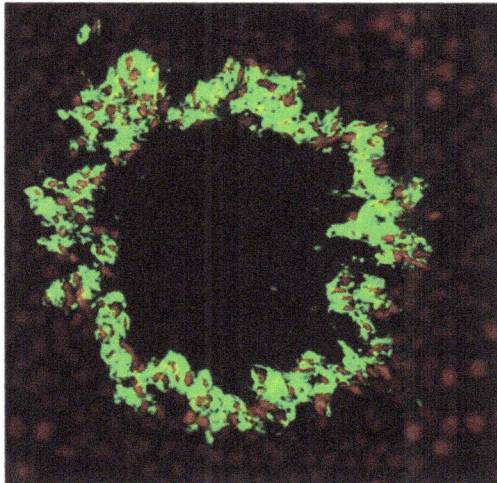

**Fig. 5.18.** A merged confocal image showing a vivid green fluorescent signal localizing a specific protein in cells after the tissue section was stained with Alexa Fluor 488 conjugated anti-Chicken IgG. Note that cell nuclei were stained red with propidium iodide.

then sectioned before proceeding to incubate with primary and fluorescent dye-conjugated secondary antibodies.

### Immunogold labelling for electron microscopy

Many plant viruses can be observed under an electron microscope after negative staining. However, the virus particles or virus-like particles may not reveal which viruses they belong to. Immunogold labelling can determine the identity of the virus if the virus particles bind to the gold-labelled secondary antibody. The procedure is similar to that for immunofluorescent microscopy except the secondary antibody is conjugated with gold. Under the electron microscope, gold particles are viewed as electronically dense dots usually surrounding a virus particle. Below is a basic protocol for immunoelectron microscopy (Gulati *et al.*, 2019). Note that the dilution of primary and secondary antibodies needs to be adjusted for optimal results. Attach a piece of parafilm (remove paper overlay and place wax-side down) to the benchtop as a working surface to hold all droplets for floating the grid. Use self-locking forceps for easily handling light-weight grids. Always hold the grid with carbon-coated side upwards and make this side come in contact with a droplet. Wick away excess liquid from the grid after each floating by gently touching the grid side with the edge of the filter paper. Incubate for 15 min in a humidified chamber or box to prevent droplets from drying.

1. Float the grid on a droplet (15–25 µl) of viral suspension.
2. Rinse grid by floating it on a droplet of distilled water.
3. Float the grid on a droplet of blocking buffer for 15–30 min.
4. Float the grid on a droplet of diluted primary antibody in blocking buffer for 1–2 h.
5. Float the grid on a droplet of wash buffer five times for 3 min each.
6. Float the grid on a droplet of gold-conjugated secondary antibody diluted in blocking buffer for 60 min.
7. Float the grid on a droplet of wash buffer five times for 3 min each and then on distilled water three times for 3 min each.
8. Float the grid on a droplet of negative staining solution for 30–60 s.
9. Remove excess liquid and air dry fully, then load the grid to an electron microscope for observation.

## ELISA

ELISA is a plate-based assay designed for detecting and quantifying pathogens and substances such as peptides or proteins. The plate used in ELISA has 96 wells arranged 8 × 12. Each well can hold approximately 300 µl of liquid. The test consists of immobilizing an antibody (against a specific antigen) on a solid surface followed by reacting the antibody with an enzyme-conjugated secondary antibody (produced against the primary antibody in a different animal species). The antigen (for example, the coat protein of a virus), if present in a sample, will bind to the primary (also called capture) antibody that has been coated on the surface of the well. After adding a substrate, the enzyme reacts with the substrate and changes the solution into a colour (Fig. 5.19). This format is called double antibody sandwich ELISA (DAS ELISA). There are other formats such as direct and indirect assays in which the antigen is immobilized by direct absorption to the plate and then reacts and binds with the enzyme-conjugated primary antibody (direct) or a paired set of primary and conjugated secondary antibodies

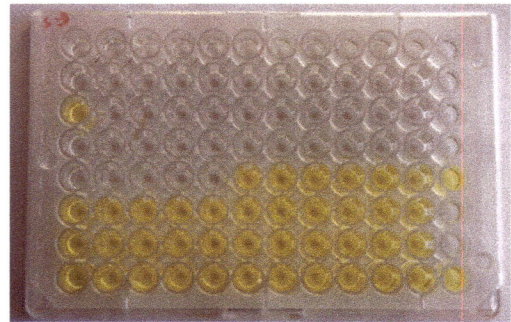

**Fig. 5.19.** A microplate showing positive *Potato virus S* (yellow-coloured wells) and negative result (clear wells). Note that wells 1C, 6E and 12H are positive controls, 1B, 6D and 12G are negative controls and 1A, 6C and 12F are blank controls.

(indirect). DAS ELISA offers a more sensitive assay than other formats and is commonly used in diagnostic labs (Cardoso *et al.*, 1998).

There are several items needed to perform an ELISA test. An 8-channel or 12-channel pipette is needed to quickly load solutions to a microplate, resulting in 8× or 12× more efficiency than a single channel pipette. A V-shaped plastic basin is used to hold the solution so it can be taken by a multichannel pipette. A microplate washer and plate reader accelerate the washing process and record the quantitative results, respectively. The flowing protocol is based on a DAS ELISA format using alkaline phosphatase and pNPP (*para*-nitrophenyl phosphate, disodium salt) detection system, but each test for a specific pathogen should follow the instructions and buffer requirements set by the manufacturer of the test kit.

1. Dilute the primary antibody (capture antibody) in 1× carbonate coating buffer and load 100 µl to each well.
2. Incubate at 4°C overnight.
3. Wash with 1× PBST 6 times with an automatic plate washer or manually by adding 300 µl of the buffer to each well and then immediately dumping it into a sink.
4. Load 100 µl of plant sap sample to each well and 100 µl positive and negative control samples to multiple wells evenly distributed in the plate as shown in Fig. 5.19.
5. Incubate the plate in a humidity box (use a Styrofoam box with a lid and line with wet paper towels) at room temperature for 2 h.

6. Wash the plate with 1× PBST 6 times.

7. Dilute the enzyme-conjugated secondary antibody (detection antibody) in 1× dilution buffer and load 100 µl to each well.

8. Incubate the plate in a humidity box at room temperature for 2 h.

9. Wash the plate with 1× PBST 6 times.

10. Dissolve pNPP in a substrate buffer and load 100 µl to each well.

11. Incubate the plate in a humidity box at room temperature for 1 h.

12. Place the plate on white copy paper and check for the colour presented in positive control wells. Negative controls and blank control wells should remain clear or have no significant colour change. Yellow colour in sample wells suggests positive samples (see Fig. 5.19).

13. Use a microplate reader to record optical density (OD) at 405 nm.

14. Determine the positive samples based on either visual assessment or OD values. A sample with OD value equal or more than three times that of the average of negative controls is considered confidently positive.

## PCR-based Detection

### Principle

Polymerase chain reaction (PCR) is a routine method used in plant diagnostic labs. It uses a DNA polymerase enzyme to amplify a piece of targeted DNA sequence in a small volume of liquid (25–50 µl) inside a plastic tube by temperature cycling. A typical PCR procedure is performed in an automated thermal cycler machine that can be programmed to perform 30–40 temperature cycles. In each cycle, there are three temperature points, each with a specific role during the DNA amplification: 94–98°C to denature DNA template, 50–65°C to anneal two DNA strands, and 72°C to elongate DNA. After 35 cycles, a small quantity (e.g. one single copy) of target DNA in a sample can yield theoretically over 34 billion copies of amplified DNA, and that amount of DNA can be visualized on an agarose gel (Fig. 5.20). This is because each cycle doubles DNA copies ($2^{35} = 3.4 \times 10^{10}$). However, exponential amplification via PCR is not solely for the visualization of a specific DNA in a sample; its product, also called amplicon, can be further sequenced, and analysed for diverse biological research purposes.

There are several different types of PCR methods. The most commonly used method is considered as conventional PCR, which uses a single set of reagents to amplify a specific DNA fragment followed by agarose gel electrophoresis. Sometimes, the targeted DNA is amplified poorly and another set of primers (designed within the amplified region of the first set of primers) is used to amplify a much shorter region of the targeted DNA, by using the amplicon generated from the first PCR reaction as a template. This succession of two PCRs is called nested PCR and is believed to offer more sensibility and specificity. However, nested PCR increases the risk of post-PCR product carry-over in the lab.

Real-time PCR or quantitative PCR is another type of PCR commonly used in plant and medical diagnostic labs. Note that real-time PCR is denoted as qPCR instead of RT-PCR, as the latter abbreviation is used for reverse transcription PCR. Unlike standard PCR, qPCR utilizes fluorescence dye to monitor DNA amplification in real time without subsequent gel electrophoresis (Holland *et al.*, 1991; Higuchi *et al.*, 1992). During each cycle, the intensity of the fluorescent signal is measured and recorded in a graph (Fig. 5.21). In initial cycles the copies of amplicon are low and the level of fluorescence is close to the background. However, with the exponential growth of amplicon, the fluorescence intensity increases above the detectable level. This point was called the threshold cycle ($C_t$), now called the quantification cycle ($C_q$) (Bustin *et al.*, 2009; Kralik and Ricchi, 2017). Because $C_q$ values corresponds proportionally to the initial copy numbers of template DNA in the sample, they are used to evaluate positive or negative detection results. A lower $C_q$ correlates with higher copies of target DNA in a sample; a higher $C_q$ indicates lower copies of target DNA; and a $C_q$ equal to zero or over 40 is considered negative. Unlike conventional PCR, qPCR employs a fluorogenic DNA probe (a fluorescent dye on the 5′ end as a reporter and a dye on the 3′ end as a quencher) that binds to the complementary sequence of a targeted region of a DNA sample; therefore, only the PCR product amplified by the primer pair and annealing with the DNA probe is detected and recorded in real time. Any non-specific amplification, if it occurs, is not detected by the system. That said, qPCR with the fluorogenic probe detection system has a much higher specificity and sensibility than conventional PCR. However, qPCR based on a chemical dye such as SYBR Green does not increase specificity, as

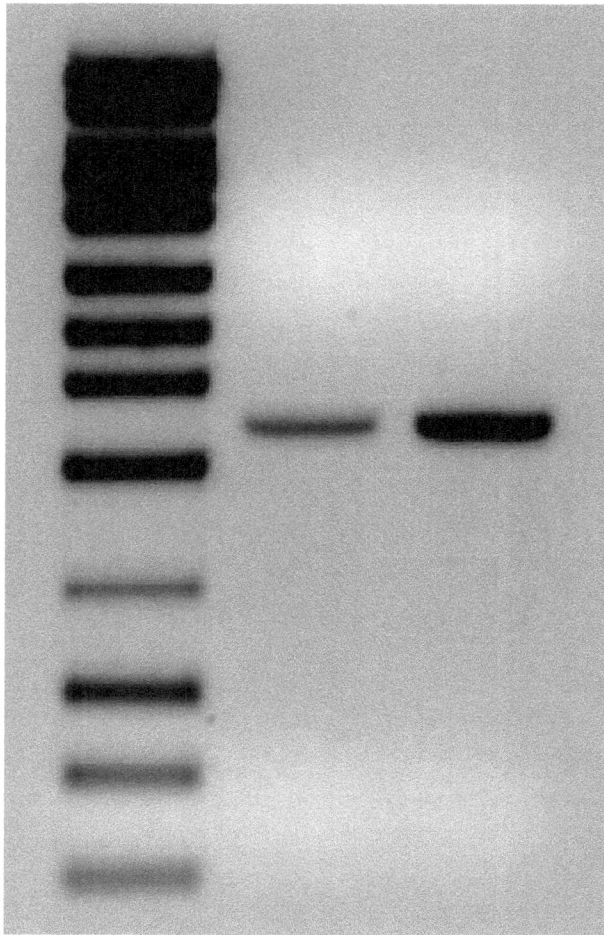

**Fig. 5.20.** PCR-amplified DNAs revealed on an agarose gel after staining with ethidium bromide. Lanes from left to right: DNA ladder, sample #1 and sample #2. Note that both samples had the same size of amplicons but sample #2 had a stronger band (higher copies of amplicons).

the dye binds any and all double-stranded DNA in a PCR reaction. Still, this dye-based qPCR is widely used, as it detects PCR products in real time.

Reverse transcription polymerase chain reaction (RT-PCR) is another type of PCR that involves reverse transcription (RT) of RNA into DNA (complementary DNA, or cDNA) followed by the amplification of specific DNA targets from the cDNA. It is primarily used to measure gene expression levels (from DNA to mRNA) in organ tissue and to detect a virus that has RNA (ribonucleic acid) as its genetic material. The step of reverse

transcription of RNA can be followed by conventional PCR or qPCR and continued with other downstream applications such as cloning and sequencing. The combination of RT and qPCR is widely used in gene expression analyses and RNA virus detections.

### Conventional PCR

A PCR reaction consists of the following ingredients: PCR buffer (provided with *Taq* polymerase enzyme), 5' primer and 3' primer, dNTP, *Taq*

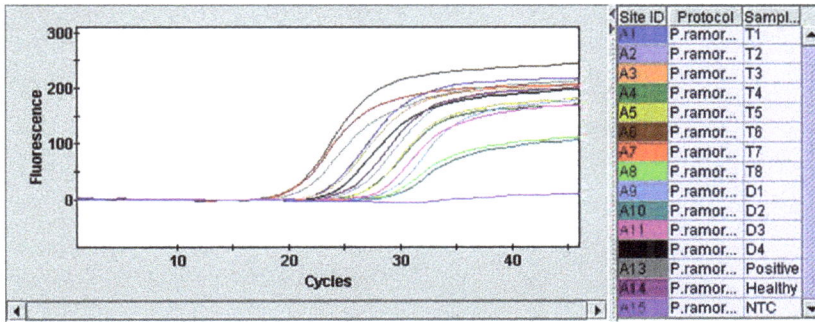

| Site ID | Protocol | Sampl... |
|---|---|---|
| A1 | P.ramor... | T1 |
| A2 | P.ramor... | T2 |
| A3 | P.ramor... | T3 |
| A4 | P.ramor... | T4 |
| A5 | P.ramor... | T5 |
| A6 | P.ramor... | T6 |
| A7 | P.ramor... | T7 |
| A8 | P.ramor... | T8 |
| A9 | P.ramor... | D1 |
| A10 | P.ramor... | D2 |
| A11 | P.ramor... | D3 |
| | P.ramor... | D4 |
| A13 | P.ramor... | Positive |
| A14 | P.ramor... | Healthy |
| A15 | P.ramor... | NTC |

**Fig. 5.21.** Amplification plots from a real-time PCR test showing the relative fluorescence intensity versus cycle number. Note that in some samples the intensity of fluorescence started to increase exponentially as soon as after 20 cycles (higher $C_q$), but in others it delayed to 25 or 30 cycles (lower $C_q$). This test was to detect plant DNA using plant cytochrome oxidase (COX)-specific primers and a Texas Red-labelled probe.

polymerase, molecular-grade water and DNA template. Depending on the *Taq* polymerase provided by the chosen vendor, a typical mix of PCR reaction can be assembled as shown in Table 5.5. When preparing multiple PCR reactions, a master mix should be prepared by adding all ingredients in the order as shown in the table to a single 1.5 ml microcentrifuge tube and thoroughly mixed. The master mix is then aliquoted to each single 0.2 ml PCR tube followed by adding sample DNA. The use of master mix increases working efficiency and saves pipette tips. Use 11× volume for every 10 reactions to ensure that there is enough master mix for all intended samples and controls. A PCR reaction can be done in 25 µl or 50 µl volume.

Here is a basic PCR protocol using Platinum *Taq* DNA polymerase.

1. Sterilize a PCR workstation by turning UV light on for 15 min.
2. Remove PCR buffer, $MgCl_2$, aliquoted dNTP, primers at working concentration (10 mM), water and DNA samples from −20°C freezer and thaw on ice. Vortex and spin down all ingredients and samples.
3. Use Table 5.5 to add all components into a microcentrifuge tube labelled as 'Master Mix' inside the PCR workstation.
4. Vortex and spin the master mix and then aliquot 49 µl of master mix into each 0.2 ml PCR tube.
5. Add 1 µl of sample DNA into each PCR tube and 1 µl of water into one or two tubes as negative control.
6. Vortex and spin down all tubes.

**Table 5.5.** A typical example of assembling a PCR test for ten samples.

| Component | Volume per reaction | Master mix for 10 reactions (×11) |
|---|---|---|
| 10× PCR buffer | 5 µl | 55 µl |
| 50 mM $MgCl_2$ | 1.5 µl | 16.5 µl |
| 10 mM dNTP mixture | 1 µl | 11 µl |
| 10 mM forward primer | 1 µl | 11 µl |
| 10 mM reverse primer | 1 µl | 11 µl |
| *Taq* polymerase | 0.2 µl | 2.2 µl |
| Molecular biology grade water | 39.3 µl | 432.3 µl |
| Sample DNA | 1 µl | |
| Total | 50 µl | 539 µl |

7. Load tubes onto a thermal cycler and run 35 cycles using the following parameters
   (a) 94°C for 1 min
   (b) 35 cycles of 94°C for 1 min, 55°C for 1 min, and 72°C for 1 min
   (c) 72°C for 10 min (or 1 min per kb of amplicon)
   (d) Hold on 4°C
8. Proceed to gel electrophoresis or store PCR product at 4°C or −20°C until use.

The above cycle parameters should be adjusted according to the primers and types of thermocyclers. The most critical parameter is the annealing temperature ($T_a$). $T_a$ is dependent on the primer melting temperature ($T_m$), which is the temperature at which one-half of the DNA duplex is dissociated to become single stranded but retains the duplex stability. A $T_a$ chosen for PCR depends on the length and composition of the primers, and an optimal $T_a$ can be calculated based on the following formula: $T_a$ (optimal) = $0.3 \times T_m^{Primer} + 0.7 \times T_m^{Product} - 14.9$ (Rychlik *et al.*, 1990). For plant diagnostics, the melting temperature of a PCR product ($T_m^{Product}$) may not be known beforehand but melting temperature of primers ($T_m^{Primer}$) can be calculated. This is okay because in a real world a $T_a$ 5°C below the $T_m$ of the primers is commonly selected for an initial test and an optimal $T_a$ can be found in several PCR try-outs by slightly increasing or decreasing $T_a$. Remember, having a $T_a$ that is too low causes one or both primers to anneal to DNA sequences other than the intended target, due to the increased tolerance to base mismatches or partial annealing. This leads to nonspecific PCR amplification and may yield multiple random products while the desired product is greatly reduced. In an agarose gel, nonspecific PCR amplification is indicated by the presence of unexpected bands and/or weak signal of the expected band. Conversely, if $T_a$ is too high, primers may have less chance to anneal with the template and amplification efficiency is reduced. That said, an optimal $T_a$ allows both specific amplification and a high yield of the correct product.

Another factor to be considered for PCR is the extension time of each cycle. The typical extension time for *Taq* DNA polymerase is 1 min/kilobase (kb), while other polymerases may be slower and require more time. Most PCRs in plant diagnostics generate amplicons of less than 1 kb, so 1 min of extension per cycle is generally sufficient. In case there are still some amplicons not fully extended after the final cycle, an extended incubation at 72°C for 10–15 min will complete synthesis to the end. The time needed for this final step also depends on the amplicon length and the enzyme. The final extension step is also critical if the amplicon is to be further cloned. Enzymes such as *Taq* DNA polymerase have terminal deoxynucleotide transferase activity, which adds extra nucleotides to the 3′ ends of the PCR products at this final step so that the amplicon has proper 3′-dA tailing for efficient PCR cloning (see cloning section of this chapter).

### Nested PCR

The procedures for nested PCR are the same as standard PCR except that there are two successive PCR runs. The first run of PCR uses a pair of primers to target a wider range of the DNA sequence, while the second run uses a different set of primers to amplify a narrower range of the sequence from the PCR product generated in the first run. The following protocol is a typical nested PCR procedure used to detect phytoplasma from any plant species.

1. Follow steps 1 to 6 as described in the section above on conventional PCR. Use P1/P7 primers (Smart *et al.*, 1996) as forward and reverse primers.
2. Load tubes onto a thermal cycler and run 35 cycles using the following parameters:
    (a) 94°C for 2 min
    (b) 35 cycles of 94°C for 1 min, 55°C for 2 min and 72°C for 3 min
    (c) 72°C for 10 min
    (d) Hold on 4°C
3. Add 1 µl of P1/P7 PCR product into 49 µl of molecular grade water in a microcentrifuge tube, vortex, and spin down.
4. Use Table 5.5 to prepare a master mix but change primers to R16F2n/R16R2 (Gundersen and Lee, 1996).
5. Vortex, spin down the master mix, and add 49 µl of master mix into each 0.2 ml PCR tube.
6. Add 1 µl of diluted P1/P7 PCR product (step 3) into each PCR tube and 1 µl of water into one or two tubes as negative control.
7. Vortex and spin down all tubes.
8. Load tubes onto a thermal cycler and run 35 cycles using the same parameters in step 2.
9. Proceed to gel electrophoresis or store the PCR product at 4°C or –20 °C until use.

### Real-time quantitative PCR (qPCR)

Real-time PCR requires specific forward and reverse primers, a fluorogenic DNA probe and a real-time PCR machine (system). The primers and probe need to be designed and then synthesized by a commercial company. In gene expression studies, researchers can easily pick a pair of primers and a probe sequence from an online software such as Primer3 (http://bioinfo.ut.ee/primer3-0.4.0/) by pasting a piece of gene sequence into the software application. This is because the DNA sequence of a specific gene in a model organism, for example house mouse (*Mus musculus*), is quite different from that of other genes in the organism. In plant diagnostics, however, there are many known and unknown microorganism species and strains that may share a high level of sequence similarity in a targeted gene or locus. Designing a unique primer and probe to detect a specific microorganism requires extensive *in silico* work and data mining. Fortunately, many researchers have accomplished this in research labs and demonstrated the usefulness of the primers and probes in peer-reviewed journals. A diagnostician can validate and optimize the methods before using them routinely for pathogen detections. The following is a general protocol adopted from the USDA-APHIS-PPQ Beltsville Lab for screening of all phytoplasma species from plants. This protocol is a multiplex qPCR with positive detection of plant DNA (18S ribosomal RNA gene) as an internal control, and it has been routinely used to successfully detect phytoplasmas from hemp crops (Schoener and Wang, 2019).

1. Order primers and probe as listed in the table below.

| Primer/probe | Sequence |
|---|---|
| JH-F1 (primer) | 5'- GGT CTC CGA ATG GGA AAA CC -3' |
| JH-F all (primer) | 5'- ATT TCC GAA TGG GGC AAC C -3' |
| JH-R (primer) | 5'-CTC GTC ACT ACT ACC RGA ATC GTT ATT AC -3' |
| JH-P uni (probe) | 5'- /56FAM/AAC TGA AAT ATC TAA GTA AC /MGBNFQ/ -3' |
| 18S F | 5'- GAC TAC GTC CCT GCC CTT AG -3' |
| 18S R | 5'-AAC ACT TCA CCG GAC CAT TGA -3' |
| 18S P | 5'- /5TET/ACA CAC CGC CCG TCG CTC C /3IABkFQ/ -3' |

2. Sterilize a PCR workstation by turning UV light on for 15 min.
3. Remove PCR buffer, MgCl₂, aliquoted dNTP, primers at working concentration, probes at working concentration, water and DNA samples from –20°C freezer and thaw on ice. Vortex and spin down all ingredients and samples.
4. Add all components into a microcentrifuge tube labelled as 'Master Mix' inside the PCR workstation following the table below.

| Reagent | 1 reaction (µl) | Master Mix for 16 reactions (×17) (µl) |
|---|---|---|
| Molecular grade water | 15.15 | 257.55 |
| Invitrogen 10× Platinum *Taq* buffer only | 2.5 | 42.5 |
| Invitrogen 50 mM MgCl₂ | 3.0 | 51 |

| Reagent | 1 reaction (µl) | Master Mix for 16 reactions (×17) (µl) |
|---|---|---|
| 10 mM dNTPs | 0.5 | 8.5 |
| 10 µM Primer Mix JH-F 1/ JH-F all/JH-R | 0.75 | 12.75 |
| 10 µM JH-P uni probe | 0.25 | 4.25 |
| 2 µM primer mix 18S F/18S R | 0.5 | 8.5 |
| 2 µM 18S P probe | 0.25 | 4.25 |
| Invitrogen Platinum *Taq* polymerase | 0.1 | 1.7 |
| Total | 23.00 | 391 |

5. Vortex, spin down the master mix, and then add 23 µl of master mix into each PCR tube or each well of a qPCR plate.
6. Add 2 µl of sample DNA into each PCR tube or well and 2 µl of water into one or two tubes or wells as negative control. Vortex and spin down.
7. Load tubes or plates onto a qPCR thermal cycler using the following parameters:
   Stage 1: 95°C for 1 min with optics off
   Stage 2: 40 cycles of 95°C for 5 s with optics off and 58°C for 35 s with optics on.
8. Enter sample information into the computer, select appropriate dye set and start run.

## Reverse transcription PCR (RT-PCR)

RT-PCR is used to detect plant viruses from plant tissue. Most plant viruses contain only coat protein and an RNA genome. They rarely have an envelope. Virus RNA genomes are small and usually single-stranded (ss), but some viruses have double-stranded RNA (dsRNA), ssDNA or dsDNA. To detect a virus from plant tissue, total RNAs are extracted using a commercial RNA extraction kit. The RNA molecules are then converted to single-stranded DNA, commonly called cDNA, by using a DNA reverse transcription kit and following the manufacturer's protocol. The cDNA can then be subjected to conventional PCR using a pair of virus-specific primers (Chauhan *et al.*, 2015). One-step RT-PCR was also developed to detect plant viruses (Li and Hartung, 2007). After RT-PCR, the PCR amplicons are visualized by electrophoresis in a 1% agarose gel containing ethidium bromide in TAE buffer and viewed under UV light. The cDNA can also proceed to qPCR if a specific probe is available to detect a specific virus. This procedure (RT followed by qPCR) is a rapid, sensitive, reliable and often cost-effective method if it is used for screening a large number of samples. However, for some labs that do not have access to PCR-based methods, ELISA and immunostrip tests are still useful and practical for detecting many viruses from plant tissue.

## Prevention of post-PCR product contamination

Post-PCR products can become a nuisance when PCR tests are routinely performed in a lab. This is especially troublesome when a pair of primers is frequently used to amplify a DNA sequence that is subsequently purified from a gel. Because of handling, working benches, pipettes, or reagents may become contaminated by the post-PCR product. If this happens, the negative control of a PCR test may become positive, resulting in the failure of quality control. Note that the chance of post-PCR product contamination is much higher than that of sample DNA template contamination or cross-contamination, because the number of copies of template DNA is much lower than that of amplicons. The following is a general guideline to prevent post-PCR production in the lab.

1. Use a specific area or room for all post-PCR product handlings such as electrophoresis, gel imaging, gel cutting and post-PCR product purification. Gels and wastes should be properly contained before being transferred to other areas of the lab. Keep the door closed.
2. Designate a separate room or area for pre-PCR procedures such as making a master mix and adding reagents for PCR. All items used for these procedures should remain in this room, including icebox, gloves, pipettes, tips, racks and trays. Ideally, a refrigerator may be placed in this room to store all reagents used for PCR tests. Keep the room door closed.
3. Place all tubes, pipettes, tube openers, trays and racks inside the PCR workstation and UV-irradiate the workstation for at least 30 min before each use.
4. Be aware where the post-PCR products are when handling lab procedures. If there is a spill or contamination of a PCR product, immediately clean the area and items with bleach solution and dispose of contaminated gloves and tips. The percentage of active ingredients (sodium hypochlorite) in the bleach solution and the time period of treatment is critical. It has been demonstrated that 3.0% (w/v) sodium hypochlorite, approximately equal to

the concentration of commercial bleach (3–6%), can effectively eliminate surface DNA contamination from bone after immersion for at least 15 min (Kemp and Smith, 2005). In general lab practice, however, 10% Clorox solution is commonly used to eliminate unwanted DNA in the lab, and such concentration has been shown to be effective in eliminating all ethidium bromide-stainable DNA (Prince and Andrus, 1992).
5. Similar care should be taken with the plasmid that contains the PCR product after cloning.

## Resuspending and diluting PCR primers and probes

Primers and probes are both oligonucleotides (oligos) synthesized by commercial companies. When they arrive, follow the instructions of the manufacturer to resuspend them in either buffer or water if they are lyophilized. Some companies offer a service to resuspend them for a small fee. Here is the common procedure to resuspend lyophilized primers and probes.

1. Upon receiving them, store primers and probes at –20°C or –80°C if not ready to resuspend.
2. Centrifuge the tubes containing the lyophilized primer or probe for 1 min at 8000 rpm.
3. Under the PCR workstation, add to the tube an amount of 1× TE buffer (4°C) that is 10 times (in μl) the oligo yield (in nmol) to reach 100 μM concentration. For example, an amount of 35.3 nmol would be resuspended with 353 μl of 1× TE buffer. Oligo yield information can be found on the tube label or specification sheet.
4. Vortex, spin down and store the concentrated primers or probes at –20 °C or –80 °C. Probes are light-sensitive and should be stored in an amber tube.
5. To make a working solution at 10 μM, dilute the stock primer solution 10 times with molecular-grade water. For sequencing, further diluting it to 2 μM or other desired concentrations may be needed. The working primer/probe solutions should be aliquoted in small amounts for each use to avoid multiple freeze/thaw cycles.

## Restriction Enzyme Digestion

Before DNA sequencing became a routine diagnostic tool, digesting a PCR product to see the different pattern and sizes of fragments in an agarose gel was used to differentiate different species or strains of pathogens. Theoretically, this approach is still a sequence-based approach, as the different patterns of digested DNA products reflect the sequence differences. However, this method is much less sensitive and less reliable than completely sequencing a PCR product, because it only targets a specific restriction site. There are many restriction enzymes discovered and commercially available and they are commonly used to cut a large DNA into smaller pieces for use in the construction of a recombinant DNA. A sequenced PCR product can be virtually cut using online tools such as NEBcutter V.2.0 (http://nc2.neb.com/NEBcutter2/) if such a cutting is still useful for a diagnosis.

## DNA Sequence-based Identification

This section describes a series of procedures used to obtain a piece of a diagnostic DNA sequence for organism identification. This is known as DNA barcoding and has been proven to be a powerful technique for identifying unknown organisms, including plants (CBOL Plant Working Group, 2009), metazoan invertebrates (Folmer *et al.*, 1994)

and nematodes (Derycke *et al.*, 2010). The whole process includes DNA extraction, conventional PCR, amplicon purification, sequencing directly or after cloning into a plasmid vector, sequence analysis, and BLAST search (Fig. 5.22).

### Purification of organism or culture

The first step for DNA sequence-based identification is to obtain a clean individual organism or a clone of an organism, such as fungus, bacterium, nematode, insect, or mite, originated from a single plant specimen. The starting material must be free from contamination by other organisms. This is because most primers used to amplify a specific section of DNA are considered universal and a mix of two different organisms or species may complicate the identification process. For the plant, a small piece of clean leaf tissue (100–200 mg) is sufficient. For arthropods, an individual large insect of any life stage or a collection of tiny mites can be used to extract total DNA. For nematodes, both a single specimen and a pool of hand-picked nematodes of the same species can be used to extract DNA. Most fungi and bacteria are culturable and a pure culture can be easily obtained

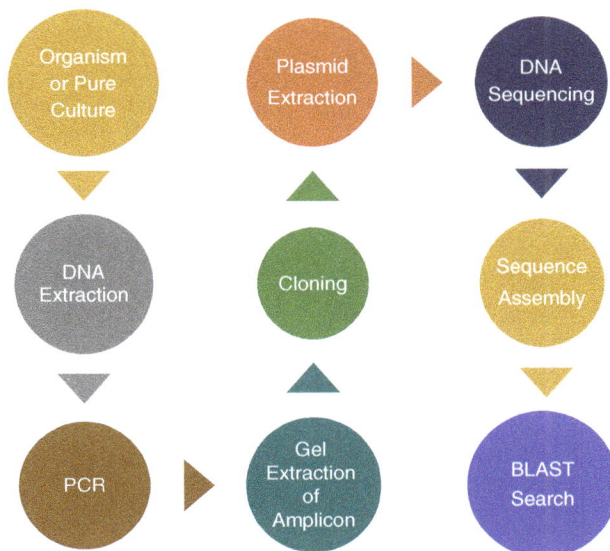

**Fig. 5.22.** A DNA barcoding scheme for identifying unknown organisms. Note that if the PCR product only contains a single amplicon less than 600 bp, the amplicon can be purified using a PCR purification kit and then submitted for sequencing. For a large amplicon or multiple amplicons, it is better to perform gel cutting and extraction followed by cloning before submitting for sequencing (see explanations in text).

via single spore isolation or single colony purification. The viruses can be purified through a series of mechanical inoculations on a local lesion indicator plant using a single necrotic lesion.

Some pathogenic organisms, such as phytoplasmas and other fastidious bacteria, are difficult to culture on a synthetic medium, nor can they be purified. In this case, total DNAs from the host plant, the pathogen and other associated microorganisms are used to amplify a targeted pathogen sequence using universal primers that are also relatively specific to a group of pathogens. Often, the DNA from other endophytes or other associated bacteria is also amplified. After sequencing the PCR product, untargeted bacteria may be identified. Non-specific amplification of the DNA of untargeted bacteria is a common issue in fastidious bacterial disease diagnosis and the final determination must be based on amplicon sequence instead of the presence of amplification. Some primers published in journals have been shown to specifically amplify targeted pathogens in the research, but when they are used in real-world plant diagnostics, those primers amplify many epiphytic and endophytic organisms' DNA as well (Garneni and Wang, 2009).

Most plant diagnostic labs mainly deal with fungal, oomycete and bacterial pathogens. It is easy to obtain a pure culture of a bacterial isolate from a single colony through the streaking plate method (Fig. 5.13). The following protocols are used to obtain a pure culture of fungal and oomycete isolates for DNA sequence-based identification.

### Hyphal tipping

This protocol uses *Phytophthora* as an example to obtain a pure culture from hyphal tipping.

1. Transfer a plug of culture to be purified into a new cornmeal agar (CMA) plate, wrap with parafilm and let it incubate upside down in an incubator for 2–5 days, depending on the growth rate.
2. After the colony grows over about one-third of the plate, place the plate under a dissecting microscope with the transmitted light on, or under a compound microscope at lower magnification. Observe the edge of the colony and locate a single outward-growing hypha.
3. Use a flame-sterilized sharp insect needle to cut a tiny piece of agar containing a single hyphal tip, then transfer it into the centre of a new CMA plate. Flame-sterilize the insect needle before picking

another hyphal tip. Seal all CMA plates with parafilm and place them upside down in an incubator at 22°C. Note that to prevent potential contamination, a PARP plate can be used as it inhibits the growth of fungi and bacteria. For fungal species, PDA amended with streptomycin (100 µg/ml) or 2% water agar can be used.
4. After a few days of growth, a nice colony will be seen.

### Single spore isolation

A pure isolate grown from a single spore can be obtained from a fungal or oomycete culture plate. The following protocol uses *Verticillium* or *Fusarium* species as an example, but it can be applied to all spore-producing fungi or oomycetes. It can also be used to obtain single microsclerotium isolates.

1. Under a compound microscope, check the isolate for the stage of spore production.
2. Inside the biosafety cabinet, use a heat-sterilized tool to transfer a 1 cm × 1 cm mycelium mass into a 1.5 ml microcentrifuge tube containing 1 ml of sterilized water. Shake vigorously to suspend spores in water. Transfer 100 µl of spore solution into a new microcentrifuge tube; this is the stock solution.
3. Inside the biosafety cabinet, prepare three microcentrifuge tubes, each labelled as 10x, 100x and 1000x, respectively. Add 90 µl of sterile water into each tube. Perform serial dilution by adding 10 µl of stock spore solution into the 10x tube, mixing, then transferring 10 µl from 10x to 100x tube, and so on.
4. Add 1 µl of each dilution and 50 µl of sterilized water to a 2% water agar plate. Add three sterilized glass beads. Stack all plates together and gentle swirl to spread spores evenly with beads. Dispose of beads, seal plates and incubate at 22°C for 24–48 h.
5. Use a compound microscope to observe spore germination at lower magnification (4x). Select an area containing only a single germinated spore and use a flame-sterilized sharp insect needle to cut the agar piece. Transfer it to a new culture plate such as PDA[+strep].
6. Incubate plates at 22°C until fully grown.

### Single sclerotium isolation

Some fungi do not produce spores in culture media. Instead, they produce abundant sclerotia, which are structures containing a compact mass of hardened fungal mycelium. A pure culture can be obtained through hyphal tipping or single sclerotium isolation.

The following protocol uses *Sclerotium cepivorum*, a pathogen causing white rot of *Allium* crops, as an example. All procedures are done inside a biosafety cabinet. The protocol can be used to isolate other sclerotium-producing fungi, such as *Rhizoctonia* and *Botrytis*.

1. Select a PDA plate that has fresh growth of *Sclerotium cepivorum* with abundant sclerotia.
2. Take a quad Petri dish (compartmentalized Petri dish), add 1% sodium hypochlorite (NaClO) to the first quarter and sterilized water into the remaining quarters.
3. Use sharp heat-sterilized forceps to pick individual sclerotium from the plate into the first quarter of the quad Petri dish and immerse all sclerotia into the solution for 30–60 s. The forceps should be sterilized by dipping in 95% ethanol and placing in the glass bead sterilizer for 15 s before and after each use.
4. Rinse the sclerotia by transferring them into the next quarter of the Petri dish filled with sterilized water. Allow the sclerotia to sit in water for 30–60 s by using forceps to submerge sclerotia intermittently.
5. Repeat step 4 until the sclerotia have been transferred to and rinsed in the fourth quarter, remove an individual sclerotium from the Petri dish, lightly touch a piece of autoclaved filter paper to absorb extra water and place the sclerotium at the centre of a potato dextrose agar (PDA) plate.
6. Wrap the plates with parafilm and place them upside down in a 22°C incubator.

## DNA extraction

The extraction of total DNA can be done using a commercial DNA extraction kit. The selection of a kit should be based on the type of tissue or medium from which DNA is to be extracted. Most diagnostics require the extraction of DNA from plant tissue that harbours pathogens such as fungi, oomycetes and bacteria. DNA extracted from infected plant tissue should be considered as total DNA that contains DNA from the plant and the pathogen, as well as endophytic and epiphytic microorganisms. The total DNA is then subjected to downstream testing such as PCR detection for a specific pathogen. For arthropods, there are commercial kits such as the DNeasy® blood and tissue kit available to extract DNA from an insect specimen. For nematode specimens, DNA extraction can be simply performed by squashing a single nematode in a droplet of water, and the resulting DNA content is sufficient to perform regular PCR-based identification (Powers and Harris, 1993; Harris *et al.*, 2020). Commercial genomic DNA isolation reagents and kits are available for extracting a high quality and quantity of DNA from a population of purified nematodes or cysts (Wang *et al.*, 2001). For purified microorganisms such as fungi, oomycetes and bacteria, there are DNA extraction kits optimized specifically for microbes. To extract total DNA from soil, choose a kit developed for soil DNA extraction. In general, there is no need to use a traditional phenol–chloroform extraction method, as many ready-to-use DNA extraction kits are available in the market.

### From plant tissue

The following protocol is for extracting DNA from plant tissue for phytoplasma analysis, based on the Qiagen DNeasy Plant Mini Kit with slight modifications. Users should always follow current manufacturer's instructions as certain buffer names may have been changed. Extraction kits from any other vendors can be used according to the user's preference or experience. Using this protocol, a diagnostician can perform up to 12 samples together during a 2 h period. To increase efficiency, a set of tubes and columns can be prelabelled numerically and prearranged in numerical order. Final DNA products can then be labelled in detail with a printed label.

1. On a clean benchtop or diagnostic table, spray surface with 70% ethanol and wipe down.
2. On a small piece of parafilm (3" × 4", or 7.5 cm × 10 cm), use a new razor blade to cut out the midribs with the petioles of multiple leaves that are symptomatic, chop the midrib and petioles into fine pieces, and transfer them into a new weigh boat. Use a new set of parafilm, razor blade and weigh boat for each sample.
3. Weigh out 200 mg of chopped midrib/petiole and pour the 200 mg of plant tissue into a labelled lysing matrix A tube (orange cap, garnet beads). Proceed to DNA extraction or store at –20 °C until use.
4. Add 800 µl buffer AP1 and 8 µl RNase A (100 mg/ml) to each lysing matrix A tube.
5. Homogenize samples using the Qbiogene FastPrep 24 at speed 6.0 for 40 s or using any available homogenizer.
6. Vortex the tubes for 15–30 s on setting seven to mix and then spin down.

**7.** Incubate the tubes in the pre-heated thermo-mixer (65°C) for 10 min with gentle shaking.
**8.** Centrifuge the tubes briefly (15–30 s) at 5000 rpm to remove liquid from the cap.
**9.** Add 260 µl of buffer AP2, vortex for 5–10 s and incubate on ice for 5 min
**10.** Centrifuge the tubes for 10–13 min at 14,000 rpm.
**11.** Pipette the lysate into a labelled QIAshredder (lilac) column placed on top of a 2 ml collection tube.
**12.** Centrifuge the column for 2 min at 14,000 rpm.
**13.** Pipette the flow-through into a new 1.5-ml microcentrifuge tube and calculate total amount of liquid recovered.
**14.** Add 1.5× the volume of the liquid from step 13 of buffer AW1 and gently mix by pipetting.
**15.** Pipette 650 µl of the mixture into the DNeasy mini spin column (white) and centrifuge for one minute at 8000 rpm.
**16.** Discard the flow-through into a hazardous-waste container and reuse the column. Add any remaining mixture from step 14 and repeat step 15.
**17.** Place the spin column into a new 2 ml collection tube.
**18.** Add 500 µl of AW2 buffer and centrifuge for 1 min at 8000 rpm
**19.** Discard the flow-through and add another 500 µl of AW2 buffer; centrifuge for 2 min at 14,000 rpm.
**20.** Transfer the spin column to a new 1.5 ml microcentrifuge tube and pipette 100 µl of elution buffer AE into the center of the column membrane. Incubate the tubes for 5 min at room temperature.
**21.** Centrifuge the tubes for 1 min at 8000 rpm.
**22.** Discard the column, save the flow-through (final DNA), measure DNA concentrations, label the DNA samples and store at –20°C until use.

### From fungal or oomycete culture

To extract DNA from cultured fungi or oomycetes, both plant mini-kits and microbe kits can be used. A lab can compare the usage and purpose of different kits to determine which one is more appropriate to use. Below is a protocol on how to collect fungal mycelium from a culture plate. After a sufficient amount of mycelium is obtained, DNA extraction can be performed using the protocol described above or using a microbe DNA extraction kit.

**1.** Inside the biosafety cabinet, open a pure culture plate with a fully grown fungal colony. Use a cell scraper to harvest 100 mg mycelium and spores by scraping the colony and rolling mycelium up. Cell scrapers are individually wrapped, sterile and disposable polyethylene blades. They are useful for harvesting cells and can be used to harvest fungal spores and mycelium. For some fungal species that mostly grow into the agar, a razor blade or forceps may be used, but try to minimize the amount of agar picked up.
**2.** Transfer mycelium pieces into a pre-weighed lysing matrix A tube. Weigh the tube again to calculate the net weight of mycelium (approximately 100 mg).
**3.** Label tubes on the lids as well as the sides with sample ID and proceed to DNA extraction procedures.

### From bacterial culture

A pure bacterial culture can be obtained from a single colony and then proceeded to DNA extraction. The following protocol combines the procedures to obtain the bacterial cell pellet and the general steps of DNA extraction outlined in the DNeasy® UltraClean® Microbial Kit. Users can choose any kits based on their preference and experience and should always follow current protocols provided with kits.

**1.** Inside a biosafety cabinet, use an inoculating loop to streak a bacterial culture on a new LB or other nutrient agar plate (Fig. 5.13). Incubate the plate (upside down) at 28–30°C until single colonies appear on the fourth streak.
**2.** Between 3–5pm, use a sterile pipette tip to scrape a single colony from the plate and dip into a 3 ml LB broth or other liquid medium in a 15 ml sterile plastic culture tube. Pipette several times to release bacterial cells into the medium. Close it with the cap.
**3.** Place culture tubes in the incubator at 37°C (or other temperatures) and shake at 220 rpm for 16 h.
**4.** Shake the tube and pour 1.5 ml of bacterial suspension into a 1.5 ml microcentrifuge tube.
**5.** Centrifuge at 10,000 rpm for 2 min to pellet bacterial cells. Remove the supernatant with a pipette tip.
**6.** Resuspend the cell pellet in 300 µl of PowerBead Solution, vortex, and then transfer to a lysing matrix A tube.
**7.** Add 50 µl of Solution SL into the tube and homogenize samples using the Qbiogene FastPrep 24 at speed 6.0 for 40 s or using any available homogenizer.
**8.** Centrifuge the tubes at 14,000 rpm for 30 s.
**9.** Transfer the supernatant to a clean 2 ml collection tube.

10. Add 100 µl of Solution IRS to the supernatant and vortex for 5 s. Incubate at 4°C for 5 min.

11. Centrifuge the tubes at 14,000 rpm for 1 min.

12. Transfer the supernatant to a clean 2 ml collection tube.

13. Add 900 µl of Solution SB to the supernatant and vortex for 5 s.

14. Load about 700 µl into an MB Spin Column and centrifuge at 14,000 rpm for 30 s. Discard the flow-through, add the remaining supernatant to the MB Spin Column and centrifuge again at 14,000 rpm for 30 s.

15. Add 300 µl of Solution CB and centrifuge at 14,000 rpm for 30 s.

16. Discard the flow-through and centrifuge at 14,000 rpm for 1 min.

17. Place the MB Spin Column in a new microcentrifuge tube.

18. Add 50 µl of Solution EB to the centre of the white filter membrane and centrifuge at 14,000 rpm for 30 s.

19. Discard the MB Spin Column and save the flow-through as the final DNA solution. Label and save the DNA at –20°C for future downstream applications.

### From nematodes

Nematodes extracted from soil comprise populations of various types including bacteria-feeding, fungus-feeding, predatory and omnivorous nematodes and root-feeding nematodes. Nematodes extracted from plant tissue often contain bacterial or fungal feeder types. Considering that nematodes are typically 50 µm in diameter and 1 mm in length, it is difficult to hand-pick a meaningful number of targeted nematodes under a stereo microscope for DNA extraction. However, for PCR-based identification, an individual vermiform nematode can be hand-picked and placed in a 15 µl drop of sterile water on a cover slip. The nematode can then be ruptured by gentle pressing, using a flat-tipped micropipette tip, and the nematode lysate can be collected and processed immediately for PCR or frozen for future analysis (Powers and Harris, 1993). For cysts or enlarged females, which can be extracted and purified from root tissue, DNA extraction can be performed using any kits designed for nematode tissue. The following protocol is an example of a procedure for extracting genomic DNA from purified cyst nematodes using the DNAzol™ genomic DNA isolation reagent (Wang et al., 2001).

1. Inoculate a single cyst to a susceptible host plant such as soybean (*Glycine max*) or lespedeza (*Lespedeza* sp.) and grow the plant in greenhouse for 4–6 weeks.

2. Extract cysts and mature females from the roots by the rubbing and sieving method (Riggs and Schmitt, 1991).

3. Transfer cyst/female collection into a 50 ml centrifuge tube and spin at 1130 ×g for 6.5 min to pellet. Remove water and then resuspend in 125% sucrose solution (5 g sugar to 4 ml water). Centrifuge at 1130 ×g for 1 min, pour sugar solution with floating nematodes into a 60-mesh sieve and collect them into a beaker.

4. Load collected cyst nematodes onto the top of 40 ml sucrose gradients in a 50 ml centrifuge tube consisting of 8 ml each in the order, top to bottom, of 30, 50, 70, 125 and 150% (sugar:water/W:W). Centrifuge at 900 ×g for 3 min.

5. Collect cysts floating on the top of the gradient into a counting dish using a glass pipette, further clean the cysts by hand-picking as necessary to remove all debris (under a stereo microscope), rinse in sterilized water three times by suspension and decanting and freeze the cysts at –80°C.

6. Grind approximately 500 mg of frozen cysts in a mortar with a pestle and homogenize in 2 ml of DNAzol reagent. Transfer the homogenate into two 1.5 ml microcentrifuge tubes.

7. Centrifuge for 10 min at 14,000 ×g at 4°C.

8. Transfer the resulting viscous supernatant to new microcentrifuge tubes. Add 500 µl of 100% ethanol and gently mix by inverting the tubes. A cloudy precipitation of DNA should be visible at this step.

9. Use pipette tip to swirl the DNA precipitate and transfer it to a new microcentrifuge tube. If there is no visible DNA precipitation, centrifuge at 4000 ×g for 1–2 min to pellet the DNA.

10. Add 1 ml of 95% ethanol to suspend the DNA, invert the tubes 3–6 times and then store the tubes vertically for 1 min to allow the DNA to settle to the bottom. Remove ethanol by pipetting.

11. Repeat step 10 to wash the DNA one more time.

12. Open the tubes to airdry the DNA pellet for 15 s at room temperature and then add 200 µl of 8 mM NaOH to solubilize the DNA.

13. Label and store the DNA at –20°C.

### From soil

There are commercial extraction kits available for extracting DNA from soil. By following the instructions of the manufacturer, isolating DNA from soil can be performed easily. For example, the PowerSoil Kit can isolate high-quality DNA from up to 250 mg samples in just 30 min. Since soil contains a variety of organisms such as fungi, bacteria, nematodes, algae, protozoa and arthropods, DNA extracted from the soil can be used to detect many targeted organisms by using PCR analyses.

### From arthropod tissue

Insects or mites can be hand-picked under a dissecting microscope and transferred into a microcentrifuge tube for DNA extraction. Make sure the specimens are from the same population without mixing with other species. For a large insect, a single specimen can be used. Collected specimens can be stored at 4°C for a few days or –20°C until ready for DNA extraction. The following protocol is used to extract DNA from insects based on the Qiagen blood and tissue kit. Other kits or methods can be utilized based on user's preference and experience.

1. Under a dissecting microscope, collect insects or mites into a pre-weighed lysing matrix A tube. Weigh the tube again to calculate the net weight of the specimen (50–100 mg).
2. Add 180 µl PBS into the tube and homogenize samples using the Qbiogene FastPrep 24 at speed 6.0 for 40 s, or using any available homogenizer.
3. Spin down contents to the bottom of tubes.
4. Add 20 µl proteinase K and 200 µl buffer AL. Vortex and incubate at 56°C for 10 min.
5. Add 200 µl ethanol (96–100%) to the sample and vortex to mix thoroughly.
6. Pipette the mixture from step 5 (including any precipitate) into the DNeasy Mini spin column placed in a 2 ml collection tube. Centrifuge at ≥ 6000 ×g (8000 rpm) for 1 min, then discard flow-through and collection tube.
7. Place the DNeasy mini spin column in a new 2 ml collection tube, add 500 µl buffer AW1 and centrifuge for 1 min at ≥ 6000 ×g (8000 rpm). Discard flow-through and collection tube.
8. Place the DNeasy Mini spin column in a new 2 ml collection tube, add 500 µl buffer AW2 and centrifuge for 3 min at 20,000 ×g (14,000 rpm). Discard flow-through and collection tube.

9. Place the DNeasy mini spin column in a new 1.5 ml microcentrifuge tube and pipette 100–200 µl buffer AE directly onto the column membrane. Incubate at room temperature for 1 min and then centrifuge for 1 min at ≥ 6000 ×g (8000 rpm) to elute.
10. Label and store DNA at –20°C.

### Measurement of DNA concentration

The DNA extracted from a tissue sample needs to be measured to determine the purity and quantity. This can be done easily using a spectrophotometer which measures the A260 value of a DNA sample. The A260 value is used as a quantitative measure for nucleic acids including both DNA and RNA. A double-stranded DNA sample with an A260 value of 1 defines a concentration of 50 ng/µl. Similarly, a single-stranded RNA sample with 1 unit of A260 has a concentration of 40 ng/µl, while an ssDNA sample would have a concentration of 33 ng/µl. The value of A280 obtained from a sample reflects the protein component. Therefore, the ratio of A260/280 is commonly used to assess the purity of DNA. A pure dsDNA (in a buffered solution) generally has an A260/280 value of 1.85–1.88. The value for a pure RNA sample is around 2.1. A DNA sample with an A260/280 value lower than 1.8 may suggest protein contamination; and higher than 1.88 may indicate the presence of RNA.

A spectrophotometer can be purchased from any vendor to fit the needs of a laboratory. Some can be used to measure the optical density (OD) value of DNA, RNA, protein and bacterial cell solutions. Each piece of equipment may have specific procedures to be followed. Below is an example protocol to measure DNA concentration using a biophotometer.

1. Turn on biophotometer, select dsDNA and enter dilution factors.
2. Add 50 µl of $H_2O$ in a cuvette and press 'blank'. The measurement value should be zero.
3. Add 1 µl of sample DNA into 49 µl of $H_2O$ in a 1.5 ml microcentrifuge tube, mix and spin down.
4. Load the total diluted (1:50) DNA solution into the cuvette, tap down to ensure solution settles into the bottom without bubbles. Place the cuvette into the sample slot of the biophotometer. Do not touch the clear sides of the cuvette.
5. Press 'sample' button to measure the sample. DNA concentration (µg/ml= ng/µl) and the values of A260/280 and A260/230 are printed or shown on screen.

6. Print a label with DNA sample ID, type (genomic, amplicon, plasmid, etc.), concentration (ng/µl) and date. Affix the label to the tube containing original sample DNA.

Unlike conventional photometers, NanoDrop One is a spectrophotometer that makes DNA measurement much simpler. It does not require a cuvette for DNA or RNA measurement, nor does it require dilution of the original sample DNA. The detection limit for dsDNA ranges from 2.0 ng/µl to 27,500 ng/µl by using a pedestal (Thermo Scientific, 2016), which covers almost all possible concentrations of DNA samples that a plant diagnostic lab usually handles. Once DNA extraction is completed, one microlitre of the final DNA product can be loaded directly onto the lower pedestal for measurement. The following procedures are based on a NanoDrop One instrument, but users should refer to instructions written specifically for each model.

1. From the home screen, select the 'Nucleic Acids' tab and select dsDNA or other types of nucleic acids to be measured.
2. Pipette 1 µl of blanking solution (usually the same buffer or $H_2O$ used to resuspend DNA) onto the lower pedestal and then lower the arm to blank (with auto-blank on). Alternatively, tap 'Blank' (with auto-blank off).
3. Lift the arm and clean both upper and lower pedestals with a clean Kimwipe, add 1 µl of DNA sample onto the lower pedestal, lower the arm to measure (with auto-measure on).
4. When the measurements for all samples are completed, save reported values, lift the arm and clean both pedestals with a clean Kimwipe.

### Amplification of a DNA fragment using universal primers

The use of a short region of a conservative or housekeeping gene of an organism as the reference sequence to aid species identification is known as DNA barcoding (Ratnasingham and Hebert, 2007). With genomic DNA extracted from an organism, culture, or soil, a pair or set of universal primers can be used to amplify a conservative region and then analyse the DNA sequence to determine the identity of an organism. For example, a 648 bp region of the cytochrome c oxidase I (COI) gene is the primary barcode sequence for identifying species in the animal kingdom (Hebert *et al.*, 2003;

Savolainen *et al.*, 2005). In plant species identification, two sets of primers that amplify chloroplast DNA regions (*matk+rbcl*) are widely used (Wattoo *et al.*, 2016) and they are the core DNA barcode for ferns (Li *et al.*, 2011). For pathogen identification, the regions of the rRNA gene are popular loci for species identification and discrimination (White *et al.*, 1990; Jeng *et al.*, 2001).

Phytoplasmas are an important group of bacterial pathogens, especially for hemp crops. A commonly used universal primers pair is P1/P7 (Deng and Hiruki, 1991; Schneider *et al.*, 1995). If P1/P7 amplicon (1.8 kb) is weak, dilute it 1:10 and amplify again using another universal primers pair R16F2n/R16R2 (Gundersen and Lee, 1996). This nest PCR generates an amplicon of 1.4 kb. In case both primer pairs fail to detect any potential phytoplasma from plant samples, another universal primer pair P1/Tint may amplify a fragment of 1.6 kb. Makarova *et al.* (2012) designed five primer cocktails containing 13 primers from the elongation factor *Tu* gene to amplify a fragment as a barcode region for identifying phytoplasmas. However, this method involves five primer cocktails that require synthesis of 13 primers, which may be too complicated to use in a routine diagnosis.

With more barcoding primers developed by researchers for different types of organisms, a plant diagnostic lab may only need to select a few sets of universal primers for routine barcoding-based pathogen and pest identification. Table 5.6 lists important primers that can be used to amplify the barcode region of plants, metazoan invertebrates, nematodes, fungi, bacteria and fastidious bacteria such as phytoplasmas.

Once a pair of universal primers are selected, conventional PCR can be performed followed by electrophoresis on a 1% (large amplicon ≥ 1kb) or 2% (amplicon ≤ 1 kb) agarose gel. The primer pairs listed in Table 5.6 usually generate a single sharp band that is suitable for gel cutting and PCR product extraction (Fig. 5.23). However, a PCR product with a single band (most likely a single amplicon) can be directly purified using any PCR purification kit without gel cutting and extraction. This is true when ITS1/ITS4 primers are used to amplify the region of the rRNA gene from purified fungal or oomycete cultures. However, if a PCR using universal primers is based on total DNA extracted from an infected plant sample or soil, a single band on a gel does not necessarily indicate a single amplicon. Further cloning of PCR products into a vector and selecting multiple clones for sequencing may be needed (see cloning section of this chapter).

**Table 5.6.** Universal primer pairs (the same colour block) commonly used for DNA barcode-based organism identification.

| Primer name | Primer sequence | Reference |
|---|---|---|
| | Insects, Mites and Nematodes | |
| LCO1490 (Forward) | 5'-GGTCAACAAATCATAAAGATATTGG-3' | Folmer *et al.*, 1994 |
| HCO2198 (Reverse) | 5'-TAAACTTCAGGGTGACCAAAAAATCA-3' | Folmer *et al.*, 1994 |
| | Plants[a] | |
| *mat*K-390F (Forward) | 5'-CGATCTATTCATTCAATATTTC-3' | Wattoo *et al.*, 2016 |
| *mat*K-1326R (Reverse | 5'-TCTAGCACACGAAAGTCGAAGT-3' | Wattoo *et al.*, 2016 |
| *rbc*L-BF (Forward) | 5'-ATGTCACCACAAACAGAAAC-3' | Wattoo *et al.*, 2016 |
| *rbc*L-724R (Reverse) | 5'-TCGCATGTACCTGCAGTAGC-3' | Wattoo *et al.*, 2016 |
| | Fungi and oomycetes | |
| ITS1 (Forward) | 5'-TCCGTAGGTGAACCTGCGG-3' | White *et al.*, 1990 |
| ITS4 (Reverse) | 5'-TCCTCCGCTTATTGATATGC-3' | White *et al.*, 1990 |
| | Bacteria | |
| 16SF (Forward) | 5'-GAGTTTGATCATGGCTCAG-3' | Jeng *et al.*, 2001 |
| 16SR (Reverser) | 5'-ACGGTTACCTTGTTACGAC-3' | Jeng *et al.*, 2001 |
| | Phytoplasmas | |
| P1 (Forward) | 5'-AAGAGTTTGATCCTGGCTCAGGATT-3 | Deng and Hiruki, 1991 |
| P7 (Reverse) | 5'-CGTCCTTCATCGGCTCTT-3' | Schneider *et al.*, 1995 |
| R16F2n (Forward) | 5'-GAAACGACTGCTAAGACTGG-3' | Gundersen and Lee, 1996 |
| R16R2 (Reverse) | 5'-TGACGGGCGGTGTGTACAAACCCCG-3' | Gundersen and Lee, 1996 |
| P1 (Forward) | 5'-AAGAGTTTGATCCTGGCTCAGGATT-3 | Deng and Hiruki, 1991 |
| Tint (Reverse) | 5'-TCAGGCGTGTGCTCTAACCAGC-3' | Smart *et al.*, 1996 |

[a]Note that a combination of a nuclear locus, for example the internal transcribed spacer (ITS) region, with two plastid loci, *rbc*L and *mat*K, can better solve the species complex problem (Li *et al.*, 2011). The ITS1/ITS4 primer pair, popularly used for fungal identification, can successfully amplify the ITS region of rDNA from many plant species (Fig. 5.23). If a plant specimen is infected with a fungus, a PCR with ITS1/ITS4 may generate two amplicons: one is from plant DNA and the other from the fungal DNA.

## Direct PCR from single bacterial colony

A single colony can be picked into a TE buffer and proceeded to PCR. This quick method is useful for amplifying a targeted region of DNA without further culturing and DNA extraction.

1. Inside a biosafety cabinet, add 5 µl of 1× TE buffer onto a single colony, pipette several times and then transfer bacterial suspension into a 0.2 ml PCR tube. May repeat one more time to collect all bacterial cells into the tube. Add additional 1× TE buffer to reach total volume of 15 µl and mix.
2. Transfer 5 µl of bacterial cells from step 1 into 9 µl of molecular grade water in a new 0.2 ml PCR tube. Mix by gently pipetting and then heat the solution in a thermocycler at 99–100°C for 10 min.
3. Remove the tube from the thermocycler and spin down briefly. The final volume will be approximately 12.5 µl, due to evaporation.
4. Add 10× PCR buffer, $MgCl_2$, primers, dNTP, *Taq* polymerase and water (use water to bring total reaction volume to 25 µl).

5. Run PCR cycles (94°C for 90 s, 30–35 cycles of 94°C for 30 s, 55°C for 1 min and 72°C for 1 min, and final extension at 72°C for 5–10 min). Adjust PCR cycling parameters to optimize amplification.

## Direct PCR from fungal spores

Many obligate fungi such as powdery mildew and rust pathogens cannot be cultured on a medium to produce a mass of material for DNA extraction, nor can they be purified through culturing. In this situation, the direct collection of spores from plant specimens can be performed for routine diagnosis. The following protocol applies to rust, powdery mildew and other fungi that produce massive spores on the surface of plant tissue (Schoener and Wang, 2018).

1. Under a stereo microscope with reflection light on, place a leaf infected with powdery mildew or rust fungus and select an area where a high density of spores/mycelium are present without any other fungi observed.

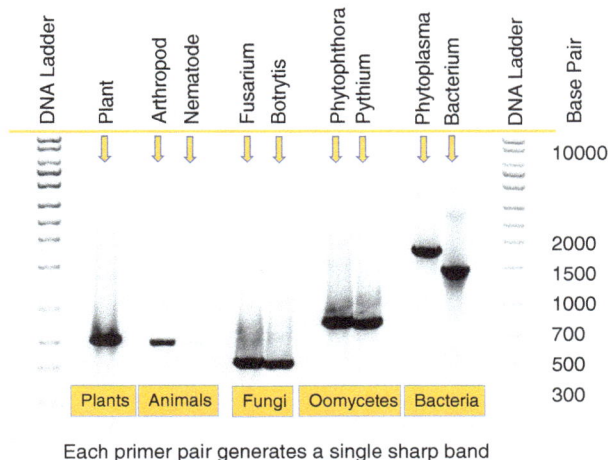

Each primer pair generates a single sharp band

**Fig. 5.23.** The size of amplicons generated from different types of organisms by the universal primers listed in Table 5.6. **Plant:** Foothill deathcamas (*Zigadenus paniculatus*) with primers ITS1/ITS4 (≈700 bp); **Arthropod:** Cannabis aphid (*Phorodon cannabis*) with primers LCO1490/ HCO2198 (≈709 bp); **Nematode:** *Ditylenchus dipsaci* with primers LCO1490/ HCO2198 (≈700 bp); **Fusarium:** *F. oxysporum* with primers ITS1/ITS4 (≈550 bp); **Botrytis:** *B. porri* with primers ITS1/ITS4 (≈550 bp); **Phytophthora:** *P. nicotianae* with primers ITS1/ITS4 (≈890 bp); **Pythium:** *P. aphanidermatum* with primers ITS1/ITS4 (≈860 bp); **Phytoplasmas:** Clover proliferation group with primers P1/P7 (≈1.8 kb); **Bacteria:** *Enterobacter* sp. with primers 16SF/16SR (≈1.5 kb).

2. Apply 15 µl of 1× TE buffer to the area where a dense mycelial/spore colony is present. Repetitively pipette to suspend spores and mycelia in the TE buffer and transfer into a 1.5 ml microcentrifuge tube.

3. Take 5 µl of fungal spores and mycelia from step 2 into 9 µl of molecular grade water in a 0.2 ml PCR tube. Mix by gently pipetting and then heat in a thermocycler at 99–100°C for 10 min.

4. Take the tube out from the thermocycler and spin down briefly. The final volume will be approximately 12.5 µl, due to evaporation.

5. Add 10× PCR buffer, MgCl$_2$, primers, dNTP, *Taq* polymerase and water (use water to bring total reaction volume to 25 µl).

6. Run PCR cycles (94°C for 90 s, 30–35 cycles of 94°C for 30 s, 55°C for 1 min and 72°C for 1 min, and final extension at 72°C for 5–10 min). Adjust PCR cycling parameters to optimize amplification.

### Gel electrophoresis

#### TAE buffer

TAE buffer is a commonly used solution to run electrophoresis. It is usually made as a 50× TAE stock solution. To make TAE buffer, add 242 g of Tris base, 57.1 ml of glacial acetic acid and 100 ml 0.5 M EDTA (pH 8.0) into a large beaker. Add water to 1000 ml and mix using a stirrer. This concentrated TAE buffer can be stored on the shelf at room temperature. To use the buffer for electrophoresis, dilute it to 1× TAE and add ethidium bromide before using it to make a gel and run electrophoresis. Here is a simple protocol for making 5 l of 1× TAE buffer with ethidium bromide (EB) incorporated for daily use.

1. Make a 50× TAE stock solution as described above (Sambrook and Russell, 2001).

2. In a big container (5–6 l), add 4900 ml nanopure water, 100 ml of 50× TAE stock solution, and 100 µl of 1% ethidium bromide.

3. Close the lid, shake vigorously to mix, and store at room temperature.

#### Prepare, load and run a gel

After PCR, the amplicon can be visualized on an agarose gel through electrophoresis. The first step is to prepare a fresh gel. The following protocol is used to make a medium-sized 1% agarose gel with EB. Gels containing EB and run in an EB-containing 1× TAE buffer can be directly placed under UV light to visualize the DNA band. Otherwise, the gel must be soaked in EB solution to stain the DNA followed by rinsing in water.

1. Add 1 g of agarose powder into 100 ml of 1× TAE buffer (containing EB) into a 500 ml beaker. Mix to suspend agarose in the buffer.
2. Microwave until the solution boils for 15–30 s and becomes clear. Let it cool briefly inside the microwave.
3. Assemble a gel casting tray with a comb inserted, pour agarose solution into the tray and let it solidify at room temperature. To avoid exposure to EB vapour, prepare gel in a fume hood.
4. Remove the comb and place the tray into the horizontal electrophoresis apparatus. Fill with 1× TAE buffer containing EB until gel is fully immersed.
5. Load DNA samples into the gel wells. The following is an example to load five PCR products into wells without additional tubes or changing the pipette volume for each sample: place a small piece of parafilm on the bench, set a 20 µl pipette to 10 µl and add five tiny drops of 6× DNA loading dye buffer in a line on the parafilm (approximately 2 µl per drop); change tip and then pipette 10 µl of the PCR product into one drop of dye, mix, depress the pipette all the way down, uptake all liquid (12 µl), then load it into the well.
6. Load DNA ladder on the first and/or last wells.
7. Place the lid on top of the apparatus and connect it to power supply.
8. Set the voltage (5V/cm) and run the gel. In general, 120V is applied for a large gel and 60–70V for a small gel.
9. When the last dye (there are usually three dyes in the loading buffer) reaches the bottom of the gel, disconnect from power and take the gel to a gel documentation system for observation or to take an image.
10. Properly dispose of the gel, as it contains EB.

### Staining gel with ethidium bromide

This procedure is to be used only when the gel and running 1× TAE buffer do not contain EB.

1. Add 300 ml of 1× TAE buffer (amount based on the size of gel and container) into a glass baking dish and then add 1.5 µl of 10 mg/ml ethidium bromide solution. Gently mix.

2. Slide the gel into the dish and ensure the gel is immersed in the solution.
3. Gently shake (on a shaker) for 30–45 min.
4. Transfer the gel into another glass dish and add water to immerse the gel.
5. Gently shake for 20 min at room temperature.
6. Observe DNA bands under a gel imaging system.

### Decontamination of EB-containing TAE buffer

EB is a mutagen and so EB-containing solutions should be handled with care. In the case of a spill, any contaminated surfaces should be treated with activated charcoal. Since EB is only used in electrophoresis buffers in a very diluted concentration (≤ 0.5 µg/ml), the following protocol can be used to decontaminate a used buffer before disposal (Bensaude, 1988; Sambrook and Russell, 2001).

1. Add 100 mg of powdered activated charcoal to each 100 ml of solution.
2. Store the solution for at least an hour, shaking it intermittently.
3. Slowly pour the charcoal solution through a Whatman No. 1 filter paper.
4. Place the filter paper with activated charcoal into a sealable plastic bag and dispose it into the hazardous waste container. Discard the filtrate.

### PCR product purification

After PCR, the amplicon needs to be isolated or purified for downstream applications such as direct sequencing, cloning or restriction enzyme digestion. There are two commonly used purification methods: PCR purification and gel extraction. A PCR purification kit generally removes primers, nucleotides, enzymes, salts and other impurities from PCR reactions and it does not separate amplicons. In diagnostic PCRs, some primers generate one sharp band with some weak or invisible bands. In this case, a gel cutting followed by extraction is the best way to purify the targeted amplicon, even if only a sharp band is observed. Gel extraction procedures can isolate a single amplicon from a PCR reaction that generates multiple bands, so it is commonly used to purify a specific amplicon even when only a single band is visible. If the amplicon is to be cloned into a vector, the gel extraction method is highly recommended.

## DNA extraction from gel

Running a gel after PCR has two purposes: (i) to check if the expected DNA fragment is successfully amplified; and (ii) to cut a single band from gel for DNA extraction. To get a strong band with a high concentration of the PCR product, a PCR reaction can be performed in 50 µl volume. Use 10 µl to run a gel to check the amplification and, if successful, load the remaining 40 µl of the PCR reaction into a combined well (use clear tape to wrap three comb prongs to create a wide well in the gel). After electrophoresis, the gel can be placed on a UV box for cutting. Wear face shield and eye protection as well as gloves to protect from UV exposure. The following protocol for DNA gel extraction is based on a Qiagen kit, but users can choose any kit from any vendor based on their preference and experience. All centrifuge steps are done at 13,000 rpm (17,900 ×g) in a conventional tabletop microcentrifuge at room temperature unless otherwise noted.

1. Prepare a 1% agarose gel with one or two wide wells.
2. Add 8 µl of 6× loading dye into 40 µl of post-PCR reaction, pipette to mix and then load into the well of the gel. Load a DNA ladder for size reference.
3. Run the gel at 5V/cm until loading dye migrates two-thirds of the gel length.
4. Attach a piece of plastic wrap on the top of a UV transilluminator and then place the gel on it.
5. Prepare a pre-weighed colourless 1.5 ml microcentrifuge tube, a clean sharp scalpel and forceps. Wear protective face shield, gloves, long sleeves, etc., before turning on the UV light.
6. Use a scalpel to cut the band off the gel precisely. Remove excess agarose.
7. Turn off UV light and place the gel slice into the pre-weighed tube. Weigh again and calculate the net weight of the gel. Depending on the extraction kit, a 400 mg gel slice is the maximum.
8. Multiply gel weight (mg) by 3 to get a volume (µl). Add this volume of buffer QG to the tube containing the gel slice.
9. Incubate at 50°C, inverting/shaking every 2–3 min, for 10 min or until the gel slice has dissolved.
10. Add 1× gel volume (1 µl per 1 mg) of isopropanol to the tube and invert to mix.
11. Add sample to the QIAquick® column and centrifuge for 1 min. Discard the flow-through.
12. Add 500 µl of buffer QG to the QIAquick® column and centrifuge for 1 min. Discard flow-through.
13. Add 750 µl of buffer PE to the QIAquick® column and centrifuge for 1 min. Discard flow-through.
14. Centrifuge again at full speed for 1 min.
15. To elute DNA, place QIAquick® column into a new 1.5 ml microcentrifuge tube, add 10 µl of buffer EB to the centre of the column membrane, let it stand for 1 min and then centrifuge for 1 min at 12,000 rpm. The flow-through is the final DNA.
16. Measure the DNA concentration, label the product and store at –20°C.

## PCR clean up

There are commercial kits available to clean up a PCR product right after a PCR reaction. This procedure is commonly used in plant diagnostics, especially when there is a large number of samples (with a single PCR product) to be analysed by sequencing. As the kit is designed to remove some non-amplicon components in a PCR reaction mix, it should not be used to purify a PCR reaction with multiple amplicons for sequencing purposes. Below is a protocol based on a Qiagen PCR purification kit. All centrifuge steps are done at 13,000 rpm (10,000 ×g) in a conventional tabletop microcentrifuge at room temperature unless otherwise noted.

1. Add five volumes of buffer PBI to one volume of PCR reaction and mix by pipetting.
2. Pipette the sample into the MinElute column placed in a 2 ml collection tube.
3. Centrifuge for 1 min and discard flow-through.
4. Add 750 µl of buffer PE to MinElute column and centrifuge for 1 min. Discard the flow-through.
5. Centrifuge for an additional 1 min at maximum speed (14,000 rpm).
6. Place MinElute column into a new 1.5 ml microcentrifuge tube, add 10 µl of buffer EB to the centre of the column membrane, let it stand for 1 min and then centrifuge for 1 min at 12,000 rpm. The flow-through is the final DNA.
7. Measure the DNA concentration, label the product and store at –20°C.

## Direct sequencing of PCR products

PCR products purified by gel extraction or directly cleaned up can then be submitted for DNA sequencing. PCR products are generally sequenced by the Sanger sequencing method (Sanger *et al.*, 1977). This method reads up to 800–1200 bp, a size range for most PCR products generated in plant diagnostics. Sanger sequencing is available in most genomic centres and costs as low as US$2 per sample. The turnaround time may be less than 24 h. The low price and quick service allow many diagnostic labs to delegate this task to commercial services to avoid purchasing expensive sequencing equipment and spending unnecessary time on sequencing. Although next-generation sequencing is a new and powerful technique, it is mainly used for large-scale genomic research projects with in-depth bioinformatics analyses. However, it may eventually evolve to the point where diagnostic labs can use it in routine diagnosis for only a tiny fraction of the current costs.

To prepare a DNA sample for Sanger sequencing, follow the instructions of the genomic centres. There are two key issues to be considered when submitting DNA samples: (i) optimal amount of DNA template; and (ii) the amount of primer. In dye-terminator Sanger sequencing, only one primer is needed for each sample. The primer is either one of the PCR primers used for amplification or a primer designed from the vector (e.g. plasmid DNA) into which a PCR amplicon is inserted. An optimal sequencing requires the appropriate number of copies of the primer and DNA template, instead of the actual weight. For example, 40 ng is sufficient for sequencing a 1.5 kb PCR amplicon, but it is too much for a 150 bp amplicon, as 40 ng of 150 bp has far more copies than 40 ng of 1.5 kb. Conversely, 4 ng is optimal for 150 bp but falls short for 1.5 kb because there is a very low number of molecules in 4 ng of 1.5 kb. Therefore, when calculating the amount of DNA needed for Sanger sequencing, the 'divide by 20' rule is applied for plasmid DNA and the 'divide by 50' rule is used for PCR amplicons. Table 5.7 lists the common types and sizes of templates a plant diagnostic lab usually generates and the amount of each needed for dye-terminator Sanger sequencing at the Nevada Genomic Center.

To submit a DNA sample for sequencing, add the required amount of DNA (in ng) and primer into a 0.2 ml PCR tub, and, if applicable, add molecular grade $H_2O$ to bring the total volume to 6.5 µl. If the concentration of DNA is low, the total volume can be brought to a maximum of 25 µl. An easy way to calculate the required volume of a DNA template is to divide the total amount of template needed (e.g. 3800 bp plasmid DNA, 250 ng) by the template DNA concentration (e.g. 500 ng/µl), which equals 0.5 µl in this example.

**Table 5.7.** Template and primer amounts needed for Sanger sequencing sample submission (Nevada Genomic Center, 2020).

| Type and size of template | Amount of template | Amount of primer |
|---|---|---|
| Plasmid, 3000–5000 bp | 250 ng | 1 µl of 2 µM primer |
| Plasmid, 5000–10,000 bp | 400 ng | 1 µl of 10 µM primer |
| Amplicon, 500–999 bp | 20 ng | 1 µl of 2 µM primer |
| Amplicon, 1000–1999 bp | 40 ng | 1 µl of 10 µM primer |
| Amplicon, more than 2000 bp | 50 ng | 1 µl of 10 µM primer |

A mixed sample can be frozen at –20°C until ready for submission. It is important to note that the DNA product should be free of ethanol. During the last few steps of the DNA extraction, ethanol residue must be removed completely by additional full-speed centrifuge after the washing step. When eluting the DNA, centrifugation should be performed at a lower speed or in soft mode. An ethanol-containing DNA sample may result in a sequencing failure.

## Cloning of a PCR product

### Rational and principle

Many diagnostic PCR products are single amplicons around 500 bp, especially for fungal pathogens (Fig. 5.23). In this case, the PCR product can be purified and then sequenced. However, the sizes of PCR products from many species are larger than 700 bp. One-time Sanger sequencing can only read a partial sequence of the amplicon with the first 50 bp, centre portion, and last 50 bp missing or inaccurate (Fig. 5.24). To obtain a full sequence of a PCR product, one strategy is to insert the amplicon molecule into a plasmid vector and transform the recombinant plasmid into *Escherichia coli* competent cells. Each cell will produce many copies of the amplicon-containing plasmid during reproduction (Fig. 5.25). This procedure is called molecular cloning (Sambrook and Russell, 2001). The amplicon-containing plasmid can then be extracted from the bacterial cells and submitted for sequencing. By using forward/reverse primers designed from the vector, which flank the inserted amplicon, both ends of the amplicon can be read accurately. When the amplicon is over 1 kb (which is too long to be read in a single run of Sanger sequencing), sequence data from both strands can be used to design a new set of primers to perform another run of sequencing. The primers can be easily picked by using an online tool called primer 3 (http://primer3.ut.ee/). Repeat this process until a full coverage of the amplicon sequence is obtained with minimal redundancy (Fig. 5.26). This strategy is known as primer walking (Sterky and Lundeberg, 2000) and it can be used to sequence large DNA fragments up to 7 kb. However, in plant diagnostics, most PCR products

or inserts are less than 2 kb and one-time primer walking and two runs of sequencing are sufficient to obtain their complete sequences.

Another benefit of cloning PCR products is the separation and identification of different amplicon molecules from a single PCR reaction or band. Some believe that a sharp band on an agarose gel represents a single amplicon molecule. However, this may not be true, especially when a PCR is run from total DNA extracted from an infected plant tissue. For example, in a survey for phytoplasma infection on many ash trees exhibiting leaf yellow and dieback symptoms, we used primers designed for detecting mycoplasma-like organisms (Ahrens

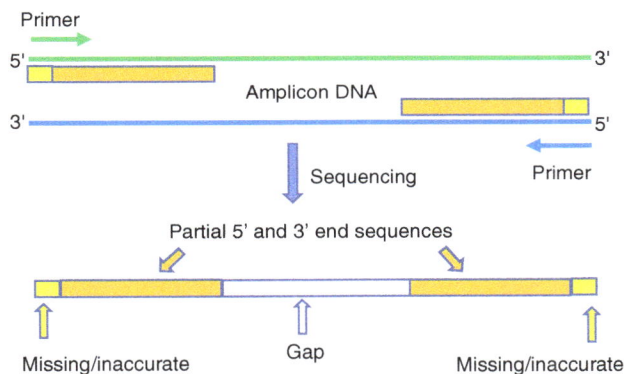

**Fig. 5.24.** A typical sequence gap in the middle and inaccurate sequence read at the 5'- and 3'- ends of a large PCR amplicon sequenced with PCR primers.

**Fig. 5.25.** A typical cloning scheme to insert a PCR product into a pGEM®-T vector. Note that the inserted PCR amplicon can be sequenced using primers T7 and SP6 designed from the vector sequence.

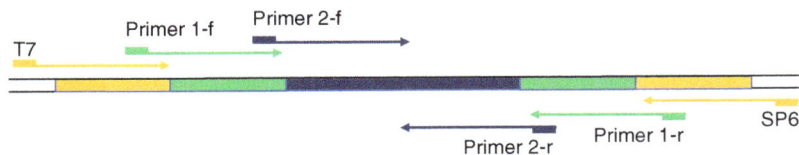

**Fig. 5.26.** The principle of primer walking to sequence the entire PCR amplicon. Note that most PCR products (less than 2 kb) cloned into pGEM®-T vectors only need one additional primer pair after sequencing with T7/SP6 primers.

and Seemüller, 1992) to amplify phytoplasma DNA from the ash trees and then examined the PCR products (566 bp) by gel cutting, cloning and sequencing. Surprisingly, the 566 bp band contained different DNA molecules amplified from diverse environmental and endophytic bacterial species (Garneni and Wang, 2009). Even with a pure DNA template for PCR, there are still many errors introduced during PCR, including editing errors, misincorporation of bases, PCR-mediated recombination and DNA damage (Pienaar *et al.*, 2006; Potapov and Ong, 2017). That being said, a single-band PCR product from a diagnostic sample is likely to be a mix of DNA molecules. Through cloning, each individual molecule is inserted into a vector that is transformed into a single bacterial cell. If a number of colonies from a transformation plate are picked and the plasmids are extracted and sequenced, multiple distinct amplicon sequences may be identified. This strategy is especially important for universal detection when universal primers are used. For example, cloning PCR products generated from universal phytoplasma primers followed by examining sequences of multiple clones led to the detection of multiple members of the clover proliferation phytoplasma group as well as the bacterium *Spiroplasma citri* (Schoener and Wang, 2019).

### Procedures

The procedure described here is based on the Promega pGEM®-T-vector system (Promega, 2018) and the QIAprep Spin Miniprep kit. Other kits can be selected at the user's discretion. To perform, set a 3-day time window as some steps require overnight incubation, but the total time spent on the procedures from ligation to submission for sequencing may be less than 8 h.

### Day 1: Ligation
1. Thaw the reagents listed below on ice, vortex and then spin down.
2. Add reagents in the order listed below into a 0.2 ml PCR tube and mix by pipetting. Note that the volume of a PCR product is based on its concentration. Typically, a 3:1 molar ratio of the PCR product to the vector is used. To calculate what amount of PCR product is needed, use the formula: (50 ng of vector × kb size of PCR product × 3) ÷ 3.0 (kb of vector) = $x$ ng of PCR product needed.

| Reagent | Amount per reaction |
|---|---|
| 2× rapid ligation buffer for T4 DNA ligase[a] | 5 µl |
| pGEM-T or pGEM-T easy vector DNA (50 ng) | 1 µl |
| PCR product (cleaned or gel extracted) | $x$ µl |
| T4 DNA ligase (3 Weiss units/µl) | 1 µl |
| Molecular grade water | To final volume of 10 µl |

[a]Note: Always use the buffer provided with the ligase. Aliquot the buffer to single-use amount to avoid frequent freeze–thaw.

3. Place the tube in a thermocycler programmed to 24°C for 1 h, then incubate in 4°C overnight.

### Day 2: Transformation and Culture
4. Thaw aliquoted SOC medium (in 1.5 ml tube) and IPTG solution. Add 20 µl of X-gal and 100 µl of IPTG to each LB agar plus ampicillin plate (LB[+amp]). Add three sterilized glass beads and swirl the plate to evenly distribute solution on the plate. Place the plates (LB[+ ampicillin/IPTG/X-gal]) into 37°C incubator (upside down) for 1 h.
5. Remove JM109 high efficiency competent cells from –80°C and thaw on ice for 5 min. Gently mix by flicking the tube. Aliquot the cells to 50 µl in each 1.5 ml microcentrifuge tube. Use only one aliquot for each PCR product cloning and freeze unused aliquots immediately.
6. Add 2 µl of ligation reaction to the tube with 50 µl competent cells and gently pipette to mix, then place the tube in the ice for 20 min.
7. Heat-shock the cells for 45–50 s by placing the tube into a water bath at exactly 42°C.
8. Immediately return the tubes to ice (pushing tube into the ice) for 2 min.
9. Add 950 µl room-temperature SOC medium to the tube, mix by pipetting and then transfer cell solution to a sterile culture tube.
10. Incubate the sterile culture tube at 37°C with shaking at 150 rpm for 90 min.
11. Take 3 LB[+ ampicillin/IPTG/X-gal] plates from the 37°C incubator and add 30 µl, 70 µl and 100 µl of cell solutions onto the three plates, respectively. These three volumes are designed to have one plate with optimal density for colony pick-up. Add three autoclaved glass beads to each plate and swirl gently to distribute the cell solution evenly.
12. Incubate plates upside down at 37°C overnight (16–24 h) or until the colony sizes are optimal for pick-up (1–2 mm in diameter). Transformed

colonies with the PCR amplicon insert are white and those without the insert are blue. Some white colonies may have a blue colour in the centre and they should not be picked up.

**13.** At 4:00 pm, inside the biosafety cabinet, arrange several culture tubes as desired (e.g. three to five per cloning sample) on a rack, pipette 3 ml of Super Broth (or LB broth) and 3 µl of 100 mg/ml ampicillin solution to each tube. Use a 2-20 µl pipette to pick an individual white colony by gently scraping the colony with the pipette tip and then dipping the tip into the medium and releasing bacterial cells by gently pipetting. Discard the tip into a beaker containing bleach solution.

**14.** Place the culture tubes in the incubator at 37°C with shaking at 220 rpm for 16 h.

### Day 3. Plasmid Extraction and Preparation for Sequencing

**15.** Remove tubes from the shaking incubator and pour each cell suspension into a 1.5 ml microcentrifuge tube and centrifuge for 3 min at 8000 rpm to pellet the bacterial cells. Save the remaining (1.5 ml) cell culture at 4°C.

**16.** Pour the liquid out into a beaker that contains bleach (to kill the bacteria if any). Use a 20-200 pipette to remove the remaining liquid without dislodging the pelleted bacteria and eject the tip into the bleach solution.

**17.** Add 250 µl of buffer P1 to each tube and resuspend the bacterial cells by quickly scraping the tube bottom across the wells of an 80-well microcentrifuge tube holder.

**18.** Add 250 µl of buffer P2 and mix thoroughly by inverting the tube 4–6 times.

**19.** Add 350 µl of buffer N3 and mix immediately and thoroughly by inverting the tube 4–6 times.

**20.** Centrifuge for 10 min at 13,000 rpm in a tabletop centrifuge.

**21.** Pipette or pour supernatant into the QIAprep spin column.

**22.** Centrifuge for 1 min at 13,000 rpm and discard the flow-through.

**23.** Add 500 µl of buffer PB and centrifuge for one minute at 13,000 rpm. Discard the flow-through.

**24.** Add 750 µl of buffer PE and centrifuge for 1 min at 13,000 rpm. Discard the flow-through.

**25.** Centrifuge again for 1 min at 14,000 rpm to remove ethanol residue. Discard the flow-through.

**26.** Place the QIAprep column in a clean and labelled 1.5 ml microcentrifuge tube.

**27.** Add 50 µl of buffer EB to the centre of the column. Let it sit for 5 min and centrifuge for 1 min at 13,000 rpm.

**28.** Measure plasmid DNA concentration and store at –20°C.

**29.** To submit for sequencing, follow Table 5.7 or the instructions of selected genomic service centre for determining the appropriate amount of template and primers. Most diagnostic PCR products are less than 1 kb, so the total size of a plasmid with an insert is less than 4 kb. To sequence from both DNA strands, prepare two PCR tubes and add a volume of plasmid DNA (250 ng) into each tube. Add 1 µl of 2 µM T7 promoter primer (TAATACGACTCACTATAGGG) to one tube and 1 µl of 2 µM SP6 upstream primer (ATTTAGGTGACACTATAG) to another tube. Spin down and store at –20°C until ready to submit.

**30.** If needed, the remaining bacterial cells can be preserved in glycerol at –80°C using the procedures described below.

### Storing transformed cells in glycerol

After sequencing results are available and suggest that the bacterial clone contains a vector successfully inserted with an expected PCR product, the remaining bacterial cells cultured from each colony can be preserved. Whenever the plasmid is needed, a loop of bacterial cells can be streaked on an LB+amp plate or directly cultured in LB broth (amended with ampicillin) to extract more plasmids. The plasmid with the targeted PCR product can be used as a positive control in certain diagnostic tests, but extra caution should be taken for high-copy plasmids that may become a source of cross-contamination. The following procedure is used to store bacterial cells in 15% glycerol solution at –80°C.

**1.** Add 75 ml of glycerol and 25 ml of nanopure water to a 100 ml screw-top bottle, mix, autoclave, cool down and store at 4°C.

**2.** In the biosafety cabinet, add 200 µl of autoclaved 75% glycerol (4°C) into a 1.5 ml microcentrifuge tube.

**3.** Pipette 800 µl of bacterial cell culture into the microcentrifuge tube and mix with glycerol by pipetting.

**4.** Store the tube at –80°C.

### Sequence assembling

Once sequencing data are available, sequence assembling is needed to obtain a full amplicon sequence. As mentioned earlier in this chapter, a large amplicon may need additional sequencing by using the primer walking strategy. Assembling Sanger sequence data can be easily done by any commercial DNA software. For example, a Lasergene application called SeqMan Pro can assemble the 5'- sequence and 3'- sequence data into one contig in just minutes. Users simply need to save the 5'- and 3'- DNA sequence data in one Lasergene application file and add the sequence files into SeqMan Pro application. By clicking the 'assemble' button, all sequence pieces are assembled into one contig. The contig can be shown as a graph detailing the depth of coverage. If a PCR product is cloned into plasmid vector, multiple plasmid DNA samples can be sequenced and subjected to assembling to get a consensus sequence. SeqMan Pro can trim sequence ends, scan for vector sequence, remove contaminant sequences and perform many other functions.

Not all diagnostic labs have DNA sequence analysing software. In this case, manual assembling can be performed using Microsoft Word. It is a bit time-consuming but doable for a small number of assembling projects. Below is an example to assemble a PCR product sequence that is amplified by ITS1/ITS4, cloned into the T-vector, and sequenced by T7 promoter/SP6 upstream primers.

**1.** Paste sequence data from primer T7 promoter (T7 sequencing read) and data from primer SP6 upstream (SP6 sequencing read) in a Word document.
**2.** Use the online reverse complement tool (https://www.bioinformatics.org/sms/rev_comp.html) to reverse SP6 sequencing read. Paste the reverse complement into the Word document and delete the original SP6 sequencing read from the document.
**3.** Paste PCR primers (e.g. ITS1/ITS4) into the top of the Word document, reverse complement one of them and paste it to the document.
**4.** Select all documents, change the font to Courier New and choose font size 8 or 10 so that there is the same number of nucleotides and space in each line.
**5.** Click 'Find' and select 'Advanced Find', copy and paste one of the primer sequences (e.g. ITS4) into the 'Find what' box, click 'Find Next'. The ITS4 primer may be found in either the T7 sequencing read, the reverse complement of SP6 sequencing read, or both (in the case of small amplicons). This also depends on the orientation of the insert in the vector. Highlight the found sequence in green. Repeat this finding process with the reverse complement sequence of another primer (e.g. ITS1) and highlight the sequence in red. Now the starting and ending sequences of the amplicon have been located by the primers.
**6.** In the T7 sequencing read, select a short sequence in the middle, run the 'Find' tool and highlight all

found sequences in yellow. If it is found in both the T7 and reversed SP6 sequencing reads, it is an overlapping sequence. Continue this process until all the overlapping sequences are found and marked yellow. Mark the end of the T7 sequencing read in the SP6 sequencing read and vice versa.
**7.** Now it is time to examine the overlapped area and find any base variations between the T7 and SP6 sequencing reads. If a piece of sequence in the T7 read cannot be found in the SP6 read in the overlapping region, this suggests a sequence variation. In this situation, consider the sequence nearer the sequencing primer to be more accurate than those further away from the primer. Alternatively, increase the coverage by sequencing plasmid DNA samples extracted from multiple clones.
**8.** After the primer and overlapping sequences have been located, delete the vector sequences (upstream PCR primers) and connect the T7 and reverse complement SP6 sequencing reads using the overlapping sequence as a bridge.
**9.** Now the assembled sequence contains the ITS1/ITS4 PCR primers at both ends and an overlapped sequence in the middle (Fig. 5.27). Do a word count to determine the total base pairs.
**10.** Copy the sequence into the NCBI website and perform a quick BLAST search.

### BLAST search

The Basic Local Alignment Search Tool, known as BLAST, is an approach for rapid sequence comparison (Altschul *et al.*, 1990). This tool has a powerful algorithm for searching nucleotide (nucleotide BLAST, or BLASTn) and protein sequences (protein BLAST, or BLASTp) in a database to find similar regions between sequences. The program compares nucleotide or protein sequence data to sequences deposited in databases and calculates the statistical significance of matches. BLAST can be used to infer the functional and evolutionary relationships between biological sequences as well as to help identify members of gene families. The Web BLAST has become a routine online tool used by millions of users to search for related nucleotide and protein sequences (https://blast.ncbi.nlm.nih.gov/Blast.cgi).

In DNA sequence-based diagnosis, BLAST is frequently used to search closely related biological species based on the DNA sequences obtained from the specimens. The following is the series of clicks to perform a BLAST search.

```
TCCTCCGCTTATTGATATGC ITS4 primer
CCGCAGGTTCACCTACGGA ITS1 primer (Reverse-complement)
```

## T7 promoter primer sequencing read

```
GGGGCGGGATTGGATTTCCTCCGCTTATTGATATGCTTAAGTTCAGCGGGTAATCTTGCCTGATATCAGGTCCAATTGAGATGCATACCGAAGTACAC
ATTAAGTTCCCAAATGGATCGACCCTCGACAACCGAAGCTGCCACCCTACTTCGCAACAGCAAAGCCAATTCAAAAGCCAAGCCACCGAGCTATGG
TTCACCAGTCCATCACGCCACAGCAGGAAAAACGTCCAATAAGTGCATTGTTCAGCCGAAGCCAACCATACCACGAATCGAACACTCTTCCATTAACG
CCGCAGCAGACAAACCAGTCGCCAATTTGCCACAATAGCAGCCTTCACAACCAGCTACGCCACCACTTTTCGAGCAAAGAGAAGTTCAGTTTAGTACA
TTTAAAAGGACTCGCAGCCGGTCCGAAGACCAATCGCAAGACACTTCACATCTGACATCTCCTCCACCGACTACACGGAAGGAAGAAAGTCAAGTTTA
ATGTACGGACACTGATACAGGCATACTTCCAGGACTAACCCGGAAGTGCAATATGCGTTCAAAATTTCGATGACTCACTGAATCCTGCAATTCGCATT
ACGTATCGCAGTTCGCAGCGTTCTTCATCGATGTGCGAGCCTAGACATCCACTGCTGAAAGTTGCTATCTAGTTAAAAGCAGAGACTTTCGTCCCCAC
AGTATATTCAGTATTAAAAGGAATGGGTTTAAAAGAAAAAAAGACTACTAAATCAGCCGAAGCCCAAACGTTCGCCTTTTTTTTGATAGGGCTCA
CTCAGCAGCAGCCG
```

## SP6 upstream primer sequencing read (reverse-complement)

```
CAATTCAAAGCCAAGCCACCGAGCTATGGTTCACCAGTCCATCACGCCACAGCAGGAAAAACGTCCAATAAGTGCATTGTTCAGCCGAAGCCAACC
ATACCACGAATCGAACACTCTTCCATTAACGCCGCAGCAGACAAACCAGTCGCCAATTTGCCACAATAGCAGCCTTCACAACCAGCTACGCCACCACT
TTTCGAGCAAAGAGAAGTTCAGTTTAGTACATTTAAAAGGACTCGCAGCCGGTCCGAAGACCAATCGCAAGACACTTCACATCTGACATCTCCTCCAC
CGACTACACGGAAGGAAGAAAGTCAAGTTTAATGTACGGACACTGATACAGGCATACTTCCAGGACTAACCCGGAAGTGCAATATGCGTTCAAAATTT
CGATGACTCACTGAATCCTGCAATTCGCATTACGTATCGCAGTTCGCAGCGTTCTTCATCGATGTGCGAGCCTAGACATCCACTGCTGAAAGTTGCTA
TCTAGTTAAAAGCAGAGACTTTCGTCCCCACAGTATATTCAGTATTAAGGAATGGGTTTAAAAGAAAAAAAGACTACTAAATCAGGCCGAAGCC
CAAACGTTCGCCTTTTTTTTGATAGGGCTCACTCAGCAGCAGCCGCCAACAATTAAGCCAGCAGCCGCCGCCAAATAAGACCCCCAACTATTGGGTTG
AAACGGTTCACGTGGAAAGTTTTTTAGGTGTGGTAATGATCCTTCCGCAGGTTCACCTACGGAAATGACTAGTGAATTCGGGCGGGGGCTGCAGGTGG
```

## Assembled Sequence, 892bp

```
TCCTCCGCTTATTGATATGCTTAAGTTCAGCGGGTAATCTTGCCTGATATCAGGTCCAATTGAGATGCATACCGAAGTACACATTAAGTTCCCAAATG
GATCGACCCTCGACAACCGAAGCTGCCACCCTACTTCGCAACAGCAAAGCCAATTCAAAAGCCAAGCCACCGAGCTATGGTTCACCAGTCCATCAC
GCCACAGCAGGAAAAACGTCCAATAAGTGCATTGTTCAGCCGAAGCCAACCATACCACGAATCGAACACTCTTCCATTAACGCCGCAGCAGACAAACC
AGTCGCCAATTTGCCACAATAGCAGCCTTCACAACCAGCTACGCCACCACTTTTCGAGCAAAGAGAAGTTCAGTTTAGTACATTTAAAAGGACTCGCA
GCCGGTCCGAAGACCAATCGCAAGACACTTCACATCTGACATCTCCTCCACCGACTACACGGAAGGAAGAAAGTCAAGTTTAATGTACGGACACTGAT
ACAGGCATACTTCCAGGACTAACCCGGAAGTGCAATATGCGTTCAAAATTTCGATGACTCACTGAATCCTGCAATTCGCATTACGTATCGCAGTTCGC
AGGCGTTCTTCATCGATGTGCGAGCCTAGACATCCACTGCTGAAAGTTGCTATCTAGTTAAAAGCAGAGACTTTCGTCCCCACAGTATATTCAGTATT
AAGGAATGGGTTTAAAAGAAAAAAAGACTACTAAATCAGGCCGAAGCCCAAACGTTCGCCTTTTTTTTGATAGGGCTCACTCAGCAGCAGCCGC
CAACAATTAAGCCAGCAGCCGCCGCCAAATAAGACCCCCAACTATTGGGTTGAAACGGTTCACGTGGAAAGTTTTTTAGGTGTGGTAATGATCCTTCC
GCAGGTTCACCTACGGA
```

**Fig. 5.27.** Sequence assembling of a DNA fragment (892 bp) amplified from *Phytophthora parasitica* infecting *Catharanthus roseus* (Wang and Buk, 2013) using ITS1/ITS4 primers and then cloned into a pGEM®-T vector followed by Sanger sequencing using the T-7 promoter and SP6 upstream primers. Note that there is a region (628 bp) highlighted in yellow shared by both the T7 and SP6 sequencing reads. This overlapping sequence allows two sequences to combine into a contig. In the overlapping region, there are three unhighlighted locations showing either the addition or the deletion of a single base (in red). If no additional coverage is available, base reads closer to the sequencing start point are considered more reliable than the base reads at the end. For example, the sequence 'AAAAGCC' is 170 bases away from the beginning of the T7 sequencing read, while 'AAAGCC' is 765 bases away from the beginning of the SP6 sequencing read; therefore, in the combined sequence, 'AAAAGCC' is selected. Similarly, 'AAGGAA' and 'GGCCGAA' are closer to the beginning of the SP6 sequencing read and are considered more reliable than 'AAAGGAA' and 'GCCGAA' obtained from the T7 sequencing read, respectively. Sequences highlighted in grey are from the vector and should be removed once the primer sequence is located. Underlined sequence indicates the end of the T7 or SP6 sequencing read.

1. In a search engine, type in NCBI or http://www.ncbi.nlm.nih.gov/.
2. On the right-hand popular resource column, click on BLAST.
3. Click Nucleotide BLAST box.
4. Copy and paste a query DNA sequence into the sequence box. Alternatively browse a sequence file for uploading.

5. For most diagnostic BLAST searches, use the default setting and select the highly similar sequences (megablast) program, and then click the BLAST button.
6. In a few seconds, a list of sequences with high similarities is generated. Navigate each accession number and its biological information to further investigate the relevance of the query DNA sequence to other deposited sequences.

7. In general, sequences with the highest scores, a zero E value and over 99% identical bases are considered to be mostly related. However, in determining the identity of an organism, a sequence from a published paper with a voucher specimen is considered more reliable than other directly deposited sequences.

### Submitting DNA sequences to NCBI

NCBI, the National Center for Biotechnology Information (https://www.ncbi.nlm.nih.gov/), is the world's largest public repository of biological and scientific information. This repository, aka GenBank, functions like a bank for sequences including mRNA sequences, prokaryotic and eukaryotic genes, rRNA, viral sequences and pseudogenes. GenBank comprises the DNA DataBank of Japan (DDBJ), the European Nucleotide Archive (ENA) and the GenBank at NCBI, all of which exchange data on a daily basis to ensure worldwide coverage. The database contains publicly available nucleotide sequences for almost 260,000 formally described species and these sequences are primarily submitted by individual laboratories, large-scale sequencing projects and environmental sampling projects (Benson *et al.*, 2013). Diagnosticians can access GenBank through the NCBI website (https://www.ncbi.nlm.nih.gov/genbank/) to retrieve sequence information, perform BLAST searches, or deposit sequence data. It is valuable to submit new sequence data from uniquely identified specimens into GenBank because it allows researchers to compare and identify specimens from different ecological niches or geographical areas. To submit a DNA sequence through the NCBI website, navigate to the NCBI Home page, click the 'Submit' box to bring the page to the Submission Portal, and then follow the instructions to complete the submission. Most submissions are made using the web-based BankIt or stand-alone Sequin programs. After submission, GenBank staff review and assign the sequence an accession number and the submitter can use the accession number in publications to reference the sequence data.

### References

Agrios, G.N. (1997) *Plant Pathology*, 4th edn. Academic Press, San Diego, California.

Ahrens, U. and Seemüller, E. (1992) Detection of DNA of plant pathogenic mycoplasma like organisms by a polymerase chain reaction that amplifies a sequence of the 16S rRNA gene. *Phytopathology* 82, 828–832.

Altschul, S.F., Gish, W., Miller, W., Myers, E.W. and Lipman, D.J. (1990) Basic local alignment search tool. *Journal of Molecular Biology* 215, 403–410. doi: 10.1016/S0022-2836(05)80360-2

Bell, W.C., Young, E.S., Billings, P.E. and Grizzle, W.E. (2008) The efficient operation of the surgical pathology gross room. *Journal of Biotechnic & Histochemistry* 83, 71–82. doi: 10.1080/10520290802127610

Bensaude, O. (1988) Ethidium bromide and safety – Readers suggest alternative solutions (letter). *Trends in Genetics* 4, 89-90.

Benson, D.A., Cavanaugh, M., Clark, K., Karsch-Mizrachi, I., Lipman, D.J. *et al.* (2013) GenBank. *Nucleic Acids Research*, 41 (Database issue), D36–D42. doi: 10.1093/nar/gks1195

Browning, I.A. (2009) Bioassay for diagnosis of plant viruses. In: Burns, R. (ed.) *Methods in Molecular Biology 508, Plant Pathology*. Humana Press, Totowa, New Jersey, pp. 1–13. doi: 10.1007/978-1-59745-062-1_1

Bustin, S.A., Benes, V., Garson, J.A., Hellemans, J., Huggett, J. *et al.* (2009) The MIQE guidelines: minimum information for publication of quantitative real-time PCR experiments. *Clinical Chemistry* 55, 611–622. doi: 10.1373/clinchem.2008.112797

Cardoso, T.C., Sousa, R.L.M., Alessi, A.C., Montassier, H.J. and Pinto, A.A. (1998) A double antibody sandwich ELISA for rapid diagnosis of virus infection and to measure the humoral response against infectious bursal disease on clinical material. *Avian Pathology* 27, 450–454. doi: 10.1080/03079459808419368

CBOL Plant Working Group (2009) A DNA barcode for land plants. *Proceedings of the National Academy of Sciences of the United States of America* 106, 12794–12797. doi: 10.1073/pnas.0905845106

Chauhan, R.P., Wijayasekara, D., Webb, M.A. and Verchot, J. (2015) A reliable and rapid multiplex RT-PCR assay for detection of two potyviruses and a pararetrovirus infecting canna plants. *Plant Disease* 99, 1695–1703. doi: 10.1094/PDIS-02-15-0225-RE

Choi, J.K., Maeda, T. and Wakimoto, S. (1977) An improved method for purification of turnip mosaic virus. *Annals of the Phytopathological Society of Japan* 43, 440–448.

Dáder, B., Burckbuchler, M., Macia, J.L., Alcon, C., Curie, C. *et al.* (2019) Split green fluorescent protein as a tool to study infection with a plant pathogen, Cauliflower mosaic virus. *PLoS ONE* 14, e0213087. doi: 10.1371/journal.pone.0213087

Deng, S. and Hiruki, C. (1991) Amplification of 16S rRNA genes from culturable and nonculturable mollicutes. *Journal of Microbiological Methods* 14, 53–61. doi: 10.1016/0167-7012(91)90007-D

Derycke, S., Vanaverbeke, J., Rigaux, A., Backeljau, T. and Moens, T. (2010) Exploring the use of cytochrome oxidase c subunit 1 (COI) for DNA barcoding of free-living marine nematodes. *PLoS ONE* 5, e13716. doi: 10.1371/journal.pone.0013716

Folmer, O., Black, M., Hoeh, W., Lutz, R. and Vrijenhoek, R. (1994) DNA primers for amplification of mitochondrial cytochrome c oxidase subunit I from diverse metazoan invertebrates. *Molecular Marine Biology and Biotechnology* 3, 294–299.

Garneni, S. and Wang, S. (2009) Use of DNA sequences and bioinformatics tools as standard diagnostic procedures for regulatory or fastidious pathogens. NPDN National Meeting, Miami, Florida, December 6–10, 2009.

Gulati, N.M., Torian, U., Gallagher, J.R. and Harris, A.K. (2019) Immunoelectron microscopy of viral antigens. *Current Protocols in Microbiology* 53, e86. doi: 10.1002/cpmc.86

Gundersen, D.E. and Lee, I.M. (1996) Ultrasensitive detection of phytoplasmas by nested-PCR assays using two universal primer pairs. *Phytopathologia Mediterranea* 35, 144–151.

Hardham A.R. (2012) Confocal microscopy in plant–pathogen interactions. In: Bolton, M. and Thomma, B. (eds) *Plant Fungal Pathogens. Methods and Protocols*. Methods in Molecular Biology, Vol. 835. Humana Press, Totowa, New Jersey, pp. 295–309. doi: 10.1007/978-1-61779-501-5_18

Harris, T.H., Szalanski, A.L. and Powers, T.O. (2020) *Molecular Identification of Nematodes Manual*. Available at https://nematode.unl.edu/nemaid.pdf (accessed 07 April 2020).

Hebert, P.D., Cywinska, A., Ball, S.L. and deWaard, J.R. (2003) Biological identifications through DNA barcodes. *Proceedings of the Royal Society B Biological Sciences* 270, 313–321. doi: 10.1098/rspb.2002.2218

Higuchi, R., Dollinger, G., Walsh, P.S. and Griffith, R. (1992) Simultaneous amplification and detection of specific DNA sequences. *Biotechnology* 10, 413–417.

Holland, P.M., Abramson, R.D., Watson, R. and Gelfand, D.H. (1991) Detection of specific polymerase chain reaction product by utilizing the 5'-3' exonuclease activity of *Thermus aquaticus* DNA polymerase. *Proceedings of the National Academy of Sciences of the United States of America* 88, 7276–7280.

Janse, J.D. and Kokoskova, B. (2009) Indirect immuno-fluorescence microscopy for the detection and identification of plant pathogenic bacteria (In particular for *Ralstonia solanacearum*). In: Burns, R. (ed.) *Plant Pathology. Methods and Protocols*. Methods in Molecular Biology, Vol 508. Humana Press, Totowa, New Jersey, pp. 89–99. doi: 10.1007/978-1-59745-062-1_8

Jeng, R.S., Svircev, A.M, Myers, A.L., Beliaeva, L., Hunter, D.M, *et al.* (2001) The use of 16S and 16S-23S rDNA to easily detect and differentiate common Gram-negative orchard epiphytes. *Journal of Microbiological Methods* 44, 69–77.

Kemp, B.M. and Smith, D.J. (2005) Use of bleach to eliminate contaminating DNA from the surface of bones and teeth. *Forensic Science International* 154, 53–61. doi: 10.1016/j.forsciint.2004.11.017

Kini, K., Dossa, R., Dossou, B, Mariko, M., Koebnik, R., *et al.* (2019) A semi-selective medium to isolate and identify bacteria of the genus *Pantoea*. *Journal of General Plant Pathology* 85, 424–427. doi: 10.1007/s10327-019-00862-w

Kralik, P. and Ricchi, M. (2017) A basic guide to real time PCR in microbial diagnostics: definitions, parameters, and everything. *Frontiers in Microbiology* 8, 108. doi: 10.3389/fmicb.2017.00108

Leslie, J.F. and Summerell, B.A. (2006) *The Fusarium Laboratory Manual*. Blackwell Publishing, Oxford, UK.

Li, F.W., Kuo, L.Y., Rothfels, C.J., Ebihara, A., Chiou, W.L. *et al.* (2011) *rbcL* and *matK* earn two thumbs up as the core DNA barcode for ferns. *PLoS ONE* 6, e26597. doi: 10.1371/journal.pone.0026597

Li, R. and Hartung, J.S. (2007) Reverse transcription-polymerase chain reaction-based detection of plant viruses. *Current Protocols in Microbiology* 6, 16C.1.1–16C.1.9. doi: 10.1002/9780471729259.mc16c01s6

Makarova, O., Contaldo, N., Paltrinieri, S., Kawube, G., Bertaccini, A. *et al.* (2012) DNA barcoding for identification of 'Candidatus Phytoplasmas' using a fragment of the elongation factor Tu gene. *PLoS ONE* 7, e52092. doi: 10.1371/journal.pone.0052092

McKeen, C.D. and Thorpe, H.J. (1971) An adaptation of a moist-chamber method for isolating and identifying *Verticillium* spp. *Canadian Journal of Microbiology* 17, 1139–1141. doi: 10.1139/m71-179

Nevada Genomic Center (2020) Sanger Sequencing: Optimal Amount of Template and Primer. Available at https://www.unr.edu/genomics/guidelines/ss-template-primer (accessed 18 April 2020).

Peyret, H. (2015) A protocol for the gentle purification of virus-like particles produced in plants. *Journal of Virological Methods* 225, 59–63. doi: 10.1016/j.jviromet.2015.09.005

Pienaar, E., Theron, M., Nelson, M. and Viljoen, H. J. (2006) A quantitative model of error accumulation during PCR amplification. *Computational Biology and Chemistry* 30, 102–111. doi: 10.1016/j.compbiolchem.2005.11.002

Potapov, V. and Ong, J.L. (2017) Examining sources of error in PCR by single-molecule sequencing. *PLoS ONE* 12, e0169774. doi: 10.1371/journal.pone.0169774

Powers, T.O. and Harris, T.S. (1993) A polymerase chain reaction method for identification of five major *Meloidogyne* species. *Journal of Nematology* 25, 1–6.

---

Prince, A.M. and Andrus, L. (1992) PCR: how to kill unwanted DNA. *Biotechniques* 12, 358–360.

Promega (2018) pGEM®-T and pGEM®-T Easy Vector Systems. Available at: https://www.promega.com/-/media/files/resources/protocols/technical-manuals/0/pgem-t-and-pgem-t-easy-vector-systems-protocol.pdf?la=en (accessed 24 April 2020).

Ratnasingham, S. and Hebert, P.D. (2007) Bold: the barcode of life data system (http://www.barcodinglife.org). *Molecular Ecology Notes* 7, 355–364. doi: 10.1111/j.1471-8286.2007.01678.x

Riggs, R.D. and Schmitt, D.P. (1991) Optimization of the *Heterodera glycines* race test procedure. *Journal of Nematology* 23, 149–154.

Rufián, J.S., Macho, A.P., Corry, D.S., Mansfield, J.W., Ruiz-Albert, J. *et al.* (2018) Confocal microscopy reveals *in planta* dynamic interactions between pathogenic, avirulent and non-pathogenic *Pseudomonas syringae* strains. *Molecular Plant Pathology* 19, 537–551. doi: 10.1111/mpp.12539

Rychlik, W., Spencer, W.J. and Rhoads, R.E. (1990) Optimization of the annealing temperature for DNA amplification *in vitro*. *Nucleic Acids Research* 18, 6409–6412. doi: 10.1093/nar/18.21.6409

Sambrook, J. and Russell, D.W. (2001) *Molecular Cloning A Laboratory Manual*, 3rd edn. Cold Spring Harbor Laboratory Press, Cold Spring Harbor, New York.

Sanderson, M.J., Smith, I., Parker, I. and Bootman, M.D. (2014) Fluorescence microscopy. *Cold Spring Harbor Protocols* 2014, pdb.top071795. doi: 10.1101/pdb.top071795

Sanger, F., Nicklen, S. and Coulson, A.R. (1977) DNA sequencing with chain-terminating inhibitors. *Proceedings of the National Academy of Sciences of the United States of America* 74, 5463–5467. doi: 10.1073/pnas.74.12.5463

Savolainen, V., Cowan, R.S., Vogler, A.P., Roderick, G.K. and Lane, R. (2005) Towards writing the encyclopedia of life: an introduction to DNA barcoding. *Philosophical Transactions of the Royal Society of London. Series B, Biological Sciences* 360, 1805–1811. doi: 10.1098/rstb.2005.1730

Schneider, B., Seemüller, E., Smart, C.D. and Kirkpatrick, B.C. (1995) Phylogenetic classification of plant pathogenic mycoplasma-like organisms or Phytoplasmas. In: Razin, S. and Tully, J.G. (eds) *Molecular and Diagnostic Procedures in Mycoplasmology, Vol. I*. Academic Press, San Diego, California, pp. 369–380. doi: 10.1016/B978-012583805-4/50040-6

Schoener, J. and Wang, S. (2018) First detection of *Golovinomyces ambrosiae* causing powdery mildew on medical marijuana plants in Nevada. (Abstr.) *Phytopathology* 108(S1), 186. doi: 10.194/PHYTO-108-10- S1.186

Schoener, J. and Wang, S. (2019) First detection of a phytoplasma associated with witches' broom of industrial hemp in the United States. (Abstr.) *Phytopathology*, 109, S2.1. doi: 10.1094/PHYTO-109-10-S2.1

Schuck, S., Weinhold, A., Luu, V.T. and Baldwin, I.T. (2014) Isolating fungal pathogens from a dynamic disease outbreak in a native plant population to establish plant-pathogen bioassays for the ecological model plant *Nicotiana attenuata*. *PLoS ONE*, 9, e102915. doi: 10.1371/journal.pone.0102915

Shurtleff, M.C. and Averre, C.W. (1997) *The Plant Disease Clinic and Field Diagnosis of Abiotic Diseases*. The American Phytopathological Society, St Paul, Minnesota.

Smart, C.D., Schneider, B., Blomquist, C.L., Guerra, L.J., Harrison, N.A. *et al.* (1996) Phytoplasma-specific PCR primers based on sequences of the 16S-23S rRNA spacer region. *Applied and Environmental Microbiology* 62, 2988–2993.

Sterky, F. and Lundeberg, J. (2000) Sequence analysis of genes and genomes. *Journal of Biotechnology* 76, 1–31. doi: 10.1016/s0168-1656(99)00176-5

Thermo Scientific (2016) NanoDrop One User Guide. Available at http://tools.thermofisher.com/content/sfs/manuals/3091-NanoDrop-One-Help-UG-en.pdf (accessed 18 April 2020).

Wang, S. and Buk, J. (2013) First detection and molecular identification of *Phytophthora parasitica* from annual vinca in Nevada. *Phytopathology* 103, S2.155. doi: 10.1094/PHYTO-103-6-S2.1

Wang, S. and Gergerich, R.C. (1998) Immunofluorescent localization of tobacco ringspot nepovirus in the vector nematode *Xiphinema americanum*. *Phytopathology* 88, 885–889. doi: 10.1094/PHYTO.1998.88.9.885

Wang, S., Riggs, R.D. and Yang, Y. (2001) Grouping of populations of *Heterodera trifolii* by host preference and AFLP pattern. *Nematology* 3, 667–674. doi: 10.1163/156854101753536037

Wattoo, J.I., Saleem, M.Z., Shahzad, M.S., Arif, A., Hameed, A. *et al.* (2016) DNA barcoding: amplification and sequence analysis of *rbcl* and *matK* genome regions in three divergent plant species. *Advancements in Life Sciences* 4, 03–07.

White, T.J., Bruns, T.D., Lee, S.B. and Taylor, J.W. (1990) Amplification and direct sequencing of fungal ribosomal RNA genes for phylogenetics. In: Innis, M.A., Gelfand, D.H., Sninsky, J.J. and White, T.J. (eds) *PCR Protocols-A Guide to Methods and Applications*. Academic Press, San Diego, California, pp. 315–322.

Wright, S. (2020) Checking for Bacteria Streaming. Available at: https://wiki.bugwood.org/Checking_for_Bacterial_Streaming retrieved on 2/25/2020 (accessed 8 April 2020).

# 6 Diagnosing Seed and Bud Diseases

## Inviable Seeds

Healthy seeds are the prerequisite for healthy seedlings and crops. It is possible that a portion of seeds used for planting a crop will lose viability due to infection by pathogens, improper storage, or genetic and physiological factors. Inviable seeds may not be identified until they are sowed and fail to germinate. There is no visible difference in general between viable and inviable hemp seeds, but certain seeds may show physical damage or abnormal morphology. A lab test is needed to determine the viability and germination rate of seeds.

### Field diagnosis

#### Symptoms

In the field, the number of seedlings is less than expected and a crop exhibits uneven or lower crop stands. In the soil, seeds fail to germinate normally and become rotten or mouldy. In seed stock, affected seeds may have necrotic lesions, abnormal shapes or sizes, a light weight, or internal drying and decay.

#### Problem classification

The loss of seedlings and crop stands can be caused by soil variability in a field. However, inviable seeds generally cause the random loss of seedling stands without a defined patch in the field. The pattern of loss of stands can be scattered. Try to recover the ungerminated seeds from the soil to see what could have happened to the seeds. An intact but mouldy or rotten seed suggests it is defective and lacks viability. When seed viability is identified as the cause, an additional investigation should be conducted to determine which factors affect seed viability. In many cases, inappropriate storage temperatures and moisture levels are the two major factors that greatly impact seed longevity, vigour and viability.

### Do your own diagnosis

Growers can perform a seed germination test with the following procedure.

1. Use a clear plastic box and place two layers of paper towels at the bottom.
2. Add enough cooled boiled water to just moisten the paper towel.
3. Place a number of hemp seeds on the paper, close the lid and place the box on a table at room temperature (20–25°C).
4. Check results within a week and calculate the germination rate.

### Sampling

If seed viability is suspected to be the cause of poor crop stands, representative seed samples may be submitted to a seed lab for a viability test. Be sure to submit a sufficient number of seeds (sample size) to achieve an accurate germination rate reading (Ribeiro-Oliveira *et al.*, 2016). Call the seed lab for instructions about what sample size is needed for a viability test.

### Lab diagnosis

A professional seed lab will perform a germination test using a specialized tray and germination paper in a germinator that controls temperature, moisture and lighting (Fig. 6.1). A plant diagnostic lab can test hemp seed germination rate using a standard Petri dish following the procedure described below.

1. Take a sterile Petri dish (100 mm diameter × 15 mm height) and place a piece of filter paper that is cut to fit the dish.
2. Add a small amount of sterilized water, just enough to wet the filter paper completely.
3. Surface sterilize the seeds in 1% sodium hypochlorite (NaClO) for 1 min and then rinse them in sterile water (optional).

DOI: 10.1079/9781789246070.0006

4. Place five seeds evenly on the dampened filter paper in the Petri dish, place the lid on the dish and seal with parafilm.

5. Place the dish on a lab bench or in an incubator (18–22°C) with 12 h lighting.

6. After 7–10 days, check the seeds for germination. Seeds with a tap root grown out are considered viable. Calculate germination rate.

### Key diagnostic evidence

The germination rate of a seed lot used in planting is lower than a standard rate, or 85%, and its low germination rate corresponds to the low stands of seedlings in the field. In cases where the seed lot has a healthy germination rate, other factors should be investigated.

## Latent Infection of Seeds by *Alternaria* Species

Hemp seeds can carry or be infected by pathogenic microorganisms originating from mother plants,

**Fig. 6.1.** Germination of newly harvested hemp seeds on a piece of dampened germination paper. Note that there were three seeds that did not germinate (inviable).

without visible symptoms. The infection is often inactive, known as latent infection. When these infected seeds are planted, seedlings may show certain symptoms, including damping off, wilting, or pre-emergence death, depending on the pathogen carried. In many cases, seeds rot before germination because their viability has been destroyed. Seed-carrying pathogens can affect seed vigour, germination, seedling stage and beyond. Asymptomatic latent infections are common in seeds, but it is not rare to see some infected seeds with physical symptoms such as necrotic lesions and internal decay. These symptomatic seeds generally fail to germinate. There are other types of microorganisms associated with hemp seeds internally (endophytes) and externally (epiphytes) without significant impacts on seed and seedling health; in certain situations, these microorganisms may play a role in secondary metabolite production in cannabis plants (Taghinasab and Jabaji, 2020) or otherwise be beneficial to plant growth. That said, not all microorganisms associated with seeds are considered pathogenic.

### Field diagnosis

#### Symptoms

*Alternaria*-infected hemp seeds may not exhibit visible symptoms. Infected seeds appear to germinate normally in the germination tray, but brown to dark grey fungal mycelia may grow extensively from the seeds, impairing germination tests (Fig. 6.2). One study showed that 56% of hemp seeds from a seed batch were infected with *Alternaria infectoria* and *A. tenuissima* but the germination rate remained 86% (Schoener *et al.*, 2018). *Alternaria* species are considered seed-borne pathogens in other crops (Groves and Skolko, 1944) and they can be passed to the next crop, causing diseases such as leaf spots and foliar blight.

#### Problem classification

Latent infection of hemp seeds with *Alternaria* species is an issue of seed quality. The species infecting seeds and the rate of infection largely depend on the crop from which the seeds are produced. Certified hemp seeds generally have a lower rate of infection by seed-borne pathogens if the crop is inspected during the growing season and confirmed to have no seed-borne diseases.

### Do your own diagnosis

Some companies manufacture prepared medium plates or small medium trays ready for immediate isolation of a range of microorganisms. To verify if hemp seeds are potentially infected by *Alternaria* and other seed-borne pathogens, select the plates or trays containing PDA (potato dextrose agar) to perform a simple isolation test.

1. Soak 10–15 hemp seeds in 10 ml of 1:5 diluted house bleach solution (mix 2 ml bleach with 8 ml tap water) for 5–10 min and then rinse the seeds with cooled boiled water.
2. Use disinfected forceps to place five seeds in each plate evenly and then leave the plates at room temperature.

*Alternaria* usually grows out from the seed within 2–5 days and forms a greenish black or olive brown colony in 10 days (Fig. 6.3). To identify the isolates accurately, submit the plates to a plant diagnostic lab.

### Sampling

To test the quality of seeds and potential infections of seed-borne diseases, representative seed samples should be taken and submitted to a plant health diagnostic lab or a seed health testing lab. The sample size and the number of samples vary, depending on the quantity of seeds and the number of varieties. Consult a testing laboratory for instructions.

### Lab diagnosis

#### Isolation and colony observation

Soak up to 50 hemp seeds in 1% sodium hypochlorite for 15 min with shaking, rinse in sterilized water three times, plate on potato dextrose agar plates amended with streptomycin sulfate (PDA$^{+strep}$) and incubate at 22°C in the dark. Colonies may be visible as early as 2 days after plating. Initial colonies are white or light grey and later turn to a darker colour (Fig. 6.3). Infection rate should be counted after 10 days of incubation, as the fungus may grow slowly from some seeds. The colony is generally flat in the beginning but becomes woolly with greyish, short, aerial hyphae in older cultures. The diagnostic characteristic for *Alternaria* colonies is a greenish black or olive brown colour. The reverse side of the colony is typically brown to black due to pigment production.

#### Microscopic examination

Inside a biosafety cabinet, pick a small piece of the colony (containing hyphae and spores) using sterilized

**Fig. 6.2.** Growth of *Alternaria* mycelia (shown as dark stain) from hemp seeds on sterile filter paper during a germination test. Note that almost all seeds germinated.

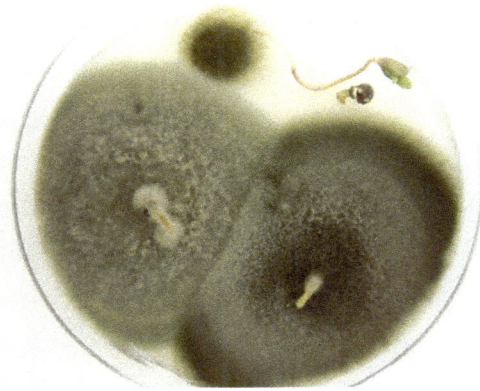

**Fig. 6.3.** Growth of *Alternaria* species associated with hemp seeds on a potato dextrose agar plate incubated in the dark at 22°C.

forceps and place it into a drop of water on a glass slide. Cover with a slip, gently press the slip to flatten the specimen and load the slide onto a compound microscope to observe the mycelia and/or spores. A determining characteristic for *Alternaria* spores is multicellular spores with divisions both vertically and horizontally (Fig. 6.4). Spores of *Alternaria* are mostly formed in chains. Hyphae are thick and dark brown. Some species may not produce abundant spores on PDA, in which case a DNA-based identification should be performed.

### DNA-based identification

If isolates do not produce the spores needed for morphology-based identification, collect 100–200 mg of mycelia and proceed to DNA extraction followed by PCR using primers ITS1/ITS4 (see Chapter 5). Amplified DNA fragments are then sequenced, assembled and used as a query sequence to search the NCBI database for closely related species. Depending on the *Alternaria* species, the size of the PCR fragment may range from 540 bp to 600 bp.

### Key diagnostic evidence

Morphological characteristics of the colony and spore should be in line with general descriptions of the *Alternaria* genus and DNA sequence data support the morphological identification. It is likely that a given seed lot is infected by multiple *Alternaria* species.

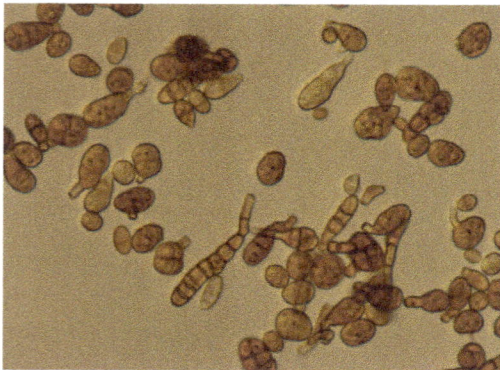

**Fig. 6.4.** Typical morphology of the conidium spores of an *Alternaria* species. Note that a variety of shapes and sizes are present due to the different development stages.

## Infection of Seeds by *Rhizopus* Species

*Rhizopus* is a genus of fungal pathogens that infect grains or seeds (Haware, 1971) and also cause postharvest diseases such as the soft rot of fruits and vegetables (Kwon *et al.*, 2011). Infected seeds may cause poor seed germination or damping-off of younger seedlings (Fig. 6.5). *Rhizopus* can infect other parts of plants too. For example, *R. oryzae* was reported to cause stem disease in sunflower (Mathew *et al.*, 2015). *Rhizopus* has been frequently detected from harvested hemp seeds and may potentially impact hemp seedling health.

### Field diagnosis

#### Symptoms

Like *Alternaria* infection, *Rhizopus* infection may not cause visible symptoms in seeds. However, *Rhizopus* causes the pre-emergence death of seeds (Howell, 2002) and post-emergence damping off (Fig. 6.5). *Rhizopus* infection can be spotted during germination tests when greyish and thick mycelia grow out from the seeds and impact the tests. If multiple infected seeds are present in a germination tray (see Fig. 2.20), the tray may become very mouldy as germination proceeds. Often, *Rhizopus* mycelia grow directly out from the newly formed tap root. The infection may cause brownish and sunken lesions (Fig. 6.6) and impair tap root health, which may eventually lead to damping-off of young seedlings.

**Fig. 6.5.** Pre-emergence death of seeds and post-emergence damping-off of microgreen caused by a seed-carrying *Rhizopus* species. Note the several patches of white mycelia growing from both germinated and ungerminated seeds.

### Problem classification

*Rhizopus* is a seed-borne pathogen and affects seed quality and seed germination. Symptoms associated with emergence and seedling health should be traced to the quality of seeds sowed and potential infection by *Rhizopus* or other pathogens.

### Do your own diagnosis

Use similar procedures to those described above in the section of latent infection of seeds by *Alternaria*. Note that *Rhizopus* grows rapidly on PDA medium and produces abundant aerial mycelium in the plate within 2–3 days. In the mass of mycelia, numerous tiny dots of brown to black colour (sporangia) may be visible to the naked eye. Hyphae are thicker than those of most pathogenic fungi and they are filament-like (see Fig. 2.20). The colour of mycelia initially looks white, then turns grey and brown with age. To identify the fungus accurately, submit the plates to a plant diagnostic lab.

### Sampling

Representative seed samples should be taken and submitted to a plant health diagnostic lab to test for *Rhizopus* and other seed-borne pathogens. Consult a testing laboratory for instructions regarding sample size and number.

### Lab diagnosis

### Visual and microscopic examination of seeds

Received seed samples should be visually examined to see if there are any physical defects and internal lesions. More detailed symptoms may be revealed by observing the seed under a dissecting microscope. This examination can identify unhealthy seeds and the nature of the damage, which helps in selecting a small set of seeds for plating.

### Isolation and colony observation

Use similar procedures to those described above in the section of latent infection of seeds by *Alternaria*. *Rhizopus* is a fast-growing fungus and can grow from one seed to the rest of plate within a few days. If multiple seeds are plated on a single plate, some seeds may be overgrown by fungus originating from other seeds, which may mess up the infection rate counting. Therefore, it is better to plate only one seed per plate. In most cases, *Rhizopus* grows out from the seed shells or newly formed tap roots. Colonies of *Rhizopus* are initially white and then become grey, brown and filament-like. A trademark characteristic of *Rhizopus* colonies is its rapid growth rate with thick fibre-like mycelia and visible black sporangia.

### Microscopic examination of Rhizopus

*Rhizopus* fungi are known to have large sporangia with a distinct dark or brown colour. The sporangia are globose or sub-globose in shape ranging from 60 μm to 180 μm in diameter (see Fig. 2.19). Sporangiophores are straight to support the sporangium and their colours range from pale brown to brown. Sporangiospores are mostly elliptical or globose (Fig. 6.7).

**Fig. 6.6.** A necrotic and sunken lesion on the tap root due to *Rhizopus oryzae* infection. Note that the lesion almost breaks the root at the middle.

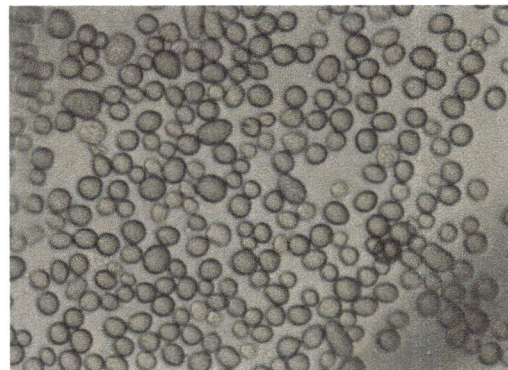

**Fig. 6.7.** Various shapes of sporangiospores of *Rhizopus* sp.

### DNA-based identification

*Rhizopus* is easily identified based on the microfibre-like colonies, distinct hyphae and sporangiophores, large and dark sporangia and massive sporangiospores. To identify an isolate to a species, DNA-based identification may be carried out using primers ITS1/ITS4 (see Chapter 5). Depending on the species, the size of the PCR fragment may range from 550 bp to 600 bp.

### Key diagnostic evidence

Pre-emergence death or post-emergence damping-off, if it occurs in the field, should have a clear link with the quality of the seed sowed. The rate of seed infection by *Rhizopus* must correspond to the scale of the problem in the field. Although other microorganisms may be found in the same seed lot, the dominance of the *Rhizopus* infection can explain a *Rhizopus*-caused emergence or damping-off disease. Accurate identification of the *Rhizopus* species can be achieved using morphological characteristics and DNA sequence data.

## Powdery Mildew of Buds

### Field diagnosis

#### Symptoms

Powdery mildew becomes very noticeable when a colony of white mould forms on a leaf or bud, but it is not evident during the early stages of infection. The typical symptom is fuzzy-looking white mycelia covering the bud tissue (Fig. 6.8). The distribution of mycelia and spores on buds may not be even and they may be denser in some areas than others, exhibiting distinct white patches. In most cases, white mycelium patches are also found on leaves and petioles. Damage to the bud tissue ranges from mild to severe, depending on the stage of infection. Infection may cause reduced growth of buds, discoloration and even cell death, leading to brown lesions.

#### Problem classification

Powdery mildew is an airborne disease affecting the leaf, stem, bud and flower. It is a mould disease frequently occurring in cool and humid environment. This disease may become a nuisance, as it affects

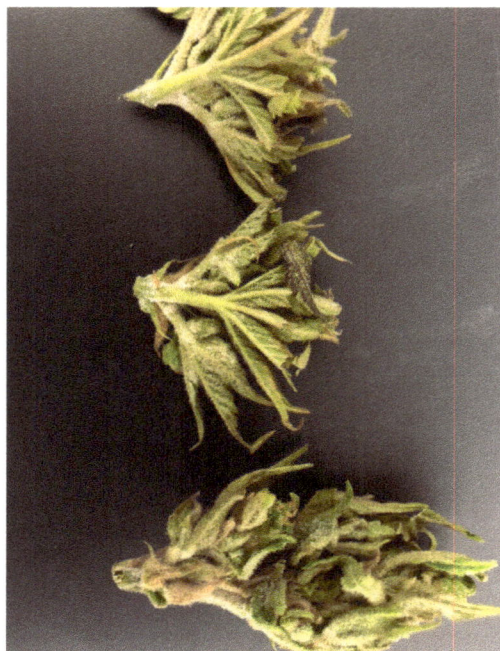

**Fig. 6.8.** Bud samples infected with a powdery mildew fungus. Note the white mycelium/spore patches and tissue damage.

the bud quality and leaf health. The pathogen, once introduced, can rapidly infect a large number of plants and be circulated inside the production facility. The 'behaviour' of this disease is different from soilborne diseases and vascular wilt diseases. While other mould diseases, such as *Botrytis* blight, may be confused with this disease, the white and flat colonies on the surface (mostly upper surface) can easily classify the problem as powdery mildew. There are several related fungi causing powdery mildew, but *Golovinomyces cichoracearum* (Punja *et al.*, 2019) and *G. ambrosiae* (Schoener and Wang, 2018) are mostly found from indoor cannabis cultivation.

#### Do your own diagnosis

Powdery mildew fungi are obligate pathogens that are not culturable, but their mouldy growth is visible on infected buds. A microscope is often needed to identify the type of mould, as other types of fungi may grow on the bud tissue, resembling powdery mildew.

### Sampling

A few (three to five) buds representing different levels of infection should be collected and individually placed in Ziploc bags. Place the sample bags in a sturdy cardboard box to protect the integrity of the bud samples. If bud samples are prohibited for submission to external labs due to restrictions, the pathogen can be blotted onto pieces of clear tape (see Chapter 5): place the sticky side of the tape on the surface of the bud tissue where white mould is visible, press gently and remove the tape. Submit bud samples or tapes to a diagnostic lab.

## Lab diagnosis

### Microscopic examination

During the growing season, only conidia (asexual spores) are produced. Conidia develop singly or in chains on specialized hyphae called conidiophores. Most hyphae grow along the surface (epiphytically), but conidiophores may grow aerially. Place a bud sample under a stereo microscope. The hyphae and spores should be visible. Diagnostic characteristics are the dense patches of white mycelium with a chain of conidia on the conidiophores (Fig. 6.9). The spores can be further blotted onto a piece of clear tape and placed on a drop of water (sticky side face down) at the centre of a glass slide. Under a compound microscope, observe the shape of the conidia. The diagnostic characteristic for conidia is their ellipsoid or almost rectangular shapes (Fig. 6.10).

At the end of the growing season, powdery mildew fungi may enter the sexual stage and produce cleistothecium, also known as chasmothecium (plural: chasmothecia). A cleistothecium is a globose and closed fruit body inside which ascospores, enclosed in a sac-like ascus (plural: asci), are produced. However, it may not be common to see the sexual stage from samples collected from growing cannabis plants.

### DNA-based identification

Powdery mildew fungi are obligate parasites. General culture-based diagnostic methods do not apply to these organisms. The genus of powdery mildew fungi is largely identified based on the cleistothecia and their appendage structures, but in almost all cases these structures are not present during the growing season. When plant samples arrive, symptom observations and microscopic examination of hyphae and conidia can often lead to the correct diagnosis of powdery mildew in general, but it is not sufficient to pinpoint the exact species unless the organism's sexual stage is present. In this case, a DNA-based identification can be performed.

**1.** Place an infected leaf under a dissecting microscope and add 15 µl of TE buffer to the leaf surface where a dense mycelial colony is present.
**2.** Suspend spores in the TE buffer droplet by repetitive pipetting and transfer the mix into a microcentrifuge tube.
**3.** Heat the spore solution at 100°C for 10 min followed by PCR using primers ITS1/ITS4.

**Fig. 6.9.** A patch of mycelia and spores on the surface of a flower bud (observed under a stereo microscope). Note that the white and dense area indicates the aerial growth of conidiophores and conidia.

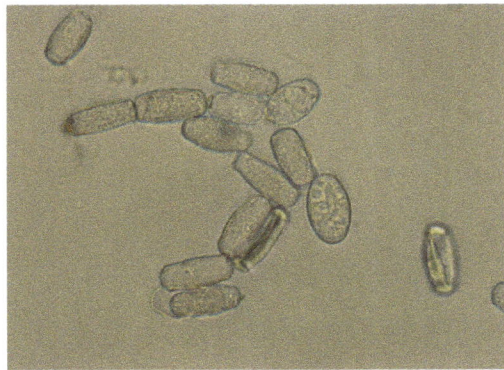

**Fig. 6.10.** Conidia of *Golovinomyces ambrosiae* produced on cannabis plants. Note the characteristic rectangular shape.

**4.** Purify the PCR product, sequence it directly and then BLASTn search (nucleotide–nucleotide basic local alignment search tool) for closely related species. The expected PCR product size is around 596 bp (Schoener and Wang, 2018).

### Key diagnostic evidence

There is no dilemma in general powdery mildew diagnosis, as visual and microscopic examination can easily determine the nature of the disease. However, the determination of species should be supported by DNA sequence data. Koch's postulates are not necessary, as powdery mildew fungi are obligate pathogens.

## *Botrytis* Bud Rot Caused by *Botrytis cinerea*

*Botrytis*, commonly known as grey mould, is a genus of fungi that infects a wide array of herbaceous annual and perennial plants and causes problems on many crops. It contains about 30 different species, many of which are plant parasitic. *Botrytis* produces heavy, grey and branching conidiophores that bear clusters of conidia (Fig. 6.11) and its intensive growth may cover the surface of a plant organ with a grey colour (Fig. 6.12). The spores produced from infected plants can be disseminated by wind and air circulation, and they can be detected in spore trap samples. *Botrytis* is often found in an indoor environment where plants are crowded and humidity is high, or in the field when

cool and rainy weather persists for a week or longer. *Botrytis cinerea* is one of the most common species, causing blight or grey moulds on many plant species, and it is a pathogen causing bud rot on cannabis plants (Punja *et al.*, 2019).

### Field diagnosis

#### *Symptoms*

Early infection of cannabis plants with *B. cinerea* results in the browning and decaying of leaves and bracts in developing inflorescence. The infection advances and may destroy the entire inflorescence. Harvested inflorescences, if infected during the growing season, may show intensive mycelium growth within the bud; and massive spores may be observed under a microscope (Fig. 6.12). Although *B. cinerea* produces sclerotia on culture media such as PDA, it may not produce sclerotia in the infected bud. Besides bud rot, infected leaves may develop individual spots or coalescent lesions (Punja *et al.*, 2019).

#### *Problem classification*

*Botrytis* bud rot is an airborne disease that may affect not only the bud but also other parts of the plant, including the leaf and stem. Similar to powdery mildew, it is a mould disease favouring cool and humid conditions. The presence of a fuzzy grey colour on the surface of buds and leaves suggests *Botrytis* bud rot. This disease affects the quality of buds and impacts

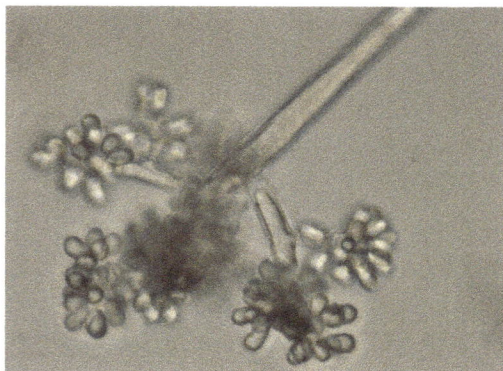

**Fig. 6.11.** Clusters of conidia on branching conidiophores of *Botrytis cinerea* observed under a compound microscope.

**Fig. 6.12.** Plant organ covered by dense clusters of spores of *Botrytis cinerea* observed under a stereo microscope.

plant growth and the pathogen may spread throughout the production facility via air circulation.

### Do your own diagnosis

This disease can be self-diagnosed by observing the presence of greyish mould covering the buds or leaves and the prolific production of cluster spores, like the symptom shown on a verbena plant when it is attacked by *Botrytis cinerea* (Fig. 6.13). Using a magnifier one can clearly see a cluster of spores on each conidiophore (Fig. 6.11). Besides visual observations of symptoms, mould can be collected on a pre-poured PDA plate with the following procedure.

1. Check leaves, flowers or buds for any sign of mould or related symptoms.
2. Use a sterile cotton swab gently to wipe across the surface of affected leaves or flower buds.
3. Open the Petri dish and streak the medium with the cotton swab in a Z-pattern.
4. Close and seal the dish and label it with the sample information.
5. Submit the Petri dish to a laboratory for further diagnosis.

**Fig. 6.13.** Prolific growth of *Botrytis cinerea* mycelia and spores on verbena flowers and leaves.

### Sampling

Use the same procedures as those described above for powdery mildew on buds.

### Lab diagnosis

### Microscopic examination

Place a whole or a portion of an infected bud that may look fuzzy or greyish under a stereo microscope to observe the mycelia and spores. Clusters of spores, conidiophores and mycelia are present on the surface of bud tissue. Hyphae grow epiphytically along the surface or aerially. Conidiophores generally point away from the surface and bear clusters of spores (Fig. 6.12). The spores can be transferred to a glass slide (by blotting with a piece of clear tape) to observe spores under a compound microscope. Spores are single-celled, mostly uniform, and oval shaped (Fig. 6.14).

### Isolation and Colony Observation

*Botrytis* can be identified easily based on its morphological characteristics of spores, conidiophores and mycelia, which are always present in infected plant samples. That said, isolation of this fungus from infected tissue may not be needed for a general diagnosis. However, a formal diagnosis should be based on the positive isolation of *Botrytis* from symptomatic plant tissue instead of the organism being directly observed from a plant specimen, since *Botrytis* can grow as a saprophyte or secondary

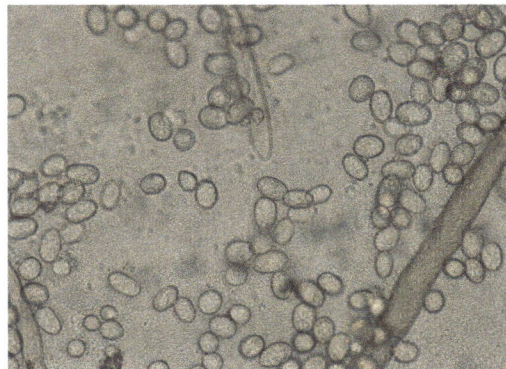

**Fig. 6.14.** Spores of *Botrytis cinerea* blotted from infected plant tissue and observed under a compound microscope.

invader on some rotted bud tissue. *B. cinerea* grows rapidly on PDA amended with streptomycin and the initial colony appears creamy white before turning to grey (Fig. 6.15). Black and irregular sclerotia are present in old cultures. The isolates of *Botrytis* can be further used to perform downstream diagnostic procedures such as morphological measurements and DNA extraction.

### DNA-based identification

Isolates of *Botrytis* may be further identified to the species level using standard DNA-based identification procedures described in Chapter 5. In brief, perform PCR using primers ITS1/ITS4, purify the PCR product, sequence it directly and then BLASTn search for closely related species. The expected PCR product size is around 550 bp.

### Key diagnostic evidence

*Botrytis* species should be found to be associated with bud rot. In the case of multiple pathogens

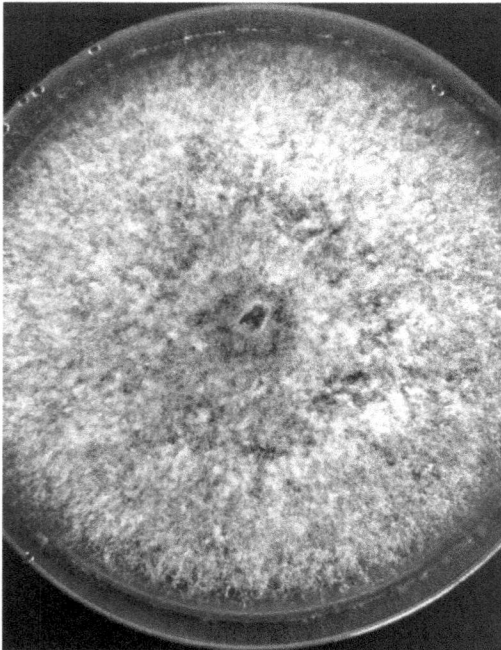

**Fig. 6.15.** A culture of *Botrytis porri* grown on a PDA[+strep] plate for 10 days. Note that sclerotia had not yet been produced.

involved in bud rot, *Botrytis* species is the one most frequently isolated from bud tissue. Both morphological and DNA sequence data support the classification of *Botrytis* species.

## Other Moulds Associated with Buds

Moulds are a group of fungi that grow on living or dead plant organs. Extensive growth of mould is detrimental to overall plant health and the quality of plant products. In some cases, mould fungi are just incidental contaminants without the onset of a symptom or bud rot. However, these fungi may start to grow on post-harvested plant products if conditions favour. There are a number of mould fungi reported to be associated with bud tissue and/ or present in air samples collected from indoor or greenhouse-grown cannabis plants (Punja *et al.*, 2019). These include *Alternaria alternata*, *Beauveria bassiana*, *Botrytis cinerea*, *Cladosporium westeerdijkieae*, *Fusarium oxysporum*, *Penicillium citrinum*, *Penicillium copticola*, *Penicillium corylophilum*, *Penicillium griseofulvum*, *Penicillium olsonii*, *Penicillium sclerotiorum*, *Penicillium simplicissimum* and *Penicillium spathulatum*. Some of them occur on both growing cannabis buds and harvested and trimmed buds, as well as in air samples; some are recovered only from buds or air samples. Research using internal transcribed spacer (ITS) analysis revealed approximately 4000 fungal taxonomic classifications from cannabis buds in dispensaries, among which *Penicillium*, *Cladosporium*, *Golovinomyces*, *Aspergillus*, *Alternaria*, *Botryotinia*, *Chaetomium* and *Fusarium* were predominantly detected (Thompson *et al.*, 2017). The occurrence of specific mould fungi, the composition of species and the colony-forming units (the concentration of a mould fungus) may depend on the growing facilities, locations, seasons and indoor/field growing environments. The varieties of hemp or marijuana may be attributed to the dynamics of microbial communities in buds.

### Field diagnosis

#### Symptoms

Mould fungi detected from buds may simply be epiphytic colonization on the surface of bud tissue if the bud does not show any visual or microscopic symptoms. Under humid conditions, certain fungi may grow on the bud tissue, turning the buds to a different colour such as black, grey, or blue, for example, if they are infected by *Alternaria*, *Botrytis*, or *Penicillium*, respectively. The colonization and growth of mould fungi can cause necrosis or decay of bud tissue. Extensive growth of moulds can form clouds of spores when disturbed.

### Problem classification

Bud mould affects the quality of harvested buds and is considered as a post-harvest disease. However, certain fungi cause bud diseases during the growing season and the problem may continue or become worse during the post-harvest stage. Most fungi identified from buds are airborne and may originate from the environment or substrates. Moulds grown from cultivation substrates can be disseminated to the inflorescences by air, and spores floating in the air can land and colonize on the substrate, buds, or exposed cutting surfaces (Punja *et al.*, 2019).

### Do your own diagnosis

Use similar procedures as described for *Botrytis* bud rot to self-diagnose bud moulds. Look closely or with the aid of a magnifier to observe the mycelia and the mould colour. For example: a white colour may suggest powdery mildew; grey suggests either *Pythium* or *Botrytis*; blue suggests *Penicillium*; black suggests either *Alternaria* or maybe *Aspergillus*; and pink suggests *Fusarium*. The presence of a unique colour in buds indicates the extensive growth of a given fungus. Alternatively, using a sterile swab to transfer potential moulds to a PDA[+strep] plate and allowing them to grow at room temperature helps visualize different fungi on the Petri dish, and the different types of colonies can then be further identified.

### Sampling

Collect a few (three to five) buds exhibiting discoloration, necrotic lesions, or fuzzy growth of mould and place each bud in a Ziploc bag. Place the sample bags in a sturdy cardboard box to protect the integrity of the bud sample during shipping and handling. Do not add any moisture to the bag. To monitor or detect the early stage of mould growth, a few asymptomatic buds can be submitted to a lab for testing. If a bud sample is prohibited from submission, blot the mould onto a piece of clear tape and submit the tape to a diagnostic lab for preliminary identification. Alternatively, use a sterile cotton swab to wipe gently across the surface of buds and streak the swab on a PDA[+strep] plate. Spores captured on a swab are transferred to the Petri dish. Cover the dish with a lid and seal tightly. To sample moulds from air, place a few PDA[+strep] plates on a growth bench for 1 h and then cover with lids and seal. Submit the plates to a laboratory for further identification.

### Lab diagnosis
### Microscopic examination

Place a whole or a portion of a bud under a stereo microscope to observe the mycelia and spores. If moulds are present, mycelia should be clearly seen under the microscope. Spores are generally present and may be in clusters. Transfer both mycelia and spores to a glass slide (by blotting with clear tape) to observe their morphology under a compound microscope. Unique characteristics of spores, conidiophore structures and hyphae branching can be used to identify the common mould fungi. For example, *Penicillium* is easily recognized by clusters of flask-shaped phialides from which chains of spores are produced. Spores are single-celled, mostly uniform and elliptical shaped (Fig. 6.16). Other mould genera can be recognized by observing their morphological characters under a microscope and by referencing a genus illustration book.

### Isolation and colony observation

Not all mould fungi can be observed directly from bud samples and identified under a microscope. It is always necessary to isolate mould fungi by plating pieces of bud tissue on PDA[+strep] plates or using a sterile swab to transfer moulds to the plates. Multiple fungal species may grow in the plates, but they vary in frequency. Certain fungi may be isolated more frequently than others, suggesting they are major moulds. Colonies can be grouped according to their colour, texture, shape, growth pattern and sporulation. Some may be easily identified to the genus level. For example, blue colonies (Fig. 6.17) and broom-like spore-bearing structures strongly support a call for *Penicillium*. Still, representative colonies of all types should be selected for further morphological and DNA-based identification.

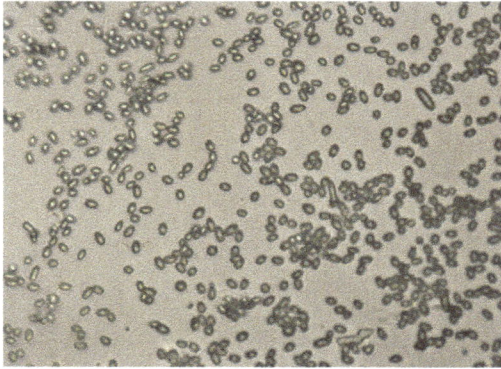

**Fig. 6.16.** Conidia (asexual spores) of *Penicillium digitatum* observed under a compound microscope.

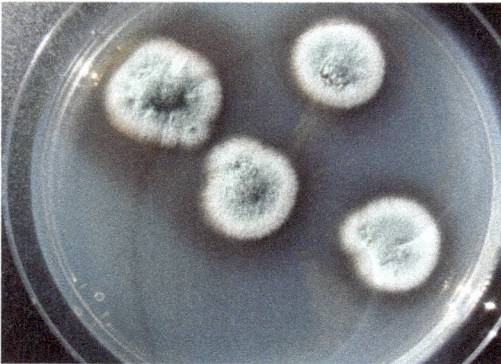

**Fig. 6.17.** Blue colonies of *Penicillium* sp. on a PDA[+strep] plate.

### DNA-based identification

A quick way to identify all encountered mould fungi to species is the standard DNA-based identification described in Chapter 5. In brief, perform PCR using primers ITS1/ITS4, purify the PCR product, sequence it directly and then BLASTn search for closely related species. The expects PCR product size is around 550 bp.

### Key diagnostic evidence

One or more mould fungi are observed or isolated from bud tissue and at least one species occurs predominantly in the samples tested. A low frequency of detection without visible symptoms may suggest an accidental epiphytic relationship. More samples may be needed to analyse the trend of each mould fungus and its level of association with the buds. Species identification of all mould fungi are supported by their DNA sequences.

## Seed and Bud Disease Management

### Using certified seeds

Seed certification is an effective programme to eliminate or minimize seed-borne diseases, as the certification process promotes seed quality and purity. Depending on the crop, the process involves meeting special field requirements, planting certified stock, inspecting the crop for genetic impurities, chimera and diseases during the growing season, analysing seeds for viability and purity, and ensuring proper labelling. That said, certified seeds not only ensure seed quality but also decrease the chance of seed-borne diseases.

### Properly storing seeds

Storing harvested seeds in proper environmental conditions maintains seed quality and viability. One research showed that storing hemp seeds at 15°C, or even 4°C or −4°C, maintained seed quality for one year (Suriyong *et al.*, 2015). The study used four hemp cultivars and stored seeds either in a sealed polypropylene plastic bag at room temperature or in aluminium foil at different temperatures (room temperature, 15°C, 4°C and −4°C) for 12 months. Seeds packed in both materials at room temperature had no significant change in germination and vigour for the first 6 months, while containing seeds in a plastic bag resulted in the most adverse effect on viability. Packing in aluminium foil and storing at 15°C, 4°C and −4°C did not significantly affect seed viability during a year of storage. The study suggests that a cold room at 15°C is the best condition for hemp seed storage.

### Reducing salt stress

Although cultivars of *Cannabis sativa* are tolerant to salts to a certain degree, recent studies show that alkaline salts, such as $Na_2CO_3$ and $NaHCO_3$, exhibit stronger toxicity to hemp seed germination compared with neutral salts such as NaCl and $Na_2SO_4$ (Hu *et al.*, 2018). The impact of salt stress on hemp seed germination varies with the type and concentration of salts and the cultivars. In Nevada, poor seed germination and lack of stands were observed in hemp fields with alkaline soil. Therefore, it is important to take representative soil samples for salt tests. If an extremely high concentration of salts is detected in a field, it may be worth avoiding planting hemp in the field. If a hemp crop is already planted in an alkaline field, soil salinity should be managed to achieve a vigorous crop.

### Adjusting growth and postharvest environment

Most mould fungi favour a high relative humidity (85–100%) and cool temperature. If possible, lower the humidity and increase the temperature to suppress mould growth. Avoid using a mist spray or overhead watering, as this will add extra moisture to the plant surface. Keep the air circulating when the facility is mould-free. However, if plants are infected with mould, air blowing may disseminate fungal spores inside the facility. Harvested buds should be stored properly to avoid excessive moisture accumulation in a closed environment that promotes mould growth.

### Inspecting and sanitizing

Early detection leads to an easy fix by just removing a few infected plants and treating the rest of the plants in good time. Therefore, periodically monitoring mould growth on plants should be a routine practice. For indoor production, infected plants, once noticed, should be removed in good time to prevent further spread and affected areas should be sanitized with protectants or disinfectants. The best way to monitor mould growth at the earliest stages is to swab a leaf or bud and then streak a PDA[+strep] plate periodically. This swab test can detect a mould fungus before any sign of mould is noticeable. Alternatively, leave a PDA[+strep] plate on a growth bench with the lid removed for 1 h and then cover it with the lid. Colonies grown on PDA[+strep] plates can be further identified to determine if a mould fungus is present inside the facility.

### Treating seeds

Completely healthy seeds are ideal for preventing seed-borne diseases such as pre-and post-emergence damping-off. However, in the real world, seeds are not always free from pathogens and thus require treatments to eliminate seed-borne pathogens.

Treating hemp seeds with an approved fungicide or disinfectant can effectively eliminate some seed-borne pathogens that notoriously affect seed germination and seedling health. This is a preventive measure proven to be effective in disease control, especially for managing damping-off diseases (Lamichhane et al., 2017). In addition to chemical treatment, physical treatments such as using hot water or hot air and screening out defective seeds have been used. As the use of pesticides is restricted in hemp crops, seeds may be treated with approved biological control agents or other natural compounds such as plant extracts, which have proved to be effective in managing seed-borne diseases in vegetable crops (Mancini and Romanazzi, 2014). These treatments may be used in combinations to control a certain type or a group of diseases.

### Coating Seeds

Seed coating with a fungicide has been used to manage seedling diseases in diverse crops. Some materials, such as aluminosilicate (natural zeolite) when coated on alfalfa seed at a rate of 0.33 g of zeolite per gram of alfalfa seed, can effectively control *Phytophthora* root rot (Samac et al., 2015). Similarly, some plant-beneficial microbes, such as plant growth-promoting rhizobacteria (PGPR), arbuscular mycorrhizal fungi and the common biocontrol agent *Trichoderma*, can be coated on seeds to improve germination and seedling establishment and to protect crops from a broad range of diseases and abiotic stresses (Rocha et al., 2019).

### Treating plants with PGPR

Plant growth-promoting rhizobacteria (PGPR) have been widely used in crop disease protection for many years. In China, at least 53 species belonging to 27 genera of PGPR have been identified, some of which have been commercialized and widely used to control crop diseases (Liu et al., 2016). PGPR have been reported to be effective in controlling various powdery mildews. For example, foliar spray of *Pseudomonas aeruginosa* can control powdery mildew caused by *Erysiphe pisi* in pea (*Pisum sativum*) (Bahadur et al., 2007). Therefore, PGPR are speculated to be applicable in managing powdery mildew in *Cannabis* crops (Lyu et al., 2019), especially when chemical-based fungicides are restricted.

## Treating plants with registered products

Using a pesticide or fungicide should be the last resort for disease control in hemp and cannabis crops. There are a few products registered and approved by the US Environmental Protection Agency (EPA) to be used on hemp (EPA, 2019). As of December 2019, nine biopesticides and one conventional pesticide had been approved. As EPA receives and reviews more applications for amending product labels for usage on hemp, it is likely that more products will become available.

## References

Bahadur, A., Singh, U., Sarnia, B., Singh, D., Singh, K. et al. (2007) Foliar application of plant growth-promoting rhizobacteria increases antifungal compounds in pea (*Pisum sativum*) against *Erysiphe pisi*. *Mycobiology* 35, 129–134. doi: 10.4489/MYCO.2007.35.3.129

EPA (2019) Pesticide products registered for use on hemp. Available at: https://www.epa.gov/pesticide-registration/pesticide-products-registered-use-hemp (accessed 7 July 2020).

Groves, J.W. and Skolko, A.J. (1944) Notes on seed-borne fungi: II. *Alternaria*. *Canadian Journal of Research* 22c, 217–234. doi: 10.1139/cjr44c-018

Haware, M.P. (1971) Fungal microflora of seeds of *Pisum sativum* L. and its control. *Mycopathologia et Mycologia Applicata* 43, 343–345. doi: 10.1007/BF02051756

Howell, C.R. (2002) Cotton seedling preemergence damping-off incited by *Rhizopus oryzae* and *Pythium* spp. and its biological control with *Trichoderma* spp. *Phytopathology* 92, 177–180. doi: 10.1094/PHYTO.2002.92.2.177

Hu, H., Liu, H. and Liu, F. (2018) Seed germination of hemp (*Cannabis sativa* L.) cultivars responds differently to the stress of salt type and concentration. *Industrial Crops and Products* 123, 254–261. doi: 10.1016/j.indcrop.2018.06.089

Kwon, J.H., Kim, J. and Kim, W.I. (2011) First report of *Rhizopus oryzae* as a postharvest pathogen of apple in Korea. *Mycobiology* 39, 140–142. doi: 10.4489/MYCO.2011.39.2.140

Lamichhane, J.R., Dürr, C., Schwanck, A.A., Robin, M., Sarthou, J. et al. (2017) Integrated management of damping-off diseases. A review. *Agronomy for Sustainable Development* 37, 10. doi: 10.1007/s13593-017-0417-y

Liu, D.-D., Li, M. and Liu, R.-J. (2016) Recent advances in the study of plant growth-promoting rhizobacteria in China. *Chinese Journal of Ecology* 35, 815–825. doi: 10.13292/j.1000-4890.201603.033

Lyu, D., Backer, R., Robinson, W.G. and Smith, D.L. (2019) Plant growth-promoting rhizobacteria for cannabis production: yield, cannabinoid profile and disease resistance. *Frontiers in Microbiology* 10, 1761. doi: 10.3389/fmicb.2019.01761

Mancini, V. and Romanazzi, G. (2014) Seed treatments to control seedborne fungal pathogens of vegetable crops. *Pest Management Science* 70, 860–868. doi: 10.1002/ps.3693

Mathew, F.M., Prasifka, J.R., Gaimari, S.D., Shi, L., Markell, S.G. et al. (2015) *Rhizopus oryzae* associated with *Melanagromyza splendida* and stem disease of sunflowers (*Helianthus annuus*) in California. *Plant Health Progress* 16, 39–42. doi: 10.1094/PHP-RS-14-0042

Punja, Z.K., Collyer, D., Scott, C., Lung, S., Holmes, J. et al. (2019) Pathogens and molds affecting production and quality of *Cannabis sativa* L. *Frontiers in Plant Science* 10, 1120. doi: 10.3389/fpls.2019.01120

Ribeiro-Oliveira, J.P., Ranal, M.A., de Santana, D.G. and Pereira, L.A. (2016) Sufficient sample size to study seed germination. *Australian Journal of Botany* 64, 308–324. doi: 10.1071/BT15254

Rocha, I., Ma, Y., Souza-Alonso, P., Vosátka, M., Freitas, H. et al. (2019) Seed coating: a tool for delivering beneficial microbes to agricultural crops. *Frontiers in Plant Science* 10, 1357. doi: 10.3389/fpls.2019.01357

Samac, D.A., Schraber, S. and Barclay, S. (2015) A mineral seed coating for control of seedling diseases of alfalfa suitable for organic production systems. *Plant Disease* 99, 614–620. doi: 10.1094/PDIS-03-14-0240-RE

Schoener, J. and Wang, S. (2018) First detection of *Golovinomyces ambrosiae* causing powdery mildew on medical marijuana plants in Nevada. *(Abstr.) Phytopathology* 108, S1.186. doi: 10.194/PHYTO-108-10- S1.186

Schoener, J., Wilhelm, R. and Wang, S. (2018) Molecular identification of *Alternaria* species associated with imported industrial hemp seed. *(Abstr.) Phytopathology* 108, S1.185. doi: 10.194/PHYTO-108-10- S1.185

Suriyong, S., Krittigamas, N., Pinmanee, S., Punyalue, A. and Vearasilp, S. (2015) Influence of storage conditions on change of hemp seed quality. *Agriculture and Agricultural Science Procedia* 5, 170–176. doi: 10.1016/j.aaspro.2015.08.026

Taghinasab, M. and Jabaji, S. (2020) Cannabis microbiome and the role of endophytes in modulating the production of secondary metabolites: an overview. *Microorganisms* 8, 355. doi: 10.3390/microorganisms8030355

Thompson, G.R. III, Tuscano, J.M., Dennis, M., Singapuri, A., Libertini, S., Gaudino, R. et al. (2017). A microbiome assessment of medical marijuana. *Clinical Microbiology and Infection* 23, 269–270. doi: 10.1016/j.cmi.2016.12.001

# 7    Diagnosing Foliar Diseases

Foliar diseases in this chapter refer to those diseases mainly affecting leaves with a pathogen confined in the leaf tissue. Diseases that may cause obvious foliar symptoms but affect more than one plant organ or behave systemically are included in the following chapters. Foliar diseases are commonly caused by fungal or bacterial pathogens. Under favourable environmental conditions, especially with extended wet weather, an initial infection may progress into visible symptoms ranging from a tiny necrotic lesion to a sizeable leaf spot or leaf blight. In most cases, these symptoms stay on the leaves.

## Hemp Leaf Spot Caused by *Alternaria* Species

*Alternaria* is a genus of fungal pathogen commonly causing leaf spot or blight on many plant species. On hemp, *Alternaria* species were detected from harvested seeds (see Chapter 6) but it is not known whether the species latent in seeds (Schoener *et al.*, 2018) actually cause the leaf spot disease. The fungus survives in plant debris as the primary source of inoculum to infect plants in early season and then produces abundant spores during the middle or late season that may spread among plants via wind, rain, or indoor air circulation. Warmer temperatures and wet weather may increase the disease incidence and severity. Species associated with leaf spot disease may differ by geographical locations, but *Alternaria alternata* is commonly diagnosed from hemp leaf samples. Other species, such as *A. infectoria* and *A. malorum*, are also detected from hemp leaf spot tissue.

### Field diagnosis

#### Symptoms

Conidia, the asexual spores of *Alternaria* spp., may be dispersed either by water or by air movement and land on a leaf surface. When environmental conditions are favourable, the spores germinate, penetrate the leaf cells and further spread their growth within the leaf tissue. The initial symptom is necrotic lesions that may expand as the infection continues. In the middle of disease development, various sizes and shapes of leaf spots may be seen on old leaves (Fig. 7.1). Lesions may coalesce into a large necrotic area where the surrounding leaf tissue turns yellow. These lesions are typically dark brown to black. In humid conditions, conidia-producing conidiophores may be produced from fully developed lesions, making the lesion appear fuzzy as a result. Spores produced from lesions serve as secondary inoculum to infect other leaves or plants. The infection cycle may repeat multiple times before harvest. Heavy infections can impact overall plant health and reduce yields.

### Problem classification

Many factors, both biotic and abiotic, can cause leaf spot. It is important to rule out chemical injury and other abiotic causes before considering any infectious agents. When it comes to an infectious disease, leaf spots can be a symptom of other diseases originating from a root, stem, or systemic infection. In general, however, *Alternaria* leaf spot does not cause direct symptoms on the root and stem, nor is there any sign of *Alternaria* presence in those plant parts. *Alternaria* leaf spot behaves as an airborne disease that mainly damages leaf tissue.

### Do your own diagnosis

Growers can systemically examine several symptomatic plants to determine if the damage is restricted to leaves. To do so, examine the surface and internal tissue of the roots, crown and stem by using a pocketknife and assess if other diseases are present. Most likely, the majority of the affected plant shows foliar symptoms only, without any evidence of root, crown, or stem diseases. Examine several symptomatic leaves to see if they share similar characteristics of leaf spot: brown and necrotic (Fig. 7.1). Use a handheld magnifying glass to examine several necrotic spots for the fuzzy look of fungal conidiophores and spores (Fig. 7.2). Irregular brown to dark and sometimes fuzzy lesions suggest *Alternaria* leaf spot. Besides visual observations of symptoms, *Alternaria* can be transferred onto pre-poured PDA plates:

**1.** Use a sterile cotton swab to wipe gently across the surface of necrotic lesions.
**2.** Open the Petri dish and streak the medium with the cotton swab in a Z-pattern.
**3.** Close and seal the dish and label it with the sample information.
**4.** Submit the Petri dish to a laboratory for further diagnosis.

### Sampling

Collect a few (three to five) leaflets exhibiting necrotic lesions and place each leaflet in a Ziploc bag. A whole plant may be submitted for comprehensive diagnosis if a field assessment cannot determine the problem. Place the sample bags in a sturdy cardboard box to protect the integrity of the sample during shipment. Do not add any moisture to the bag.

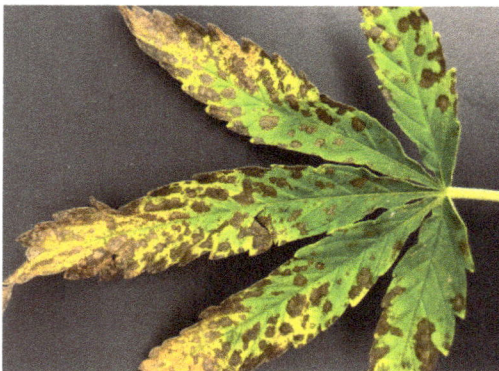

**Fig. 7.1.** Hemp leaf spot caused by *Alternaria* sp.

### Lab diagnosis

#### Microscopic examination

Place a leaflet under a stereo microscope to observe the conidiophores and spores on the surface of lesions. In humid growing conditions, conidiophores and spores are often present. If the sample has no sign of a fungal pathogen, it can be incubated in a Petri dish with added moisture (see Chapter 5) to induce the production of spores. The spores can then be transferred to a glass slide by blotting with a piece of clear tape and observed under a compound microscope. Spores are multicellular, variable in shape and brown in colour (Fig. 7.3) (also see Fig. 6.4).

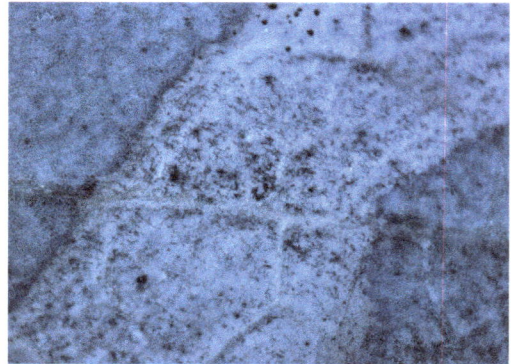

**Fig. 7.2.** Black conidiophores and spores grown out from the necrotic leaf tissue (observed under a stereo microscope).

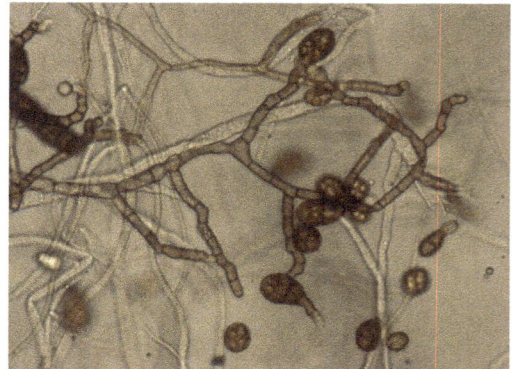

**Fig. 7.3.** Spores and hyphae of *Alternaria alternata* observed under a compound microscope.

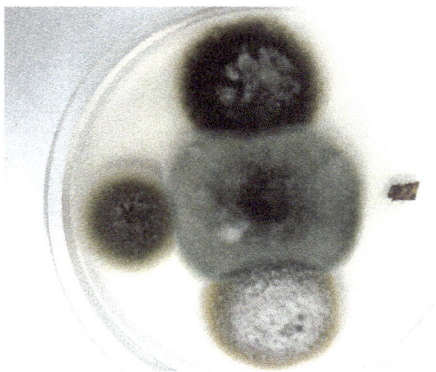

**Fig. 7.4.** Multiple species of *Alternaria* isolated from hemp leaf spot tissue.

### Isolation and colony observation

A standard diagnosis method is to isolate the pathogen from the infected leaf tissue using PDA$^{+strep}$ plates. *Alternaria* readily grows on the medium and forms characteristic thick and dark colonies (Fig. 7.4). It is not uncommon to isolate multiple species of *Alternaria* from a single infected leaf, as *Alternaria* is considered a ubiquitous fungal genus that is not only pathogenic but can also be saprophytic or endophytic (Woudenberg *et al.*, 2013). In Fig. 7.4, three different colonies are obtained from a hemp leaf sample and represent three different species based on their DNA sequence of the ITS region of rRNA gene. *A. alternata* readily produces abundant spores in the medium that can be identified based on spore morphology and morphometrics. However, some species may not produce spores in certain conditions. In this case, fungal isolates can be proceeded to DNA-based identification using the universal DNA barcode marker for fungi (Schoch *et al.*, 2012).

### DNA-based identification

Isolates of *Alternaria* may be further identified to the species level using standard DNA-based identification procedures described in Chapter 5. In brief, perform PCR using primers ITS1/ITS4, purify the PCR product, sequence it directly and then BLASTn search for closely related species. The expected PCR product size ranges from 542 to 600 bp.

### Pathogenicity test

Koch's postulates may be performed to determine the pathogenicity of obtained isolates.

**1.** Culture each isolate on a PDA$^{+strep}$ plate for 6–10 days or until sporulation. A single spore isolate (SSI) should be used in this test if possible (see Chapter 5).
**2.** Add 10–15 ml of sterilized deionized water into the plate and suspend spores from the colony by using a sterile cell scraper.
**3.** Transfer the spore solution (with some hyphae) into a sterilized 250 ml glass flask.
**4.** Use a haemocytometer to determine the spore concentration and then adjust the spore suspension (inoculum) to a given volume with a designed concentration, e.g. $10^6$ spores/ml.
**5.** Using a spray bottle, apply the inoculum onto the leaves of a group of young hemp plants grown in a greenhouse or growth chamber. Inoculated plants should remain in humid conditions for the first 2 days after inoculation.

Leaf symptoms can be evaluated as soon as 14 days post-inoculation, but the disease progress should be assessed throughout the experiment. A group of healthy plants sprayed with sterilized water should be grown under the same experimental conditions as a control. The appearance of necrotic lesions on inoculated leaves confirms the pathogenicity of the isolate.

### Key diagnostic evidence

Brown lesions always occur, especially on older or early infected leaves. Under greenhouse or humid conditions, the fungus grows out from the lesions and sporulates, making the lesion darker. *Alternaria* is frequently isolated from necrotic leaf tissue and causes similar symptoms on inoculated plants.

## Hemp Leaf Blight Caused by *Alternaria* Species

Leaf blight is a more severe condition compared with leaf spot. It is defined as the rapid death of a leaf. Although both diseases may look similar, leaf spot is a symptom where the majority of spots are discretely separated by green tissue, while leaf blight is a disease that causes the sudden death of a portion of the leaf or the merging of multiple spots into a large diseased area. One may speculate that a leaf spot disease may progress into a leaf blight if conditions favour disease development.

## Field diagnosis

### Symptoms

Infected leaves exhibit leaf spots, yellowing and dieback (Fig. 7.5). Lesions may coalesce into a large necrotic area where surrounding leaf tissue turns yellow. Lesions may also become darker, due to the growth of fungal mycelia or spores. The infection may extend to the leaf petiole and cause dark lesions or death. The life cycle and disease development are similar to *Alternaria* leaf spot.

### Problem classification

Similar to leaf spot disease, leaf blight can be caused by many factors. Initial assessment should rule out abiotic factors and other diseases associated with the stem and root. Perform a preliminary diagnosis to narrow the problem down to a fungal disease mainly affecting leaf tissue.

### Do your own diagnosis

Use the same strategy described for leaf spot disease to confirm that the disease is limited to foliage. Then, use a handheld magnifying glass to examine several leaves for any signs of fungal growth. A dark colour on an older leaf or matured lesion suggests the presence of fungal spores (Fig. 7.6).

**Fig. 7.5.** Hemp leaf blight caused by *Alternaria* sp. Note the yellowing and death of a large area of the leaf instead of segregated lesions as shown in Fig. 7.1.

Perform similar procedures described for leaf spot disease to transfer spores onto a pre-poured PDA plate. The plate can sit on a table for several days to allow the fungus to grow. A dark or green colony suggests the presence of *Alternaria* species. Alternatively, submit the Petri dish to a lab for further diagnosis.

### Sampling

Collect a few (three to five) leaflets exhibiting necrotic lesions or blight and place each leaflet in a Ziploc bag. Follow sampling guidelines described for leaf spot disease.

## Lab diagnosis

### Microscopic examination

Place a leaflet under a stereo microscope to observe if there is anything on the leaves, including tiny insects, eggs, mites and mould. This step is used to rule out arthropod causes and other diseases. Then examine the lesion or blighted area for any growth of conidiophores and spores. In most cases, typical conidiophores and spores can be observed directly

**Fig. 7.6.** Black conidiophores and spores grown out from blighted leaf tissue.

from a leaf sample. If the sample is collected during an early stage of disease, there may not be any signs of fungal growth. In this case, the leaf can be incubated in a Petri dish with added moisture to induce fungal growth. The mycelia and spores can be transferred to a glass slide by blotting and then observed under a compound microscope (Fig. 7.7).

### Isolation and colony observation

Use a standard isolation method (see Chapter 5) and procedures described above for leaf spot disease.

### DNA-based identification

See hemp leaf spot disease, above.

### Pathogenicity Test

See hemp leaf spot disease, above.

### Key diagnostic evidence

Leaf blight symptoms may not be very diagnostic in general, as they may resemble abiotic injuries and other foliar diseases. Characteristic leaf spots may not be available at the time of diagnosis. Therefore, the key evidence is that *Alternaria* is frequently and dominantly isolated from the blighted leaf but not from healthy leaves and that the fungal isolate causes typical leaf blight on inoculated test plants.

## Sooty Mould Caused by Aphid Infestation and *Cladosporium* Species

Sooty mould occurs when nutrients on the leaf surface are available for some opportunistic or saprophytic fungi to grow. These fungi do not infect plants but grow on the surface, giving the leaf the appearance of being covered with a layer of black 'soot' (mycelia and spores). The source of nutrients promoting sooty fungal growth generally comes from a sugary exudate produced by the host plant or honeydew secreted by insects. Honeydew appears to be a major cause of sooty mould, as it is a sweet and sticky liquid that can support the growth of a number of fungi. Sooty mould occurs on many plant species, but there is no specific relationship between the plant host and fungal species. Its occurrence depends on the plant surface nutrients and humidity as well as the presence of sooty-mould fungi in the environment. *Cladosporium* and *Alternaria* spp. are commonly observed in hemp sooty mould samples. Plants with leaves covered with sooty mould may grow poorly or become weakened as the total leaf area available for photosynthesis is reduced.

### Field diagnosis

### Symptoms

Affected leaves exhibit patches of black mould, mostly on the upper leaf surface (Fig. 7.8). Aphids at different development stages, including nymphs,

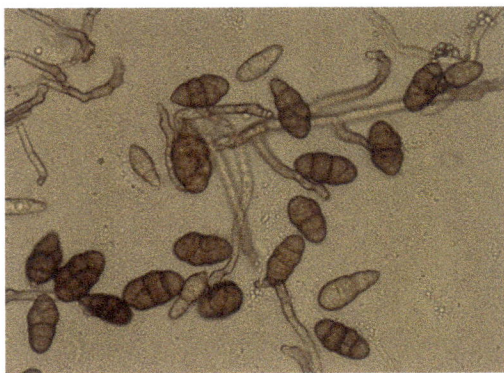

**Fig. 7.7.** Spores and hyphae of *Alternaria* species blotted from a hemp leaf onto tape and observed under a compound microscope.

**Fig. 7.8.** Visible patches of black mould on the upper side of a hemp leaf due to the abundance of sugar-rich sticky honeydew secreted by aphids. Note several aphids feeding on the leaf.

cast skins (exuviae) and adults, may be seen on both sides of leaves (Fig. 7.9). Leaves look wet and sticky due to the sugar-rich honeydew secreted by aphids. Leaves may also be infested by other insects in the order Hemiptera that produce honeydew. Usually, multiple fungal species use the honeydew as a nutrient to grow mycelia and spores on the wet surface, forming a cluster of black colonies. Affected leaves become ragged and stained, due to the rampant insect feeding and fungal growth.

### Problem classification

Sooty mould is a problem of cosmetic damage due to the superficial growth of certain fungi that use the honeydew as nutrients. It is associated with insect feeding activities and does not commonly affect plant parts beyond the leaves. The root cause of sooty mould is the honeydew produced by certain insects, with sooty mould considered as secondary problem. Without adequate nutrient support, sooty mould fungi rarely grow on healthy leaves or infect leaf tissue.

### Do your own diagnosis

First, inspect plants in the field to identify the pattern and percentage of plants affected. Then locate which parts of plants are affected with sooty mould. Use a handheld magnifying glass to look for signs of active insects or cast skins. Even without an active insect present at the time of examination, the cast skins or damage symptoms can provide

evidence of a previous infestation. Leaves will appear to be wet and sticky. The mould can be scraped easily with a hard stick and leaf tissue underneath the mould may still be green. The coexistence of honeydew, insect infestation and the black mould suggests honeydew-induced sooty mould.

### Lab diagnosis

#### Microscopic examination

Place a leaflet with sooty mould under a stereo microscope to observe everything on the leaves, including tiny insects, eggs, cast skins, mites and mould. Aphids and patches of black mould are commonly seen on the specimen (Fig. 7.10). Use a piece of clear tape to blot the mould and mount it on a slide to observe the fungal mycelia and spores under a compound microscope. Spores morphologically resembling *Cladosporium* species are frequently observed (Fig. 7.11). Spores of *Alternaria* species may also be present, as they are ubiquitous.

#### DNA-based identification

Sooty mould spores can be collected from a leaf surface, diluted, then plated on 2% water agar to obtain a single spore colony (see Chapter 5). Each single spore isolate can be further identified by its rDNA-ITS sequence, as described in Chapter 5. It is common to have multiple species or genera identified

**Fig. 7.9.** Aphids and their cast skin on the lower side of a hemp leaf. Note that the upper side was covered with sooty mould (not shown).

**Fig. 7.10.** Small patches of black mould and an aphid on the surface of a hemp leaf observed under a stereo microscope.

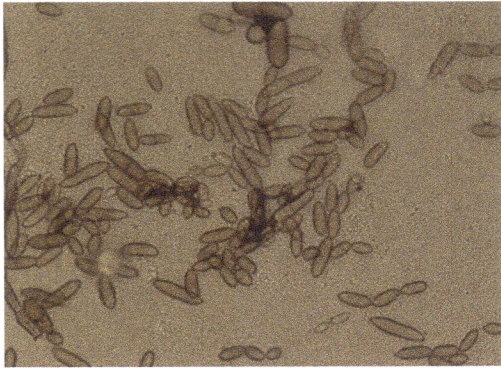

**Fig. 7.11.** One- or two-celled spores *in situ* blotted from a hemp leaf with sooty mould.

**Fig. 7.12.** A typical powdery mildew symptom on the adaxial side of a cannabis plant leaflet (right). Note that there was no visible mildew on the abaxial side.

from a sooty mould plant sample. Isolation of fungi from superficially disinfected leaf tissue on a PDA$^{+strep}$ plate is not helpful for sooty mould diagnosis, as these fungi grow epiphytically on leaves.

### Key diagnostic evidence

Sooty mould can be easily diagnosed in both the field and the lab. Unless a specific fungal species identification is needed, a general microscopic examination can conclude the diagnosis. The key evidence of sooty mould is the presence of a wet and sticky liquid, black mould covering and a pre-existing or active aphid infestation on the leaves.

## Powdery Mildew Caused by *Golovinomyces* Species

Both hemp and indoor cannabis crops are susceptible to powdery mildew disease. It may become problematic if the disease cannot be eradicated in a cultivation facility. The fungi causing powdery mildew may be from several genera of related fungi, but in *Cannabis* plants, only the genus *Golovinomyces* is reported. In Nevada, *Golovinomyces ambrosiae* was identified from indoor cannabis plants (Schoener and Wang, 2018); in Kentucky, *Golovinomyces spadiceus* was reported from greenhouse-grown hemp crops (Szarka *et al.*, 2019); and in Canada, *Golovinomyces cichoracearum sensu lato* was identified from indoor-grown *Cannabis* plants with a disease incidence ranging from 20% to 35% (Pépin *et al.*, 2018).

### Field diagnosis

#### Symptoms

Affected leaves exhibit patches of whitish dust-like powder on the adaxial surface, but rarely on the abaxial surface (Fig. 7.12). The initial symptom appears as a tiny and inconspicuous white spot, but the fungus quickly grows into a large patch. The entire leaf surface can be covered by the mycelia and spores. Spores produced from infected leaves can readily spread to other plants. Although powdery mildew occurs mostly on the upper leaves, it may also infect petioles and the nearby stalk and buds (see Chapter 6).

#### Problem classification

Powdery mildew is an airborne leaf disease that has mild to moderate impacts on leaf health. The pathogen is obligate, reproduces on live plant tissue and spreads easily from plant to plant inside the greenhouse and indoor cultivation facility.

#### Do your own diagnosis

Powdery mildew can be easily recognized and diagnosed, because the symptoms are quite distinctive. Based on the symptoms alone, one can confidently diagnose the problem as powdery mildew. However, determining the exact fungal species causing the disease requires submission of a leaf sample to a laboratory for diagnosis.

### Sampling

A few (three to five) leaflets representing different levels of infection can be collected and each placed individually in a Ziploc bag. For further sampling and submission guidelines, see the section describing powdery mildew on buds in Chapter 6.

### Lab diagnosis

See the section on powdery mildew of buds in Chapter 6.

## Hemp Leaf Spot and Blight Caused by *Bipolaris* Species

*Bipolaris* is a genus containing many plant pathogens that may cause leaf spot, leaf blight, melting out and root rot in many plant species, including high-value crops such as rice, maize, wheat and sorghum (Manamgoda *et al.*, 2014). *Bipolaris* was found to be responsible for hemp leaf spot disease in Kentucky hemp crops (Szarka *et al.*, 2020). *Bipolaris* has also been isolated from hemp and diagnosed as a pathogen of hemp leaf blight in other parts of the USA.

### Field diagnosis

#### Symptoms

Symptoms may vary significantly and depend on environmental conditions. In Kentucky, infected leaves exhibited an initial symptom of scattered specks or a tiny spot that further developed into larger spots with diameters ranging from 1 to 2 mm in 2 weeks (Szarka *et al.*, 2020). As the infection proceeds, spots extend from the upper sides of leaves to the lower sides and necrotic spots are visible from both sides. Most spots are dark brown with a noticeably darker border, but some spots are lighter without a visibly darker margin. Under favourable conditions, spots may coalesce into a large necrotic area, resembling leaf blight (Fig. 7.13). Conidiophores, the structures for bearing spores, may be visible from both sides of leaves under a magnifier. On an infected plant, leaf spot or blight can occur on both new and developed leaves. In the field, symptomatic plants are scattered.

#### Problem classification

*Bipolaris* leaf spot is a foliar disease that causes moderate to severe damage to leaves. There is no

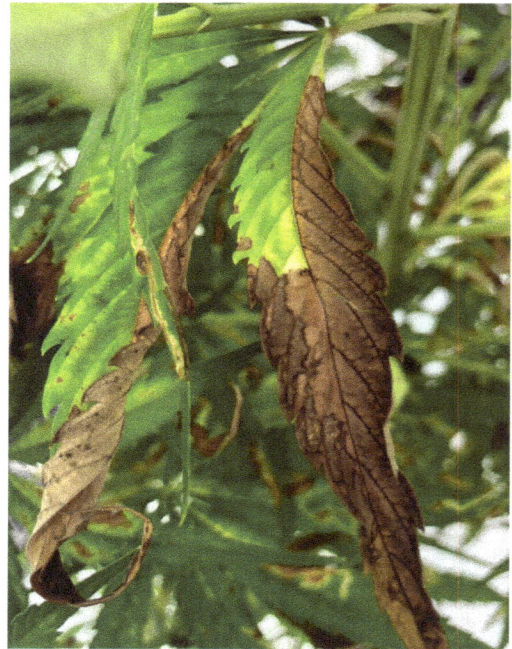

**Fig. 7.13.** Hemp leaf blight caused by a *Bipolaris* species. Note both isolated spots and large blighted areas.

evidence of its infection on other parts of plants. The pathogen overwinters in infected plant tissue as a primary inoculum and spreads among plants during the growing season by rain splash and air circulation.

#### Do your own diagnosis

By examining leaves and confirming the presence of scattered brown spots in many plants throughout the field, one can suspect the disease to be leaf spot. However, determining if it is a fungal leaf spot or if it is even caused by this specific fungus genus (*Bipolaris*) requires a lab diagnosis. Growers can conduct a preliminary assessment as to when the symptom started, the specific environmental conditions when the symptom showed up, the percentage of plants affected, the severity, pattern and time frame of symptom development, and recent foliar applications of any products. A thorough assessment helps to determine whether the leaf spot is caused by an abiotic factor or a fungal pathogen.

### Sampling

Five to ten leaflets showing mild to severe leaf spot or blight should be collected from different areas of the field and each placed individually in a Ziploc bag. Do not add any extra moisture. If multiple varieties are affected, take leaf samples from each variety.

## Lab diagnosis

### Microscopic examination

The leaf sample should be placed under a stereo microscope to look for any arthropods. If insects and mites are ruled out, examine the lesions for the presence of conidiophores rising from leaf tissue, from which conidia are born. *Bipolaris*, if present, produces large macroconidia with four to six septa. This is a diagnostic characteristic for *Bipolaris* leaf spot. Be cautious with the growth of some saprophytic fungi on some mailed leaf samples. If the lesions do not produce characteristic fungal signs that lead to a final diagnosis, proceed to the pathogen isolation step.

### Pathogen isolation and colony observation

Using PDA[+strep] plates, *Bipolaris* is readily isolated from infected leaves. Plate 10–20 pieces of tissue cut from the edge of a brown lesion and incubate the plates at 22°C for several days. A fungus, if it is the cause, should grow out from the majority or all of the tissue pieces plated. Transfer the most dominant fungal isolate to a fresh PDA plate for further morphological observation. Morphometric data such as the length and width of conidiophores and macroconidia are useful but may vary from case to case. In the case of the disease reported in Kentucky (Szarka *et al.*, 2020), conidiophores were dark brown with an approximate size of 192.6–365.3 × 7.1–10.7 µm, had four to six septa, and generally bore a single macroconidium. Macroconidia are hyaline, cylindrical, approximately 204.2–364.8 × 20.9–27.9 µm, and have three to six septa. Microconidia, when present, are much smaller (12.9–28.2 × 4.5–6.0 µm) and contain only one to three septa.

### DNA-based identification

*Bipolaris* isolates can be further identified to the species level using standard DNA-based identification procedures described in Chapter 5.

### Key diagnostic evidence

Leaf spot and leaf blight can be caused by a number of pathogens and abiotic factors. The key diagnostic evidence is the positive isolation of *Bipolaris* from necrotic leaf tissue and apparent association of *Bipolaris* with symptomatic plants. Pathogenicity test may be conducted if the fungal species is new or causes a disease on a new host.

## Hemp Leaf Blight Caused by *Exserohilum rostratum*

*Exserohilum* is a genus closely related to the *Bipolaris* and *Dreschlera* genera. It is considered a cosmopolitan organism found naturally in diverse habitats. Species in this genus can cause plant diseases such as bottle gourd (*Lagenaria siceraria*) leaf spot (Choudhary *et al.*, 2018) as well as human diseases (Katragkou *et al.*, 2014). *Exserohilum rostratum* was reported to cause bean blight in Egypt (Farag and Attia, 2020) and was also found to be associated with rice seed (Cardona and González, 2007). On hemp, *E. rostratum* was found to cause foliar blight in North Carolina (Thiessen and Schappe, 2019).

### Field diagnosis

#### Symptoms

During the growing season, brown or black lesions or spots develop on leaves or stems. Spots may have dark margins within which fungal spores (conidia) may be present (Thiessen and Schappe, 2019). Progression of this infection causes death of large areas of leaves or leaf blight.

### Problem classification

This disease is a fungal leaf spot/blight disease that primarily impacts foliar health. It may affect the stem when plants are severely infected under favourable conditions. The pathogen survives in the infected plant tissue as a primary inoculum and spreads among plants by rain splash.

### Do your own diagnosis

To determine if this is a fungal disease, closely examine the leaves for any sign of fungal spores present at the centre of necrotic areas. The spores may appear fuzzy and dark. Use a hand lens to see more details of the fungal structures. Older leaves or matured lesions are more likely to develop fungal spores, as they are infected earlier and ready to produce secondary inoculum (spores) for reinfection. Alternatively, use a clear tape to blot the lesion area and then place it under a compound microscope to observe fungal conidia spores (see Chapter 5). The presence of straight to slightly curved conidia with 7–12 septa suggests *Exserohilum* infection. A leaf sample should be submitted to a lab for further diagnosis.

### Sampling

See hemp leaf spot caused by *Bipolaris* spp., above.

### Lab diagnosis

#### Microscopic examination

Examine the lesions from infected leaves for the presence of conidia. The presence of spores with multiple septa (7–12) supports the diagnosis of *Exserohilum* leaf spot disease. Based on the disease described from North Carolina (Thiessen and Schappe, 2019), spores of *E. rostratum* were ellipsoidal, straight or slightly curved, olive-brown, and with a cylindrical hilum at the base. Most conidia had 7–12 septa and were about 75.64 μm long and 15.61 μm wide. If no spores or signs of insect or mite infestation are observed, samples can be proceeded to the next step: direct isolation of microorganisms from necrotic tissue.

#### Pathogen isolation and colony observation

*Exserohilum* can be readily isolated on a PDA plate amended with streptomycin (100 μg/ml). The colonies may be initially white but turn to dark brown or black after 2–3 days (Thiessen and Schappe, 2019) and may grow abundant aerial mycelia. Make several temporary slides to observe the spores and conidiophores for preliminary morphological identification.

#### DNA-based identification

Isolates of *Exserohilum* can be further identified to the species level using standard DNA-based identification procedures described in Chapter 5.

#### Key diagnostic evidence

A high frequency (≥ 90% of tissue pieces plated) of isolation of *Exserohilum* from necrotic leaf tissue and/

or observation of *Exserohilum* conidia from necrotic spots are key to supporting the diagnosis. Pathogenicity test is not needed for a general diagnosis.

## Hemp Leaf Spot Caused by *Cercospora* Species

*Cercospora* species are important fungal pathogens mainly causing leaf spots on certain crops, for example grey leaf spot of corn (*Zea mays* L.) caused by *Cercospora zeae-maydis* (Beckman and Payne, 1983) and narrow brown leaf spot of rice caused by *C. janseana* (Uppala *et al.*, 2019). The first hemp leaf spot caused by *Cercospora* species in North America was reported from an experimental plot in Mississippi and the causal organism was identified as *Cercospora cannabina*, based on its significant morphological differences from *Cercospora cannabis* previously recorded in Wisconsin and Missouri (Lentz *et al.*, 1974). Recently, *Cercospora* cf. *flagellaris* was diagnosed as the causal agent of hemp leaf spot in Kentucky (Doyle *et al.*, 2019).

### Field diagnosis

#### Symptoms

Initial symptoms may start on matured leaves in the lower canopy as visible yellow flecks or spots on the upper sides of the leaves. When environmental conditions favour disease development, the spots expand in size and spread to the rest of the canopy. The spots are generally round with dark brown to purplish borders. The centres of the spots may turn tan or white. The sizes of the spots vary and can grow to be 6 mm in diameter. Clusters of conidiophores of the pathogen may become visible at the centres of the spots, with or without using a hand lens (Doyle *et al.*, 2019). Infected leaves may curl, wilt, or drop from the plants and the disease moves progressively from older leaves to newer leaves, causing severe damage to the plants (Lentz *et al.*, 1974).

#### Problem classification

Similar to *Bipolaris* leaf spot, *Cercospora* leaf spot is a foliar fungal disease that causes moderate to severe damage to leaves. There is no evidence of its infection on other parts of hemp plants. The pathogen overwinters in infected plant tissue as a primary inoculum and spreads among plants during the growing season via rain splash.

### Do your own diagnosis

A round leaf spot with a raised border and light colour in the centre may signal *Cercospora* infection. However, a diagnosis based on the shape, size or colour of leaf spots is not reliable. Many leaf spot diseases look similar and the symptoms of each disease vary greatly due to the variety, environmental conditions and the location. Once leaf spots are noticed on many plants throughout the field, growers should perform a preliminary assessment to determine whether the problem belongs to an infectious leaf spot disease or an abiotic injury. Leaf samples should be collected and submitted to a lab for further diagnosis.

### Sampling

Use similar procedures described for *Bipolaris* leaf spot, above.

### Lab diagnosis

### Microscopic examination and moisture chamber induction

Examine the lesions from older leaves for the presence of conidiophores rising from the leaf tissue. When conidiophores are present, they are borne in fascicles of six to 12 (Lentz *et al.*, 1974). Needle-shaped spores may be present. If no pathogen signs are visible, infected leaves can be surface sterilized with 1% sodium hypochlorite (NaClO) for 1 min, rinsed with sterile water and incubated in a moist chamber. Conidiophores and conidia of *Cercospora* will form and become visible from infected lesions in 2–7 days.

### Pathogen isolation and colony observation

*Cercospora* species generally grow slowly and do not sporulate well on PDA plates (Uppala *et al.*, 2019); therefore, the PDA[+strep] plate commonly used for fungal isolation is not suitable for *Cercospora* isolation from plant tissue. Alternatively, spores can be induced from necrotic tissue under moist conditions and then needle-picked into a V8 agar plate (Doyle *et al.*, 2019). Spores can also be picked with a flamed (to sterilize) platinum needle

into a microcentrifuge tube with 30–50 µl of sterile water. Add the spore mix solution onto one V8 agar plate amended with streptomycin (100 µg/ml) and spread the spores evenly on the plate using three glass beads (see Chapter 5). Allow the spores to germinate at 22–23°C and form colonies, each originating from a single spore. The resulting single-spore colonies can be transferred onto fresh V8 plates and incubated with alternated 12 h fluorescent light and dark to induce sporulation. For *Cercospora* cf. *flagellaris*, colonies are grey to white, appear flat but grow aerial mycelia and produce abundant conidia. Conidia are hyaline, needle-shaped, straight or slightly curved and have up to 20 septa (Doyle *et al.*, 2019).

### DNA-based identification

Isolates of *Cercospora* can be further identified to species level using standard DNA-based identification procedures described in Chapter 5.

### Key diagnostic evidence

Like all other leaf spot diseases, judging whether a disease is caused by *Cercospora* species requires the positive isolation of *Cercospora* from necrotic leaf tissue and an obvious association of this organism with the majority of symptomatic plants.

## Hemp Leaf Spot Caused by *Serratia marcescens*

The genus *Serratia* was long thought to be harmless saprophytes, but some species, in particular *S. marcescens*, are now recognized as significant human pathogens (Mahlen, 2011). *Serratia*, belonging to the family Enterobacteriaceae, contains 14 recognized species, most of which are either opportunistic human pathogens or saprophytes. Recently, *S. marcescens* was first isolated from a hemp leaf exhibiting leaf spot and confirmed to be a causative agent for the leaf spot disease (Schappe *et al.*, 2019).

### Field diagnosis

### Symptoms

Infected plants develop leaf spots on leaves that are characterized by their angular shape, vein limited, small sizes ranging from 1 to 3 mm and dark brown colour. Lesions may also develop on buds or stems. As the disease progresses under favourable conditions, lesions may coalesce to kill large portions of the leaves. Severely infected plants can be killed. Bacterial ooze may be seen from infected tissue under a microscope.

### Problem classification

This disease is bacterial leaf spot primarily impacting foliar health, but the bacterium can cause necrotic spots on the stem and buds too. Note that there are both fungal pathogens and abiotic factors causing leaf spots that may look similar to or be indistinguishable from the bacterial spot. However, one key factor that can be used to classify a problem as bacterial leaf spot is the water-soaked lesions and bacterial oozing from the necrotic tissue. Leaf spots caused by fungi may look dry or even crispy and commonly have visible conidiophores or spores at the centre of lesions. Abiotic necrotic lesions are generally uniformly distributed on leaves without evident progression over time or any pathogen signs such as conidiophores or bacterial fluid. Like other bacterial pathogens, *S. marcescens* may survive in infected plant tissue as a primary inoculum and spread among plants by rain splash or direct contact during the growing season.

### Do your own diagnosis

A tiny or small angular-shaped spot can be used to suspect bacterial leaf spot. Perform bacterial streaming under a compound microscope to observe the red bacterial ooze streaming from the necrotic tissues. Nevertheless, leaf spots caused by other bacteria may look similar symptomatically, so a leaf sample should be collected and submitted to a lab for further diagnosis.

### Sampling

Collect five to ten leaflets exhibiting symptoms ranging from tiny spots to large coalescent lesions from different locations of the field. Place each individual leaflet in a Ziploc bag. Do not add any extra moisture. If multiple varieties are planted, take samples from all varieties, even those that are asymptomatic. Adequate representative sampling more accurately reflects the characteristics of the disease in the field and helps a lab diagnostician to

determine whether the crop is affected with more than one pathogen or if the infection is specific to certain varieties.

### Lab diagnosis
#### Microscopic examination and bacterial streaming observation

Perform a general microscopic examination to rule out any fungal diseases and arthropod infestations. Under a stereo microscope, fresh leaf lesions may appear to be moist with a thin layer of fluid seeping out from the tissue. If leaf samples are less fresh, use a razor blade to slice a piece of necrotic tissue into fine pieces on a glass slide, add a drop of water, place a cover slip over the slide and observe bacterial streaming (see Chapter 5).

#### Pathogen isolation and colony observation

*Serratia marcescens* grows well on potato dextrose agar (PDA) and forms raised and dark-pink coloured colonies (Schappe *et al.*, 2019). PDA can be used as a basic and all-purpose medium for isolating various bacteria (see Chapter 5 for bacterial isolation procedures). A defined agar medium, also known as CT agar (caprylate-thallous agar), was reported to be highly selective for the isolation of all *Serratia* species and suppressed the growth of most non-*Serratia* species in the same habitat (Starr *et al.*, 1976). This medium contains 0.01% yeast extract, 0.1% caprylic (octanoic) acid and 0.025% thallous sulfate. Although it was originally developed for use in the isolation of *Serratia* from urine, faeces and sputum specimens, it may be useful for isolating this bacterium from plant tissue.

#### DNA-based identification

Isolates of *Serratia* can be further identified to the species level using standard DNA-based identification procedures described in Chapter 5. Briefly, grow isolates from a single colony in either nutrient broth (Lysyk *et al.*, 2002) or LB broth (Wei and Chen, 2005) and extract the DNA from the bacterial cells using any commercially available microbial DNA extraction kit. Use the 16SF and 16SR primers for conventional PCR followed by sequencing.

## Key diagnostic evidence

Diagnosing a disease as being caused by any species of *Serratia* requires consistent isolation of this particular bacterium from necrotic leaf tissue and a strong link between the bacterium and the symptoms. If other bacterial species are also isolated from symptomatic tissue, inoculation tests must be done to confirm the pathogenicity of each species.

## Hemp Leaf Spot Caused by *Xanthomonas campestris* pv. *cannabis*

*Xanthomonas* is a genus of Gram-negative, short rod-shaped bacteria in the family Pseudomonadaceae. Some can survive and grow as epiphytes, but many cause spots and blights on the leaves, stems and fruits of diverse monocot and dicot plant species, including commercially important crops such as rice and citrus. Some pathogenic species show high degrees of specificity and can be further classified into multiple pathovars, a subspecies designation for strains that only attack a specific crop species. Hemp leaf spot caused by *Xanthomonas campestris* pv. *cannabis* has been reported in Romania (Severin, 1978) and Japan (Netsu *et al.*, 2014).

### Field diagnosis

#### Symptoms

Leaf spots appear brown, necrotic, water-soaked, round or irregular, 1–2 mm in diameter and often surrounded by yellow halos 2–3 mm wide. The spots may be delineated by the leaf veins and scattered all over the leaves. The bacterium may infect the shoot apex and cause bud blight. However, it does not infect stems (Netsu *et al.*, 2014).

#### Problem classification

This disease is similar to other bacterial leaf spot diseases that primarily impact foliar health. It can cause bud blight if plants are severely infected under favourable conditions. The pathogen survives in infected plant tissue as a primary inoculum and spreads among plants by rain splash or direct contact during the growing season.

#### Do your own diagnosis

Small (1–2 mm), brown and round lesions with yellow halos 2–3 mm wide are a defining characteristic used to suspect *Xanthomonas* leaf spot. A bacterial streaming test can further verify the bacterial infection. However, always take a leaf sample and submit it to a lab for further diagnosis, as many leaf spot diseases look similar.

### Sampling

See hemp leaf spot caused by *Serratia marcescens*, above.

### Lab diagnosis

#### Microscopic examination and bacterial streaming observation

Perform similar procedures to those described above for hemp leaf spot caused by *Serratia marcescens*.

#### Pathogen isolation and colony observation

While some media are developed for the selective isolation of a targeted *Xanthomonas* species (Juhnke and des Jardin, 1989) or for the semi-selective isolation of a specific pathovar (Pruvost *et al.*, 2005), they may not be suitable for a general diagnosis. That said, a general bacterial culture medium like LB agar is more appropriate for obtaining any *Xanthomonas* species or other pathogenic bacteria genera from diseased tissue. On yeast dextrose calcium carbonate (YDC) medium or sucrose peptone agar (SPA), colonies appear mucoid and yellow. Most *Xanthomonas* species produce a characteristic yellow pigment called xanthomonadin (Poplawsky *et al.*, 2000) and this pigment can be used as a diagnostic marker of this genus. On nutrient agar plates, *Xanthomonas campestris* pv. *cannabis* forms circular, raised, translucent, smooth and yellow colonies with shiny surfaces (Netsu *et al.*, 2014). Bacterial cells are short and rod-shaped and have one polar flagellum.

#### DNA-based identification

Isolates of *Xanthomonas* or other bacteria can be further identified to the species level using the standard DNA-based identification procedures described in Chapter 5.

#### Pathogenicity test

A pathogenicity test may be needed to confirm whether the obtained isolates are epiphytes or pathogens. Prepare a set of hemp seedlings grown in a greenhouse or growth chamber and a bacterial suspension (dilute bacterial culture with water) at the concentration of $10^7$–$10^8$ colony forming units (CFUs) per millilitre. Spray the suspension onto the leaves of the seedlings, keep the plants under moist conditions overnight and examine the inoculated plants for water-soaked lesions after 7–10 days.

### Key diagnostic evidence

A definitive diagnosis requires evidence of a strong association of a *Xanthomonas* isolate with leaf spot, a positive pathogenicity test and an accurate identification of the isolate either by conventional or by DNA-based methods.

### Other Foliar Diseases

In addition to the formally reported diseases described above, many other diseases have been diagnosed from hemp. For example, the fungal pathogens *Stemphylium* sp. and *Septoria* sp. were diagnosed from hemp exhibiting leaf spot and blight, respectively. For all leaf spot or blight diseases, an accurate diagnosis can be achieved by an initial microscopic examination for pathogen signs and subsequent isolation of causative agents followed by DNA-based organism identification. Symptom-based diagnosis is generally not reliable.

Hemp rust caused by *Uredo kriegeriana* was recorded in Germany and the former Soviet Union (Farr and Rossman, 2019), but it is not known how prevalent *Uredo* species are in hemp production. There are many posts on the internet talking about hemp rust, but some pictures posted more closely resemble regular leaf spots rather than rust. To diagnose a rust disease, a microscopic examination of uredospores or teliospores is the first step. For example, rust caused by the genus *Puccinia*, the most common rust fungus, always produces notable pustules (a small raised spot) with massive uredospores inside during the growing season. The pustules may be scattered or densely formed mostly on the upper sides of leaves (Fig. 7.14). To the naked eye, the uredospores are orange in colour like rust (see Plate 2.2), but under a stereo microscope, uredospores appear granular, have a shining orange colour and are often dispersed to areas around the pustules (Fig. 7.15). Positive observation of pustules and freshly produced uredospores in infected leaves is the key to differentiating rust

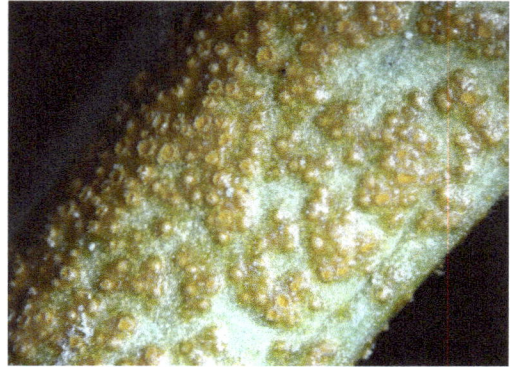

**Fig. 7.14.** Orange-coloured pustules formed on the leaf of foothill deathcamas plant (*Zigadenus paniculatus*) infected by *Puccinia* sp.

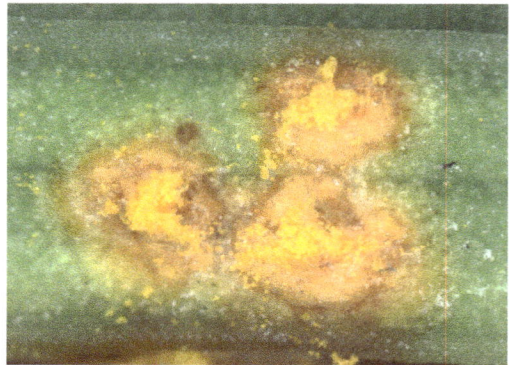

**Fig. 7.15.** Shiny orange uredospores (tiny round particles) released from pustules and dispersed on the leaf surface. Picture taken from a daylily plant (*Hemerocallis* sp.) infected by *Puccinia hemerocallidis*.

from brown or orange colour leaf spots. Uredospores can be blotted by a piece of clear tape and mounted on a glass slide to observe their morphological details under a compound microscope. In many *Puccinia* species, uredospores are nearly round with a rough surface (Fig. 7.16).

### Foliar Disease Management

In general, foliar pathogens survive in dead plant tissue in soil and crop residue, serving as a primary inoculum. They may be splashed onto plant foliage by raindrops and splashing water and initiate infection on newly grown leaves. Some pathogens may

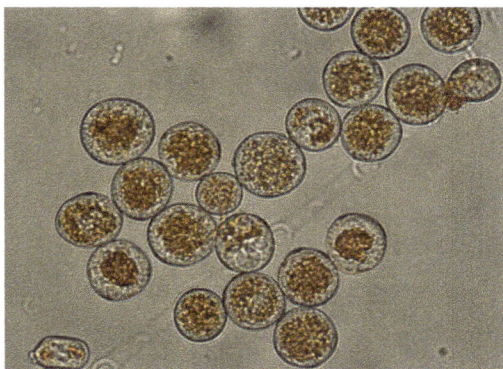

**Fig. 7.16.** Uredospores observed under a compound microscope. Picture taken from a garlic plant (*Allium sativum*) infected by *Puccinia allii*.

overwinter on certain weed species and then spread to a crop. Infected seeds are also one of the factors leading to crop foliar diseases. During the growing season, infections can spread among plants via rain splash or even air circulation. For outdoor-growing hemp, persistent rainy weather often incites leaf spot diseases if a given pathogen exists in the field. In semi-arid regions, overhead sprinklers are the main system to irrigate crops, which sometimes can worsen certain foliar diseases.

### Using resistant varieties

Use of resistant varieties is the primary and probably most effective method for combating foliar diseases on many crops, but information on hemp resistance to currently known foliar diseases is scarce. Future research is needed to characterize major diseases associated with *Cannabis* crops and explore potential resistance existing in both wild and cultivated *Cannabis* cultivars.

### Treating seeds

Depending on the pathogens diagnosed from hemp crops, a seed-borne pathogen that incites a foliar disease can be eliminated through seed treatments with a product labelled for use in hemp (see Chapter 6).

### Rotating crops

Crop rotation is an effective practice to manage soilborne diseases, but it may be used to control certain foliar diseases. Rotating with a non-host crop can significantly reduce the primary inoculum in the field and therefore reduce the incidence of disease in subsequent hemp crops. Practising rotation requires an accurate diagnosis of the diseases and knowledge of the specific host range of the pathogen so that certain non-host rotating crop can be identified.

### Reducing primary inoculum

As many pathogens survive in crop residues, removing and burning plant debris is an effective cultural practice in foliar disease management. Infected leaves that harbour plant pathogens should be removed from the field or growing facility if possible. Weeds, if they are hosts of a foliar pathogen, should be monitored and controlled.

### Spacing plants

Crowded growth generally promotes plant-to-plant transmission of foliar diseases. Thus, crowding should be avoided, especially when overhead watering is used or when rainy weather is a norm. Increased space between plants helps air circulation and allows the leaf surface to dry, creating less favourable conditions for fungal and bacterial diseases to develop.

### Using drip irrigation

Most foliar diseases can be significantly suppressed by a lack of free moisture on the leaf's surface. Drip irrigation is ideal for irrigating hemp and indoor cannabis crops, as it does not provide a moist condition favourable for leaf diseases. Furthermore, a drip system delivers water directly to the plant rhizosphere, avoiding water waste from foliar evaporation. As a result, the number of weeds growing in the field may also largely decrease.

### Applying approved products

As more fungicides and bactericides are approved for use in hemp, growers can use these products to manage foliar diseases. When spraying any of these products, ensure that both sides of the leaves are treated, as pathogens are often present on both sides. It is important to note that each product is only effective for certain foliar diseases. An accurate diagnosis helps determine the type of pathogen so an appropriate product can be selected.

# References

Beckman, P.M. and Payne, G.A. (1983) Cultural techniques and conditions influencing growth and sporulation of *Cercospora zeae-maydis* and lesion development in corn. *Phytopathology* 73, 286–289.

Cardona, R. and González, M.S. (2007) First report of *Exserohilum rostratum* associated with rice seed in Venezuela. *Plant Disease* 91, 226. doi: 10.1094/PDIS-91-2-0226C

Choudhary, M., Sardana, H.R., Bhat, M.N. and Gurjar, M.S. (2018) First report of leaf spot disease caused by *Exserohilum rostratum* on bottle gourd in India. *Plant Disease* 102, 2042. doi: 10.1094/PDIS-02-18-0315-PDN

Doyle, V.P., Tonry, H.T., Amsden, B., Beale, J., Dixon, E. *et al.* (2019) First report of *Cercospora* cf. *flagellaris* on industrial hemp (*Cannabis sativa*) in Kentucky. *Plant Disease* 103, 1784. doi: 10.1094/PDIS-01-19-0135-PDN

Farag, M.F. and Attia, F.M. (2020) A first record of *Exserohilum rostratum* as a new pathogen causing bean blight in Egypt. *Journal of Plant Pathology & Microbiology* 11, 496. doi: 10.35248/2157-7471.20.11.496

Farr, D.F. and Rossman, A.Y. (2019) Fungal Databases. Available at https://nt.ars-grin.gov/fungaldatabases (accessed 29 November 2019).

Juhnke, M.E. and des Jardin, E. (1989) Selective medium for isolation of *Xanthomonas maltophilia* from soil and rhizosphere environments. *Applied and Environmental Microbiology* 55, 747–750. doi: 10.1128/AEM.55.3.747-750.1989

Katragkou, A., Pana, Z., Perlin, D.S., Kontoyiannis, D.P., Walsh, T.J. *et al.* (2014) *Exserohilum* infections: Review of 48 cases before the 2012 United States outbreak. *Medical Mycology* 52, 376–386. doi: 10.1093/mmy/myt030

Lentz, P.L., Turner, C.E., Robertson, L.W., and Gentner, W.A. (1974) First North American record for *Cercospora cannabina*, with notes on the identification of *Cercospora cannabina* and *Cercospora cannabis*. *Plant Disease Reporter* 58, 165–168.

Lysyk, T.J., Kalischuk-Tymensen, L.D. and Selinger, L.B. (2002) Comparison of selected growth media for culturing *Serratia marcescens*, *Aeromonas* sp., and *Pseudomonas aeruginosa* as pathogens of adult *Stomoxys calcitrans* (Diptera: Muscidae). *Journal of Medical Entomology* 39, 89–98. doi: 10.1603/0022-2585-39.1.89

Mahlen S.D. (2011) *Serratia* infections: from military experiments to current practice. *Clinical Microbiology Reviews* 24, 755–791. doi: 10.1128/CMR.00017-11

Manamgoda, D.S., Rossman, A.Y., Castlebury, L.A., Crous, P.W., Madrid, H. *et al.* (2014). The genus *Bipolaris*. *Studies in Mycology* 79, 221–288. doi: 10.1016/j.simyco.2014.10.002

Netsu, O., Kijima, T. and Takikawa, Y. (2014) Bacterial leaf spot of hemp caused by *Xanthomonas campestris* pv. *cannabis* in Japan. *Journal of General Plant Pathology* 80, 164–168. doi: 10.1007/s10327-013-0497-8

Pépin, N., Punja, Z.K. and Joly, D.L. (2018) Occurrence of powdery mildew caused by *Golovinomyces cichoracearum sensu lato* on *Cannabis sativa* in Canada. *Plant Disease* 102, 2644. doi: 10.1094/PDIS-04-18-0586-PDN

Poplawsky, A.R., Urban, S.C. and Chun, W. (2000) Biological role of xanthomonadin pigments in *Xanthomonas campestris* pv. *Campestris*. *Applied and Environmental Microbiology* 66, 5123–5127. doi: 10.1128/AEM.66.12.5123-5127.2000

Pruvost, O., Roumagnac, P., Gaube, C., Chiroleu, F., Gagnevin, L. *et al.* (2005) New media for the semiselective isolation and enumeration of *Xanthomonas campestris* pv. *mangiferaeindicae*, the causal agent of mango bacterial black spot. *Journal of Applied Microbiology* 99, 803–815. doi: 10.1111/j.1365-2672.2005.02681.x

Schappe, T., Ritchie, D.F. and Thiessen, L.D. (2019) First report of *Serratia marcescens* causing a leaf spot disease on industrial hemp (*Cannabis sativa* L.). *Plant Disease* 104, 1248. doi: 10.1094/PDIS-04-19-0782-PDN

Schoch, C.L., Seifert, K.A., Huhndorf, S., Robert, V., Spouge, J.L. *et al.* (2012) Nuclear ribosomal internal transcribed spacer (ITS) region as a universal DNA barcode marker for Fungi. *Proceedings of the National Academy of Sciences* 109, 6241–6246. doi: 10.1073/pnas.1117018109

Schoener, J.L. and Wang, S. (2018) First detection of *Golovinomyces ambrosiae* causing powdery mildew on medical marijuana plants in Nevada. (Abstr.) *Phytopathology* 108, S1.186. doi: 10.194/PHYTO-108-10-S1.186

Schoener, J.L., Wilhelm, R. and Wang, S. (2018) Molecular identification of *Alternaria* species associated with imported industrial hemp seed. (Abstr.) *Phytopathology* 108, S1.185. doi: 10.194/PHYTO-108-10-S1.185

Severin, V. (1978) Ein neues pathogenes Bakterium an Hanf-*Xanthomonas campestris* pathovar. *cannabis*. *Archives of Phytopathology and Plant Protection* 14, 7–15.

Starr, M.P., Grimont, P.A., Grimont, F. and Starr, P.B. (1976) Caprylate-thallous agar medium for selectively isolating *Serratia* and its utility in the clinical laboratory. *Journal of Clinical Microbiology* 4, 270–276.

Szarka, D., Amsden, B., Beale, J., Dixon, E., Schardl, C.L. *et al.* (2020) First report of hemp leaf spot caused by a *Bipolaris* species on hemp (*Cannabis sativa*) in Kentucky. *Plant Health Progress* 21, 82–84. doi: 10.1094/PHP-01-20-0004-BR

Szarka, D., Tymon, L., Amsden, B., Dixon, E., Judy, J. et al. (2019) First report of powdery mildew caused by *Golovinomyces spadiceus* on industrial hemp (Cannabis sativa) in Kentucky. *Plant Disease* 103, 1773. doi: 10.1094/PDIS-01-19-0049-PDN

Thiessen, L. and Schappe, T. (2019) First report of *Exserohilum rostratum* causing foliar blight of industrial hemp (*Cannabis sativa* L.). *Plant Disease* 103, 1414. doi: 10.1094/PDIS-08-18-1434-PDN

Uppala, S.S., Zhou, X.-G., Liu, B. and Wu, M. (2019) Plant-based culture media for improved growth and sporulation of *Cercospora janseana*. *Plant Disease* 103, 504–508. doi: 10.1094/PDIS-05-18-0814-RE

Wei, Y.H. and Chen, W.C. (2005) Enhanced production of prodigiosin-like pigment from *Serratia marcescens* SMΔR by medium improvement and oil-supplementation strategies. *Journal of Bioscience and Bioengineering* 99, 616-622. doi: 10.1263/jbb.99.616

Woudenberg, J.H., Groenewald, J.Z., Binder, M. and Crous, P.W. (2013) *Alternaria* redefined. *Studies in Mycology* 75, 171–212. doi: 10.3114/sim0015

# 8 Diagnosing Stem Diseases

The stem is the major part of a vascular plant that supports leaves and flowers and transports nutrients and water between the roots and the shoots through its vascular system. A plant's vascular system consists of the xylem and phloem. Xylem transports minerals and water from the roots to the leaves, while phloem transports food from the leaves to other parts of the plant. Hemp plants develop a strong stem structure that contains a lignified core containing xylem vessels and a cortex with bast fibres (Andre *et al.*, 2016). A healthy stem is rigid and woody (see Fig. 2.2) and supports the growth of leaves and inflorescences. Unfortunately, plant pathogens attack the stem of *Cannabis* plants and cause primary damage on stem tissue that consequently lead to secondary symptoms on leaves and flowers. Many foliar symptoms such as leaf yellowing, wilting, dieback or even tissue necrosis are directly caused by pathogen infections on the stem and these pathogens are rarely present in leaf tissue. Stem diseases are commonly caused by fungi, oomycetes, or bacteria, many of which are soilborne. In the early stages of infection, infected plants may appear asymptomatic. As the disease progresses, plants may collapse or die quickly (Fig. 8.1). This chapter refers to stem diseases that mainly affect the stem with a pathogen confined in stem tissue. Diseases like *Fusarium* that cause obvious stem symptoms but behave systemically are presented in Chapter 10.

## Hemp Southern Blight Caused by *Athelia rolfsii*

*Athelia rolfsii* is a basidiomycetous fungus and represents the sexual stage (teleomorph) of *Sclerotium rolfsii*. The sexual stage is not common in nature, but its asexual stage (anamorph), also known as *S. rolfsii*, commonly causes southern blight, southern stem blight or white mould in over 500 plant species

(Mullen, 2001). That said, the name *Sclerotium rolfsii* is more frequently used when this pathogen causes a disease. *S. rolfsii* is a soilborne fungus and usually produces abundant, visible white and coarse mycelium on infected plant tissues. Like other *Sclerotium* species, *S. rolfsii* does not produce any asexual spores during infection; rather, its hyphae are packed densely to form a hard structure called the sclerotium (plural: sclerotia). Most *Sclerotium* species produce characteristic sclerotia that may be diagnosable. For example, *S. cepivorum*, a pathogen causing *Allium* crop white rot, produces abundant shiny, spherical, black sclerotia on the surface of bulbs (Fig. 8.2). Sclerotia can survive in soil or plant debris for several years and germinate during the spring when crops are planted. *S. rolfsii* has been reported to cause southern blight on hemp in Virginia (Mersha *et al.*, 2019).

### Field diagnosis

#### Symptoms

*S. rolfsii* infects the lower stem and crown area, causing lesions and canker. The symptoms on the stem may not be noticeable in the early stages, but the infected plant may show mild foliar symptoms such as leaf yellowing. However, the leaf yellowing is a result of stem damage rather than a direct infection of leaf tissue by this fungus. Infected stems often have lesions, canker or dry rot. Lesions may expand or merge to girdle the stem, causing the plant to wilt. When humidity around the lower stem is adequate, for example grown in plastic-mulch-covered raised beds with a drip-irrigation system, the fungus may grow rapidly to cover the surface of the lower stem with fluffy white mycelia. In the late stage of disease development, brownish (immature) to black (mature) sclerotia are produced and embedded in a fluffy white fungal mycelium

DOI: 10.1079/9781789246070.0008

**Fig. 8.1.** A hemp plant that died rapidly due to a basal stem infection by *Pythium aphanidermatum*. Note that *P. aphanidermatum* only infects crown tissue but causes secondary symptoms of leaf yellowing and shoot wilt.

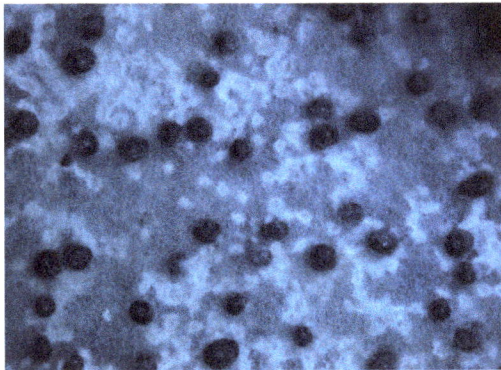

**Fig. 8.2.** Small, round and black sclerotia of *Sclerotium cepivorum* produced on PDA plate after 3 weeks of growth. Note that sclerotia are visible as tiny dots but the picture was taken under a low magnification.

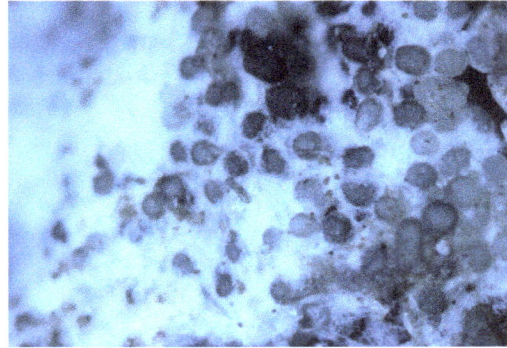

**Fig. 8.3.** An example of smooth sclerotia embedded in a mass of white mycelia observed under a stereo microscope. Note that the colour of sclerotia changes from white to light tan to brown and to black. (Picture taken from *Sclerotium cepivorum*.)

dark lesions with variable sizes in the range of 5–62 mm and the fungus produced brownish sclerotia about 0.4–1.6 mm in diameter (Mersha *et al.*, 2019).

### Problem classification

Hemp southern blight caused by *S. rolfsii* is a soil-borne disease with the actual infection located at the lower stem, or stem near the soil surface. Foliar symptoms such as yellowing, necrosis, or wilt are a result of stem infection. However, leaves should be examined to rule out any foliar diseases. In the case of plant wilt, vascular tissue should be examined to determine whether a vascular wilt disease is present. Southern blight exhibits characteristic symptoms with pathogen signs (mycelia and sclerotia) on the stem at the soil line, which can be reliably used to classify the problem as a stem canker disease.

### Do your own diagnosis

Growers can systemically examine several symptomatic plants to determine if the damage is restricted to stems. Examine the leaves for any pathogen signs and foliar disease symptoms described in Chapter 7. In the case of hemp southern blight, foliar symptoms may be similar to nutrient deficiency or drought stress. Then, examine the stem and crown area to see if any lesions or rot have occurred. The presence of

mass, similar to when *S. cepivorum* infects the lower portion of *Allium* plants (Fig. 8.3). On plants of the CBD variety 'Boax', infected lower stems exhibited fluffy white and coarse mycelia on the stem around the soil line suggests southern blight. Use a knife to scrape the fungal mycelia off the stem to reveal the lesions and cut through the stem to examine internal tissue rot. Place a magnifier close to the area with fungal mycelium mass to observe the embedded mustard-seed-like sclerotia. It is not uncommon to see numerous matured sclerotia on plant tissue with little presence of mycelia, because the fungus switches from mycelium growth to sclerotia production at the end of the season. The presence of both white mycelia and sclerotia at the lower stem in several symptomatic plants examined may confirm southern blight diagnosis.

### Sampling

Collect a few (three to five) representative symptomatic plants with roots and soil attached and place them individually in plastic bags. The submission of whole plants allows the diagnostician to examine the plants systemically and determine the problem type. If plants are too large, a few symptomatic stem pieces including the lower portion and roots should be submitted. Place the sample bags in a sturdy cardboard box to protect the integrity of the sample during shipment. Do not add any moisture to the bag.

### Lab diagnosis

### Visual and microscopic examinations

Lay the plant on a diagnostic table and inspect the leaves, stem, crown and root. Visually examine the leaf for any sign of foliar diseases. Examine the stem surface for discoloration, spots and lesions. Use a knife to cut the stem to reveal vascular tissue that may be abnormally discoloured. Examine the root for signs of black lesions, rot, or abnormal growth. In the case of hemp southern blight, a stem rot around the crown area can be easily spotted by its necrotic lesions and white mycelia. Cut the lower stem and place it under a stereo microscope to observe the sclerotia. Massive sclerotia with shiny and varied colours may be seen inside the fungal mycelium mass. The mycelium can be transferred to a glass slide by blotting with a piece of clear tape to observe the hyphae under a compound microscope. No *S. rolfsii* spores should be observed. However, spores produced by other fungi may be present.

### Isolation and colony observation

To confirm the diagnosis based on field and microscopic examinations, the pathogen must be isolated from diseased stem tissue. For a sclerotium-producing fungal pathogen, individual sclerotium can be used for pathogen isolation, which has the advantage of avoiding some of the fast-growing secondary organisms during isolation. Under a stereo microscope, use forceps to pick a sclerotium from infected tissue and place it into 1% sodium hypochlorite in a grid of a quarter plate (Petri dish with four divisions) for 30–60 s followed by three rounds of rinsing in sterilized water (see section for single sclerotium isolation in Chapter 5). Surface-sterilized sclerotia should then be placed at the centre of a PDA$^{+strep}$ plate or acidified potato dextrose agar (APDA) and incubated at 22°C or 30°C in the dark. The white mycelia of the fungus grow from the sclerotium within a few days and may completely cover the 100 mm diameter Petri dishes within a week. Sclerotia with diameters of 1.2–4.2 mm are produced after 7 days (Mersha *et al.*, 2019).

### DNA-based identification

*Sclerotium* species readily grow and produce abundant sclerotia on culture media but do not produce the spores needed for morphological identification. In this case, DNA-based identification can be performed following the procedures described in Chapter 5. Briefly, perform PCR using primers ITS1/ITS4, purify the PCR product, sequence it directly and BLASTn search for closely related species. The expected PCR product size for *Sclerotium* species may vary from 540 to 600 bp. *S. rolfsii* isolated from Virginia had a PCR product of 625 bp when ITS1/ITS4 primers were used (Mersha *et al.*, 2019). Further identification based on multiple genes or loci may be performed if necessary.

### Pathogenicity test

Performing a pathogenicity test to fulfill Koch's postulates may not be necessary unless a pathogen candidate is isolated from a *Cannabis* species for the first time. Because southern blight is a common disease, positive isolation and identification of *S. rolfsii* as well as characteristic field symptoms often warrant the diagnosis.

## Key diagnostic evidence

Infected plants show characteristic crown and basal stem rot associated with the visible growth of white mycelia and production of sclerotia typically 0.5–2 mm in diameter. DNA sequence data confirm the identity of *S. rolfsii*.

## Hemp Charcoal Rot Caused by *Macrophomina phaseolina*

*Macrophomina phaseolina* is a soilborne pathogen that infects nearly 500 cultivated and wild plant species worldwide (Khan *et al.*, 2017). It is a fungal species in the family Botryosphaeriaceae of Ascomycota and causes damping-off, seedling blight, collar rot, stem rot, charcoal rot, basal stem rot and root rot in economically important crops. The pathogen predominantly infects the stem and roots and thus impedes the normal transportation of water and nutrients throughout the plant. This results in the loss of growth vigour, plant wilting and premature death. The fungus produces abundant black microsclerotia beneath the epidermis of the lower stem and the taproot, giving them a charcoal-sprinkled appearance. The microsclerotia can survive in plant debris and soil for up to 15 years, depending on environmental conditions (Baird *et al.*, 2003; Khan *et al.*, 2017). Unlike southern blight and diseases caused by oomycetes, charcoal rot favours a higher temperature and low moisture conditions (Aegerter *et al.*, 2000; Khan *et al.*, 2017). In certain crops, drought stress may increase disease severity. On *Cannabis* crops, charcoal rot was diagnosed from several industrial and medicinal hemp varieties with disease incidences ranging from 2.7% to 37.1% in southern Spain (Casano *et al.*, 2018).

## Field diagnosis

### Symptoms

Infected plants develop leaf chlorosis or necrosis as an initial indication of the interruption of water and nutrient transportation due to fungal infection of the stem. The stem near the soil line may appear discoloured and small black sclerotia are often observed (Casano *et al.*, 2018). Roots may also be infected and exhibit necrotic vascular cambium tissue. As the disease progresses, the upper stem or stalk becomes soft and fluffy with more severe foliar symptoms. Once the fungus kills the lower stem, the upper stem tissue becomes desiccated and plants rapidly wilt and die.

### Problem classification

Similar to hemp southern blight, any foliar symptoms should be traced back to the stem or the root. A comprehensive examination of the whole plant is the best way to pinpoint the root cause and correctly classify the disease as a stem disease rather than a foliar disease. Leaf chlorosis is often a good indication that the problem originates from a stem or root infection and a combination of chlorosis and wilting warrants a holistic examination of the plant. In the case of lower-stem canker or crown rot diseases, the severity of above-ground symptoms is often proportional to the level of stem tissue disruption by the fungal infection. Inspect the stem thoroughly to reveal any abnormalities such as discoloration, soft rot, tissue disintegration, cracked or sunken areas, fuzzy growth, black dots, etc. Any abnormality found at the lower stem helps to explain foliar symptoms, especially when no sign of a foliar pathogen is present. The presence of a black colour at the lower stem of many symptomatic plants can quickly narrow down the problem to charcoal rot.

### Do your own diagnosis

Growers can follow the root-cause analysis process described above to classify the problem. Charcoal rot is an easily recognizable disease, as infected plants often show a grey to black discoloration on the taproot and lower stem and black and dusty microsclerotia are often present inside infected tissue. Use a knife to dissect the lower stem and taproot to reveal black and spherical microsclerotia. The shape and size of microsclerotia may vary significantly. Also examine the vascular tissues, as they may also turn a reddish to brown colour. In addition to symptom observation, recent environmental conditions may be assessed, because charcoal rot is also called dry-weather wilt. Chronic drought and prolonged hot weather favours disease development. Finally, the fungus has a wide host range, which suggests that previous crops such as sunflower, soybean, maize, or sorghum may have built a significant level of inoculum (microsclerotia) that unfortunately can survive several years in dry soil. With this information, charcoal rot can be easily diagnosed in the field.

## Sampling

Refer to the hemp southern blight section above for sampling instructions. If charcoal rot is suspected during field diagnosis, pieces of blackened stem with taproot can be submitted to a diagnostic laboratory.

## Lab diagnosis

### Visual and microscopic examinations

The procedures are similar to those for hemp southern blight disease. If a whole plant is submitted, visually examine the leaves, stem, crown and roots of both healthy (if submitted) and symptomatic plants to identify any abnormalities. Under a stereo microscope, examine the leaf, stem and root for lesions and fungal signs. In the case of charcoal rot, use a knife, scalpel, or razor blade to cut into the lower stem and taproot to observe the internal tissue rot, vascular discoloration and microsclerotia.

### Isolation and colony observation

Isolation of *M. phaseolina* must be performed to confirm field and microscopic diagnosis. Plate symptomatic stem tissues on potato dextrose agar or PDA amended with streptomycin (PDA+strep) and incubate the plates at 22–30°C in the dark. Be sure to cut small pieces from the junction area between the rotten and healthy tissue for plating, because that area is the front line of pathogen invasion. A fungal colony is expected to grow out from most tissue pieces plated if the stem is infected by *M. phaseolina*. Transfer each colony to a new PDA plate and continue incubating under the same conditions. *M. phaseolina* colonies are flat or slightly raised, a white to grey colour on both sides in early growth but turning black with age. Hyphae are septate, 2–4 μm wide, sub-hyaline to dark brown, while microsclerotia are black, spherical to oblong or irregular ranging from 90 μm to 180 μm (Abed-Ashtiani *et al.*, 2018). *M. phaseolina* isolated from hemp produces abundant, dark, ovoidal microsclerotia with diameters ranging from 89 μm to 141 μm at 30°C (Casano *et al.*, 2018). Unlike *S. rolfsii*, which produces sclerotia up to 4.2 mm in diameter, *M. phaseolina* produces only tiny sclerotia. Therefore, direct plating of microsclerotia on the PDA medium may not be practical.

### DNA-based identification

The accurate identification of isolates as *M. phaseolina* requires amplifying and sequencing rDNA and ITS regions. However, other genetic regions such as the *TEF-1α* region can be amplified using universal primers EF1-728F (CATCGAGAAGTT CGAGAAGG) and EF2 (GGA(G/A)GTACCAGT (G/C)ATCATGTT) (Qiao *et al.*, 2016). The calcium-modulated protein (Calmodulin) region can also be amplified using primers CAL-228F (GAGTTCAA GGAGGCCTTCTCCC) and CAL-737R (CATCTTT CTGGCCATCATGG) (Carbone and Kohn, 1999). Use the DNA sequence obtained from each genetic locus to BLAST search the NCBI GenBank database and identify a closely related species based on the DNA sequence similarity.

### Pathogenicity test

Charcoal rot is a common disease with diagnostic symptoms in plants. A positive isolation and identification of *M. phaseolina* warrants the diagnosis without a pathogenicity test.

### Key diagnostic evidence

Infected plants show characteristic crown and lower-stem rot associated with a charcoal-sprinkled appearance. Microsclerotia appear as tiny black dots underneath the epidermis of the lower stem and the taproot. Microsclerotia have a typical size of 90–180 μm in diameter. DNA sequence data confirm the identity of *M. phaseolina*.

## Hemp Stem Canker, Stem Rot and Crown Rot Caused by *Fusarium* Species

*Fusarium* is a genus of filamentous fungi that includes many species. Some of them survive in soil and plant debris as saprophytes and many cause diseases in a diverse range of crops. *Cannabis* plants are susceptible to several *Fusarium* species, especially *F. oxysporum* and *F. solani*. *Fusarium* can cause root rot, stem canker and rot (including crown rot), or vascular wilt in *Cannabis* species. However, it is not rare that infected plants have both root rot and stem canker as well as vascular wilt. In some cases, a crop can be co-infected by several *Fusarium* species, resulting in high plant mortality (Schoener *et al.*, 2017). Although *Fusarium* mainly infects hemp root and stem, it also causes flower blight. *Fusarium* species grow into white to pink colonies and produce microconidia, macroconidia and chlamydospores. Microconidia are usually common on culture medium, while macroconidia are often observed on infected plant tissue. Chlamydospores (thick-walled) are produced in response to nutrient depletion and they function as resting spores to resist unfavourable conditions. Depending on the distribution of *Fusarium* inoculum in the soil, affected plants may be clustered in a patch rather than randomly scattered in a field. Due to infected mother plants, crops grown from cuttings may exhibit scattered pattern.

### Field diagnosis

#### Symptoms

Initial infection on the lower stem may cause subtle symptoms such as minor leaf chlorosis and edge necrosis as a sign of the interruption of water and nutrient transportation. As the infection advances, foliar symptoms evolve from chlorosis to large areas of necrosis or blight. The top part of the infected plant may collapse in the late stages of infection (Fig. 8.4). The stem near the soil line may appear discoloured and turn brown (Fig. 8.5). In humid conditions, patches of white mycelia are visible on the rotted stem (Fig. 8.6). Because in most cases the infection starts at the stem near the soil line, severe rot and active fungal growth are often observed at the crown area. If the plant dies before the rot progresses upwards to the middle portion of the stem, the disease is classified as crown rot. However, a *Fusarium* infection can cause stem canker. Canker is a less severe condition than rot, as the infection mostly stays on the surface layers of the stem (Fig. 8.7). When the infection advances into internal tissue, the stem becomes completely rotten (Fig. 8.8). In the case of *Fusarium* stem canker or rot diseases, infected plants may not have root rot (Fig. 8.6).

#### Problem classification

As foliar symptoms are often noticed first, an initial examination of leaves should be performed to rule out leaf diseases. Use the same strategies described above in the southern blight and charcoal rot sections to classify a problem into a stem disease. Examine the stem of multiple symptomatic plants

**Fig. 8.4.** Foliar collapse of a hemp plant co-infected by *Fusarium oxysporum* and *F. solani*.

to reveal canker or white fungal mycelia, especially in the area close to the soil line. The presence of a white colour at the lower stem of many symptomatic plants can quickly narrow down the problem to *Fusarium* stem rot. Roots should be checked for any discoloration or decay to determine if the disease is restricted to the stem only.

### Do your own diagnosis

*Fusarium* stem rot can be recognized when the basal stem exhibits fluffy white mycelial growth. Use a knife to cut through the lower stem to reveal the extent of internal rot. Often, the canker or lesion is restricted to the outer layers of the stem and the plant may still survive. Extensive stem rot (Fig. 8.8) usually kills the plant. Continue walking through the field to determine whether foliar symptoms are mostly associated with stem canker. Such a link, if found, explains that foliar symptoms are a result of the stem disease. Still, a field observation alone is not sufficient to definitively diagnose a disease as *Fusarium*-caused stem rot. A lab diagnosis based on submitted samples is often needed.

**Fig. 8.5.** Discoloration of lower stem due to infections by *Fusarium* spp.

**Fig. 8.6.** White patches of *Fusarium* mycelia grown on the surface of infected stem.

### Sampling

Refer to the hemp southern blight section above for sampling instructions. If stem rot is suspected during field diagnosis, pieces of discoloured or decayed stem with the root system should be submitted to a diagnostic laboratory.

### Lab diagnosis

#### Visual examination and moist chamber induction of pathogen

The procedures are similar to those described above. Use a knife to dissect the stem to assess the extent of the rot. Look closely or under a stereo microscope to detect any pathogen signs such as fungal mycelia and spores. If pathogen signs are absent, the stem pieces may be incubated in a moist container to induce the growth of the suspected pathogen (see Chapter 5). This procedure is only used in a laboratory that lacks culture capability. Growers can use this method to verify if the plant is infected by a fungus.

#### Microscopic examinations

Under a stereo microscope, observe the fluffy white mycelia on the stem in greater detail if present (Fig. 8.9). Often, only the primary pathogen grows from decayed tissue, but secondary microorganisms may also be present, especially when the tissue is completely decayed. Use a clear tape to blot mycelia and spores from the stem tissue and place the tape with blotting side down on a drop of water on a glass slide. Under a compound microscope, observe both macroconidia and microconidia if present. Macroconidia are septate (three to five) and slightly curved to crescent or sickle-shaped (Fig. 8.10). Sometimes the macroconidia are referred to as boat-shaped spores. The sizes of macroconidia vary significantly, from 27–46 × 3–5 µm for three-septate macroconidia to 35–60 × 3–5 µm for five-septate macroconidia. Three-septate spores are the most common. Based on the size and shape, to some extent, macroconidia can be diagnostic for species differentiation. Microconidia are much smaller, non-septate, straight to curved, and about 5–12 × 2.2–3.5 µm in size. Depending on the species, spore

**Fig. 8.7.** Canker lesions at the middle part of stem infected by *Fusarium* species.

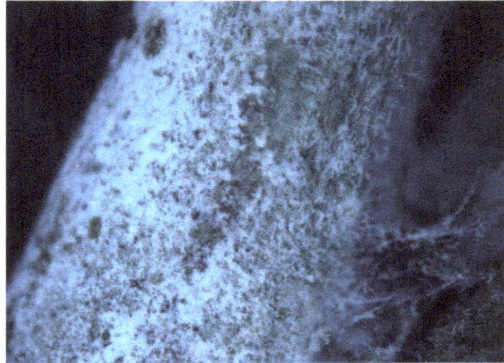

**Fig. 8.9.** White *Fusarium* mycelia observed on the surface of a hemp basal stem under a stereo microscope.

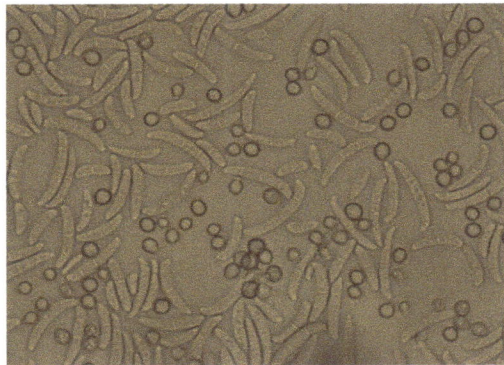

**Fig. 8.10.** Macroconidia (crescent shape) of *Fusarium oxysporum* produced on infected hemp stem tissue (observed under a compound microscope).

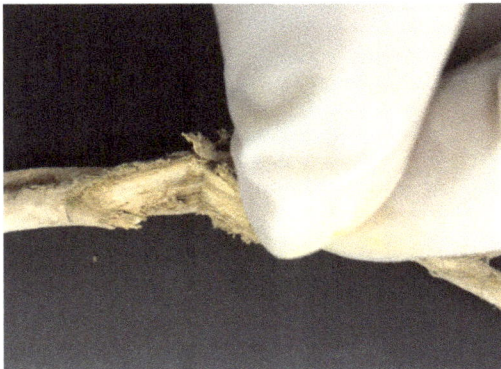

**Fig. 8.8.** Internal stem rot due to a *Fusarium* infection.

measurements vary significantly (Trabelsi *et al.*, 2017). Hyphae are observed but seldom used in species identification.

### *Isolation and colony observation*

If *Fusarium* pathogens are not directly observed from symptomatic stem tissue, isolation of a causative

agent must be performed. Plate symptomatic stem tissues on potato dextrose agar (PDA) or PDA amended with streptomycin (PDA+strep) and incubate the plates at 22°C in the dark. A white colony is expected to grow out from most tissue pieces plated if the stem is infected by *Fusarium* pathogens, e.g. *F. oxysporum* (Fig. 8.11). Transfer each representative colony to a new PDA plate and continue incubating under the same conditions to grow a full colony for morphological observation (Fig. 8.12). *F. oxysporum* colonies are cottony white but produce a dark violet to dark magenta pigment in the medium which can be observed from the underside of plates. *F. solani* has an off-white or cream colour with less pigment. Make a

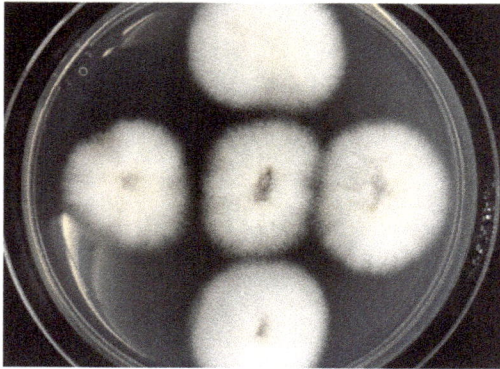

**Fig. 8.11.** *Fusarium oxysporum* colonies grown out from hemp stem tissue on a PDA[+strep] plate after 4 days of incubation at 22°C in the dark.

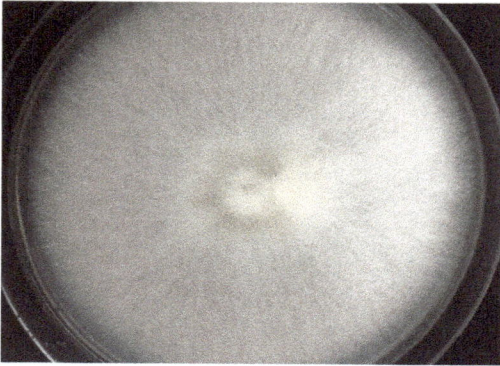

**Fig. 8.13.** Oval or ellipsoid microconidia produced on a PDA plate by *Fusarium oxysporum* isolated from hemp stem tissue. Note that some microconidia have one septum.

ITS1 and ITS4 primers followed by DNA sequencing. The expected PCR product sizes vary slightly among *Fusarium* species, but are typically 544 bp for *F. oxysporum*, 566 bp for *F. solani*, 559 bp for *F. redolens*, 563 bp for *F. tricinctum* and 546 bp for *F. equiseti* (Schoener *et al.*, 2017). Other genetic loci can be based to identify the species if necessary.

### Key diagnostic evidence

Infected plants show characteristic crown and lower-stem rot associated with fluffy white mycelia. *Fusarium* spores are observed from infected tissue and the culture of infected stem tissue results in the growth of *Fusarium* species, most likely *F. oxysporum* and *F. solani*. DNA sequence data concur with the morphological identification of *Fusarium* species.

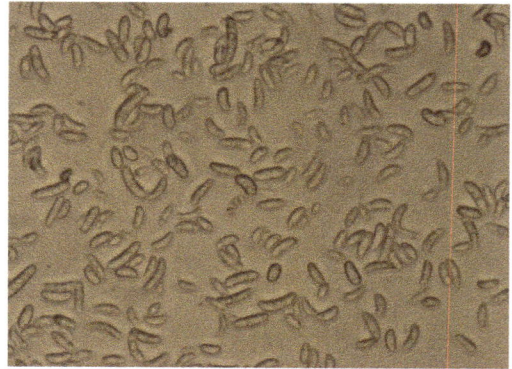

**Fig. 8.12.** A fully grown colony of *Fusarium oxysporum* on a PDA plate.

temporary slide to observe the spores under a compound microscope. For *F. oxysporum* and *F. solani*, microconidia are borne on simple phialides and are produced abundantly on a PDA plate (Fig. 8.13). While these two species are easily identified from colony and spore morphology, there are other *Fusarium* species that are often associated with diseased hemp tissue but do not produce diagnostic spores on a standard PDA medium. In this case, DNA-based identification can be performed.

### DNA-based identification

An accurate identification of all *Fusarium* isolates can be done with DNA extraction and PCR using

### Hemp Stem Canker and Crown Rot Caused by *Sclerotinia* Species

Species of *Sclerotinia*, such as *S. sclerotiorum*, *S. minor* and *S. trifoliorum*, cause the disease often known as white mould on many important crops, including legumes, sunflowers, canola and tobacco (Link and Johnson, 2007). *S. sclerotiorum* was found to cause hemp stem canker in Canada (Bains *et al.*, 2000) and recently *S. minor* was reported to cause hemp crown rot in California (Koike *et al.*, 2019). In non-*Cannabis* crops, *Sclerotinia* species can infect the upper portions of plants, causing flower or head blight and fruit rot, but more often cause crown and stem rot. However, in the case of hemp stem canker reported from Alberta, Canada (Bains *et al.*, 2000), *S. sclerotiorum* infects almost the entire stem and causes lesions up to the stem near the inflorescence. If environmental conditions are favourable, there is potential for the infection to advance to the flower heads and cause bud rot. Hemp stem and crown rot caused by *Sclerotinia* species is a monocyclic soilborne disease and favours cool and moist conditions. The primary source of infection comes from the sclerotia produced in the previous crop that have survived in the soil. When a hemp crop is planted in spring, sclerotia germinate and then infect stems near the soil line.

### Field diagnosis

#### Symptoms

The initial infection is observed as water-soaked grey lesions with distinct margins on the lower stem. The lesions expand, girdle the stem, or advance into the internal stem tissue, causing critical injury to the plant. The stem may shred and break at the lesion. However, these symptoms are often overlooked until secondary symptoms start to show up on foliage, such as yellowing, wilting and necrosis. As the infection progresses, stem tissue becomes soft with white mycelia growing from both inside and outside the tissue. Lesions can develop all the way to the stem near the inflorescence. During later stages of the disease, the mycelium is packed tightly to form a pea-sized structure that eventually develops into a hardened black sclerotium. Sclerotia are visible on the outer surface of the diseased tissue but may form inside the pith of stems. In a hemp field infected by *S. sclerotiorum*, black sclerotia were observed at the lesion surface and inside the pith cavity. The shapes of the sclerotia were round, irregular, or oblong with sizes up to 5 mm in diameter and 2–11 mm in length (Bains *et al.*, 2000). In this case, disease incidence ranged from 1% to 8%.

Hemp crown rot caused by *S. minor* shares similar symptoms with hemp stem canker. The basal stem or crown area becomes necrotic and covered with white to grey mycelia with production of small (0.5–3 mm diameter), irregular and black sclerotia. Leaves develop secondary symptoms such as abnormal colour, wilting, or drying. Infected plants may eventually collapse (Koike *et al.*, 2019). In this case, however, hemp roots and the rest of the stem appeared to be asymptomatic. Infected plants were limited to 1%.

### Problem classification

The *Sclerotinia* pathogen can attack aerial parts of the plant but mainly infects stem tissue. A complete examination of the plant for symptoms from root, crown, stem and leaf to flower bud is needed to identify which part of the plant has the active infection. Because *Sclerotinia* produces characteristic white cottony mycelia and black sclerotia on the surfaces of infected stem tissues, it is not difficult to pinpoint the problem to be a stem disease. Like many other stem diseases, foliar symptoms are often present and may show up first even before stem lesions are visible. However, foliar symptoms are secondary and should not be considered the origin of the problem. A sample of leaflets without symptomatic stem pieces will misrepresent the problem and is often misdiagnosed as nutritional or other causes.

### Do your own diagnosis

*Sclerotinia* stem canker or crown rot can be easily recognized when the infected stem exhibits the characteristic white cottony mycelia and black irregular sclerotia at the crown area. However, these diagnostic symptoms may not show up during the early stages of infection. Therefore, if the foliage symptoms are first noticed, examine the stem of a symptomatic plant and compare it with the stem of a healthy-looking plant. The surface of a healthy stem should be green and fresh white at the crown (Fig. 8.14), while the infected stem may be girdled by lesions (Fig. 8.15). Use a knife to slice through the stem to reveal the extent of lesions and internal rot. A healthy stem exhibits intact external and internal structures of epidermis, cortex, vascular bundles and pith (Fig. 8.16), while an infected stem may appear discoloured, soft, or necrotic. To better diagnose stem canker disease in the field, a randomly selected 10–20 plants showing mild to severe foliar symptoms should be examined. Characteristic signs of the pathogen such as mycelia and sclerotia are often found on some severely infected plants.

### Sampling

Refer to the hemp southern blight section above for sampling instructions. If stem canker is suspected

**Fig. 8.14.** A healthy-looking hemp stem and crown. Note the fresh white colour underneath the outer layers (peeled off) of crown.

**Fig. 8.16.** A hemp stem is sliced to expose healthy internal tissue. Note that the structure and texture of the stem are intact.

**Fig. 8.15.** A stem of young hemp plant girdled by canker due to infection by *Sclerotinia* sp. Note that the stem had sunken dry rot and the canker extended all the way up to the top portion of the stem.

during the field diagnosis, several pieces of decayed stems with mycelia or sclerotia (if present) should be submitted to a diagnostic laboratory.

## Lab diagnosis

### Visual and microscopic examinations

The procedures are similar to the *Fusarium* stem canker diseases described above. Induction of *Sclerotinia* pathogen growth in a moist chamber (see Chapter 5) may be performed by a laboratory that lacks culture capability. Growers also can use this method to verify if the stem is infected by a *Sclerotinia* species. Visual examination is used to determine the presence of lesions or canker on the stem and microscopic examination is used to determine if *Sclerotinia* mycelia and sclerotia are present. In general, *S. sclerotiorum* produces larger sclerotia ranging from 2 mm to 10 mm in diameter, while *S. minor* produces smaller sclerotia ranging from 0.5 mm to 2 mm in diameter. *S. sclerotiorum* sclerotia are generally smooth, round and less abundant, while *S. minor* sclerotia tend to be rough, angular and more abundant. Use a clear tape to blot mycelia from a stem tissue and place it with the blotting side down on a drop of water on a glass slide. Under a compound microscope, observe the mycelia and determine if their morphology is consistent with that of *Sclerotinia*. Most likely, only hyphae are observed without any spores. However, some other fungi, either saprophytic or a weak pathogen (to co-infect), may grow and produce spores. In this case, isolation of the primary pathogen from diseased tissue should be performed.

### Isolation and colony observation

If *Sclerotinia* pathogens are not directly observed from symptomatic stem tissue, isolation of a causative agent must be performed. Plate symptomatic stem tissues on PDA, or PDA amended with streptomycin (PDA+strep) and incubate the plates at 22°C in the dark. Symptomatic stem pieces can also be plated on acidified (add 2 ml of 25% lactic acid to 1 l of medium) cornmeal agar (CMA) after surface sterilizing in 0.006% NaOCl for 2 min (Koike et al., 2019). Isolates emerged from tissue pieces can be transferred to regular PDA plates. Alternatively, individual sclerotia can be picked from infected stem tissue, surface sterilized, rinsed and plated on acidified PDA. *S. sclerotiorum* produces white colonies with aerial mycelia and large numbers of sclerotia on PDA (Bains et al., 2000). Observe and measure the size of sclerotia to determine if they are consistent with characteristics of *S. sclerotiorum*. *S. minor*, if plated on acidified CMA, grows rapidly from tissue pieces and appears colourless and appressed. After being transferred to PDA, *S. minor* forms white to grey colonies that produce abundant small black sclerotia. Note that the sizes of sclerotia of *S. sclerotiorum* and *S. minor* differ significantly, so they can be used to differentiate these two species.

### DNA-based identification

*S. sclerotiorum* and *S. minor* do not produce diagnostic spores on standard PDA medium. The size and shape of sclerotia can be used as a reference but accurate identification requires DNA sequence data. Conventional PCR using ITS1 and ITS4 primers followed by direct sequencing can quickly and correctly identify the isolates to species level. For *S. minor*, the expected PCR product size is 540 bp (Koike et al., 2019). Other species may have slightly different PCR amplicon size but should be close to 540–550 bp.

### Key diagnostic evidence

Infected plants show characteristic crown and lower-stem rot associated with fluffy white mycelia as well as sclerotia. *Sclerotinia* species, most likely *S. sclerotiorum* or *S. minor*, are consistently isolated from the infected stem tissue. DNA sequence data concur with morphological identification of *Sclerotinia* species.

## Hemp Collar Rot Caused by *Plectosphaerella cucumerina*

*Plectosphaerella cucumerina* is known as a ubiquitous and polyphagous fungus associated with several different plant hosts (Pascoe et al. 1984; Palm et al. 1995). It causes root and collar rots on horticultural crops in southern Italy (Carlucci et al., 2012), cabbage (*Brassica oleracea*) root rot in China (Li et al., 2017) and wild rocket (*Diplotaxis tenuifolia*) leaf spots (Gilardi et al., 2012). *P. cucumerina* can infect hemp plants, causing collar rot (see Chapter 5).

### Field diagnosis

The symptoms of hemp collar rot caused by *P. cucumerina* are described in Chapter 5, where this disease is used as an example for illustrating a thorough and systemic examination as an important diagnostic process. Growers can follow the general guidelines presented in that section to perform a preliminary field diagnosis and collect relevant samples for a lab analysis.

### Lab diagnosis

To further isolate and identify this pathogen from hemp samples, plate several pieces (2–3 mm) of infected plant tissue on PDA+strep plates and incubate them at 22°C for 5–7 days. Colonies on PDA are buff to salmon pink and slimy. Usually there is little aerial mycelium growth. Mycelia are hyaline, septate and branched and occasionally form hyphal coils. Conidia are hyaline, smooth, thin-walled, ovoid, 0–1 septum, and $3.8$–$9.1 \times 2.5$–$4.5$ μm (Li et al., 2017). Further identification to the species level can be done by amplifying ITS-5.8S rDNA region with ITS1/ITS4 primers. Additionally, D1/D2 domain of the 28S rRNA gene can be amplified with NL-1 (5′-GCA TAT CAA TAA GCG GAG GAA AAG-3′) and NL-4 (5′-GGT CCG TGT TTC AAG ACG G-3′) primers. PCR amplicons can be then purified and directly sequenced. Sequence data from both loci can be used as query sequences to BLASTn search the NCBI GenBank to find closely related species. The amplified sequence from ITS-5.8S rDNA region is expected to be around 562 bp, while the sequence from D1/D2 domain is expected to be 608 bp (Li et al., 2017).

## Hemp Crown Rot Caused by *Pythium* Species

The *Pythium* genus along with its closely related genus *Phytophthora* are major pathogens of many important crops, including hemp. *Phytophthora* has not been reported from *Cannabis* species as of yet, but *Pythium* species were frequently diagnosed from hemp crops exhibiting root or crown rot in Nevada (Schoener *et al.*, 2018) and Indiana (Beckerman *et al.*, 2017, 2018). Both *Pythium* and *Phytophthora* have been considered 'aquatic fungi', but they are oomycetes rather than true fungi. However, they produce fungal-like hyphae and behave similarly to fungi when causing diseases. A significant characteristic is that they produce motile zoospores that survive and 'swim' in the water and infect plant roots and crown tissue. In the field, *Pythium* movement in the soil is limited and the disease is often restricted to the area where the pathogen is present. In a *Pythium*-contaminated hydroponic or aeroponic production system, however, the zoospores can move freely in the medium or nutrient solution and infect exposed roots directly, causing widespread root rot of cannabis plants (see Chapter 9). There are approximately 200 *Pythium* species described and most of them are pathogenic to plants. In *Cannabis* crops, the most prevalent species are *Pythium aphanidermatum* (Beckerman *et al.*, 2017; Schoener *et al.*, 2018), *P. myriotylum* (McGehee *et al.*, 2019) and *P. ultimum* (Beckerman *et al.*, 2018). Many *Pythium* species infect a broad range of plant species; therefore, rotation with other crops may not be effective in reducing the pathogen inoculum from the soil.

### Field diagnosis

#### Symptoms

*P. aphanidermatum* crown rot may cause significant damage to hemp crops. In one hemp field, the disease affected approximately 5–10% of plants (Schoener *et al.*, 2018). The initial sign of *P. aphanidermatum* infection may be mild leaf yellowing, curling and necrosis along the edge (Fig. 8.17). Plants may wilt temporarily when temperatures are warmer around noon. Examining the crown area of the stem may reveal necrotic lesions (Fig. 8.18) and internal discoloration made visible by cross-sectional slicing (Fig. 8.19). As the disease progresses, lesions or cankers will expand at the basal stem (Fig. 8.20) and infection may spread from the crown area upwards to the main stem and lower branches (Fig. 8.21), eventually killing the plant (Fig. 8. 22). Under moist conditions, white to grey moulds (*Pythium* mycelia) grow on the surface of the crown area (Fig. 8.23). In this case, some affected plants had a

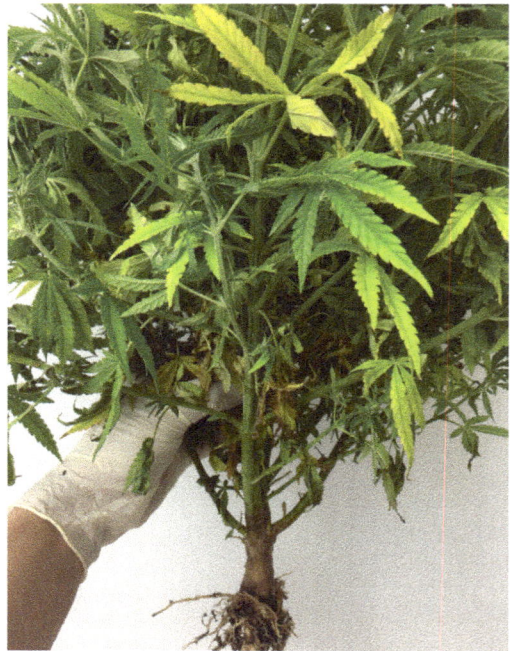

**Fig. 8.17.** Mild leaf yellowing and chlorosis on a plant infected by *Pythium aphanidermatum*. Note that the active infection was at the basal stem.

healthy root system and the infection was restricted to the crown tissue (Fig. 8.24). *P. aphanidermatum* can infect hemp seedlings and cause noticeable symptoms as soon as 13 days after sowing at growing temperatures of 25–30°C. Infected seedlings may show brown, water-soaked stem lesions with visible aerial mycelia grown on the stem surface, but the pathogen attacks the root system more severely than younger stem. Infected roots often develop brown lesions and lose feeder roots, which secondarily causes chlorotic leaves, stunted growth, wilt, or seedling collapse (Beckerman *et al.*, 2017). Similar to *P. aphanidermatum*, plants infected by *P. ultimum* exhibit stunted growth, chlorotic leaves, and crown and root rot with water-soaked brown lesions (Beckerman *et al.*, 2018).

#### Problem classification

A common mistake when diagnosing a stem disease is misclassifying the problem as a foliar disease, since infected plants almost always show noticeable secondary symptoms during the disease development.

**Fig. 8.18.** A *P. aphanidermatum*-infected hemp plant showing brown lesions at the basal stem and their progression into the middle portion of the stem. Note that the brown to dark, water-soaked lesions near the green tissue are the area that *P. aphanidermatum* actively infects and moves into the healthy tissue.

**Fig. 8.19.** Internal discoloration and rot at the basal stem of a hemp plant due to a *Pythium aphanidermatum* infection.

Such a mistake can be avoided by a complete examination of plants for symptoms from the root, crown, stem, leaf to the flower bud. For *Pythium*

**Fig. 8.20.** Crown and all basal stems of a hemp plant are killed by a *Pythium aphanidermatum* infection.

crown rot, it is not difficult to identify the active infection site from the plant, as the pathogen causes easily recognizable crown or lower-stem rot and sometimes overgrows its mycelia from rotted stem tissue. Foliar symptoms caused by crown rot are generally uniform, for example all leaves turning yellow or wilting (Fig. 8.22). However, some plants may show foliar symptoms on certain portions of the plant, for example one stalk has base rot while the rest of the stalks have a mild infection. In this case, only the branch with the basal stem killed by the pathogen develops severe foliar symptoms (Fig. 8.25). A clear linkage between stem rot and foliar symptoms can quickly classify a problem into a crown or basal stem rot disease.

### Do your own diagnosis

For disease scouting, look for water-soaked lesions and cankers around the basal stem when a plant starts to show foliar symptoms. The canker around the stem at the soil line is the key diagnostic symptom for *Pythium* crown rot. During the late season, greyish to white-coloured mould (*Pythium* mycelia) may be visible on the surface of the crown

**Fig. 8.21.** Brown to dark lesions on stalks (lateral stems) caused by *Pythium aphanidermatum* as the infection moved up from the crown area.

**Fig. 8.22.** A hemp plant killed by *Pythium aphanidermatum* crown rot. Note that adjacent plants appeared healthy.

area. Growers may preliminarily diagnose a stem disease as *Pythium* crown rot in the field based on these symptoms. However, there are other types of lower-stem rot diseases caused by other pathogens, and their symptoms and mould growth may look similar to those caused by *Pythium* crown rot. Therefore, a sample of the infected plant or its stem pieces should be collected and submitted for laboratory analysis.

Alternatively, growers can perform a handy and quick immunostrip test in the field (see Chapter 5). A *P. aphanidermatum* crown rot can be detected by Agdia's *Phytophthora* immunostrip as it cross-reacts with *P. aphanidermatum*. By grinding a small piece of rotted stem tissue in a provided plastic pouch (buffer solution inside) and then inserting the immunostrip into the solution, a positive or negative result can be read in minutes (Fig. 8.26). However, this immunostrip is designed to detect most *Phytophthora* species and cross-reacts with some of *Pythium* species. A positive detection only indicates the presence of an oomycete species, not necessarily *P. aphanidermatum*. Further lab diagnosis is still required.

### Sampling

Sampling procedures for *Pythium* crown rot are similar to those described for other stem diseases. Because *Pythium* also infects roots, a root sample should be taken along with stem samples.

### Lab diagnosis

#### Visual and microscopic examinations

The procedures are similar to those of the *Fusarium* stem canker diseases described above. Induction of *Pythium* pathogen growth in a moist chamber (see Chapter 5) may be performed by a laboratory that lacks culture capability. Visually examine the hemp sample to determine the location and extent of lesions or rot on the stem and look under a microscope for *Pythium* mycelia and sporangia that may be present.

#### Isolation and colony observation

*Pythium* species are easily isolated from the infected plant tissue by using either standard PDA plates or selective media for oomycetes such as PARP. PARP

**Fig. 8.23.** *Pythium aphanidermatum* mycelia actively grew on the surface of a rotted stem near healthy tissue. Note that mycelia were fuzzy and changed colour from off-white (newly grown) to grey (old) and that roots were not healthy.

**Fig. 8.24.** The root system of a hemp plant remained healthy despite extensive crown rot caused by *Pythium aphanidermatum*.

is a cornmeal agar (CMA) amended with pimaricin, ampicillin, rifampicin and pentachloronitrobenzene to inhibit fungal and bacterial growth (see Chapter 5). For plant samples exhibiting crown or lower stem rot, it is wise to use both PDA and PARP. This is especially important when the cause of the disease is unknown. PDA as a general-purpose medium can isolate most fungal pathogens, while PARP can selectively isolate oomycetes such as *Pythium* and *Phytophthora*. Although *Pythium* can grow on PDA, it generally outgrows other pathogens such as *Fusarium oxysporum*. In the case of a complex infection by both *Fusarium* and *Pythium* (see Chapter 13), the use of PDA alone may end up with the isolation of *Pythium* only, instead of both, and such an isolation result may lead to an inaccurate diagnosis. Isolates on PARP plates can be further transferred to fresh CMA or V8 juice agar plates for morphological observations. Isolation and subsequent maintenance of isolates should be at 22°C in the dark. Certain *Pythium* species such as

*Pythium aphanidermatum* do not grow on PDA medium amended with streptomycin (PDA$^{+strep}$), therefore medium amendment with streptomycin should be avoided in *Pythium* isolation. A *Pythium* colony is generally recognized by its fast and radiating growth on CMA and cottony aerial mycelium on PDA. For example, *P. aphanidermatum* can grow over a full plate (100 mm in diameter) on PDA medium within 24 h (Fig. 8.27). It produces oogonia, antheridia and sporangia on CMA medium (Fig. 8.28). *P. ultimum* produces spherical sporangia either terminally or intercalarily.

### DNA-based identification

There are many *Pythium* species that cause diseases or are associated with diseased tissue. It is likely that more than one *Pythium* species will be isolated from infected plants. Accurate identification of every encountered species can be achieved by obtaining the DNA sequence data from the ITS region of rDNA. Using the primer pair ITS1 and ITS4, most *Pythium* species yield a PCR product

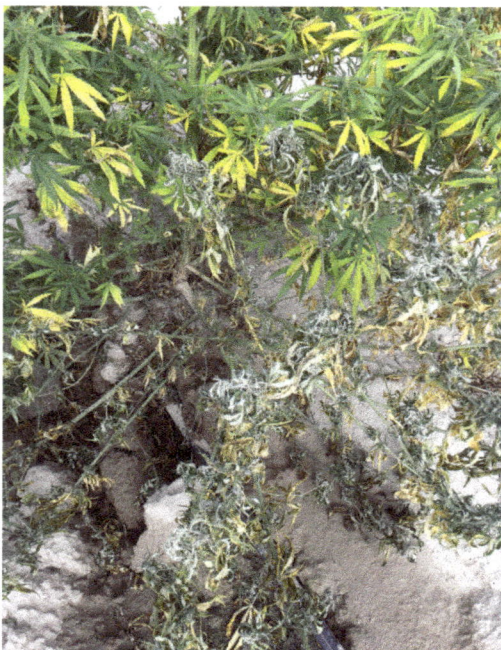

**Fig. 8.25.** One lateral branch was completely wilted due to the basal stem rot caused by *Pythium aphanidermatum*. Note that the rest of the plant only showed mild and sporadic leaf yellowing, because the main stem and lateral stems were not fully infected.

approximately 300 bp larger than that from fungal species (see Fig. 5.23). For *P. aphanidermatum*, the expected PCR product size is 866 bp (Schoener *et al.*, 2018). Note that some of the ITS1/ITS4-generated DNA sequences deposited in NCBI only represent a portion of PCR amplicon sequences, with 5' and 3' sequences missing. This is because both the 5'- and 3'- ends of the PCR amplicons are not correctly read by Sanger sequencing and therefore trimmed. To get a full PCR amplicon sequence for maximum alignment with existing DNA sequences in the GenBank, a PCR product (over 800 bp) can be cloned into the T-vector and then sequenced using primers T7 promoter and SP6 upstream (see explanation in Chapter 5, Cloning of PCR Product).

### Key diagnostic evidence

The crown, basal stem, or lower portion of the branched stalk are discoloured, have water-soaked lesions, or are completely rotten. The basal stem may be covered with white to grey fluffy mycelia. *Pythium* is consistently isolated from diseased stem tissue on PARP plates. ITS1/ITS4 primers generate a PCR product around 866 bp and the DNA sequence has over 99% similarity to that of an expected *Pythium* species based on morphological characteristics.

## Hemp Crown Soft Rot Caused by *Enterobacter* Species

*Enterobacter* is a genus in the family Enterobacteriaceae. This family contains about 41 genera, nine of which contain plant-pathogenic species (Kado, 2010). The pathogenic genera in Enterobacteriaceae commonly cause water-soaked lesions, soft rot and stem rot and may invade xylem vessels, causing systemic wilt. The *Enterobacter* species are generally considered saprophytes associated with soil and water, but some have emerged as plant pathogens. For example, *Enterobacter cloacae* causes internal decay of onions (Bishop and Davis, 1990; Schroeder *et al.*, 2009) and leaf spot of chilli pepper seedlings (García-González *et al.*, 2018). On non-*Cannabis* crops, bacterial soft rots are common diseases that affect the lower stem, crown, or underground plant organs such as the bulbs, tubers and roots, especially when the soil is saturated or overly irrigated. On hemp, one bacterium identified as *Enterobacter* sp. has been found to be associated with hemp crown soft rot (as described below) and approximately 1% of plants in the crop were affected. This case suggests that a hemp crop may develop soft rot if a pathogenic bacterial species exists in the soil.

### Field diagnosis

#### Symptoms

Infected hemp plants exhibit water-soaked lesions at the crown area. As the infection progresses, outer layers of stem tissues disintegrate and decay (Fig. 8.29). Internal tissue turns brown and wet and becomes rotten with an unpleasant odour (Fig. 8.30). Thick cream-like and off-white to pale yellow bacterial fluid may be present on rotted tissue (Fig. 8.31). Affected plants show secondary foliar symptoms ranging from mild leaf yellowing and necrosis to systemic wilt or collapse.

#### Problem classification

Hemp crown rot can be caused by several fungal or oomycete pathogens. Perform an initial assessment of

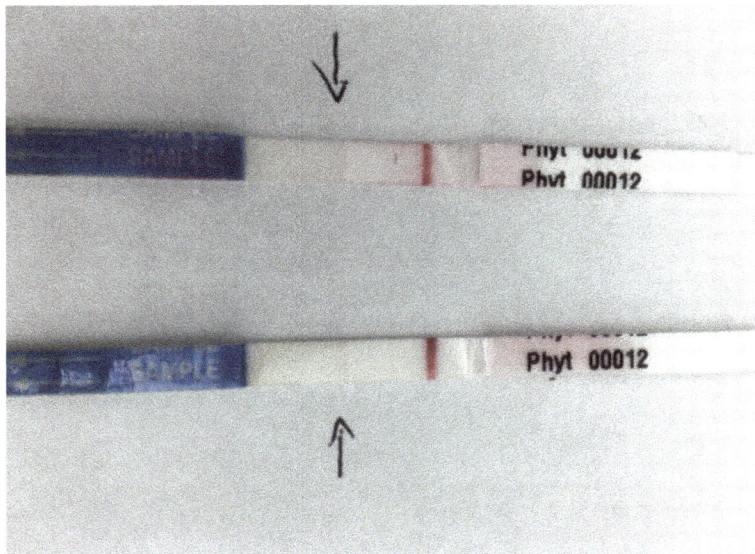

**Fig. 8.26.** Two positive detections of *Pythium aphanidermatum* from a hemp crown rot sample. Note that the left-hand pink/purple band (arrow) is a test line to determine if a targeted pathogen is present (line visible) or absent (line invisible), and the right-hand line is a control that must be present to validate the test.

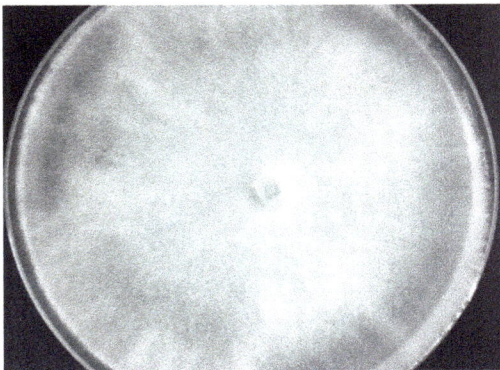

**Fig. 8.27.** A colony of *Pythium aphanidermatum* grown on PDA at 22°C in the dark for 24 h.

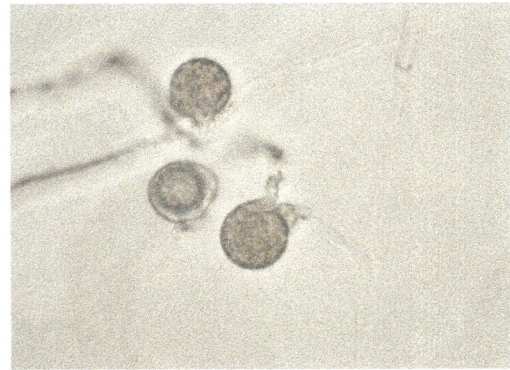

**Fig. 8.28.** Sporangia and oogonium (middle with thick wall) produced by *Pythium aphanidermatum* on CMA.

representative symptomatic plants to narrow down the problem to a stem/crown disease. Unlike fungal crown rots, where the canker or rot appears firm and dry, bacterial crown rots have a characteristic bacterial fluid that builds up in or on rotted tissue. Infected tissues are often soft and disintegrated and have a strange odour. Bacterial soft rots are often associated with excessive water in the soil. Evaluating current weather and soil conditions can help determine if the

crown rot is worsening due to recent wet weather. To summarize, a crown rot can be easily classified as bacterial soft rot if the lower stem tissue appears wet and soft with a dense bacterial fluid and odour.

### *Do your own diagnosis*

The characteristic symptoms described above can be used to preliminarily diagnose the disease as

**Fig. 8.29.** Disintegration and decay of a hemp basal stem due to the infection by an *Enterobacter* species.

**Fig. 8.30.** Internal crown tissue of hemp exhibiting a water-soaked appearance and extensive rot.

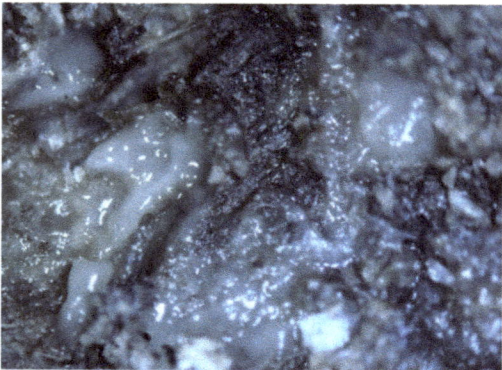

**Fig. 8.31.** Creamy, off-white, dense bacterial fluid on rotted hemp crown tissue (observed under a stereo microscope).

bacterial soft rot. There are other ways to detect signs of bacterial infection on infected tissue. For example, use a hand lens to look for a dense and cream-like fluid. The presence of fluid (containing massive bacterial cells) is a sign of bacterial infection, similar to the fungal mould (mycelium) as a sign of fungal infection. Positive detection of these signs greatly aids diagnosis and usually can classify a disease into an appropriate category. If bacterial fluid is not present, the rotted stem can be partially cut vertically at the lower end and dipped into a clear glass of water. Hold the stem pieces for a couple of minutes. A cloud of bacterial cells may move slowly down to the water, visible from the side of glass. This is a simple bacterial streaming test doable in the field or a non-laboratory setting. These procedures help to diagnose the problem as a bacterial disease, but further identification of the bacterial species requires lab testing.

### Sampling

Take three to five whole symptomatic plants with the root system intact and submit them to a laboratory for diagnosis. Each plant should be individually packed in a large plastic bag. A whole-plant sample is more informative for diagnosis than a tissue sample and, in some cases, submitting a whole diseased plant can lead to the detection of other diseases not noticed in the field. If submission of whole-plant samples is not feasible, representative stem pieces including the crown and root can be sampled from symptomatic plants and submitted.

### Lab diagnosis

#### Visual and microscopic examinations

Visually examine the submitted plants or parts systemically to determine the primary site of infection and rule out other diseases associated with the leaves or the upper portion of the plant. For bacterial stem soft rot, look at the base of stem for any lesions and liquid exuding. Use a knife to slice the stem tissue and observe the internal rot. Under a stereo microscope, observe the bacterial fluid in more detail. In certain cases, a high population of bacteria-feeding nematodes may be present inside the sticky bacterial liquid (Fig. 8.32). These nematodes consume bacteria and their active feeding suggests excessive growth of bacteria and accelerated tissue rot. Because of this, it is not rare to see

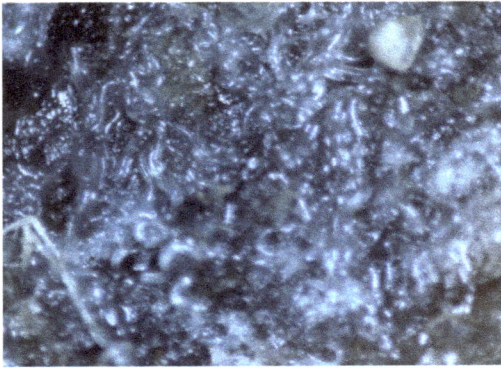

**Fig. 8.32.** A crowd of bacteria-feeding nematodes consuming bacteria cells on the surface of *Enterobacter*-infected hemp crown tissue. Note the creamy-white and worm-like nematodes.

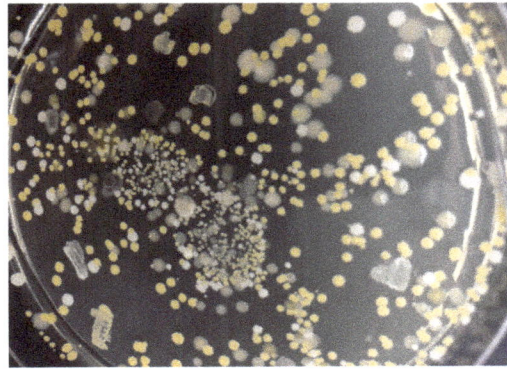

**Fig. 8.33.** Three different colonies isolated from diseased hemp crown tissue. Note that creamy and opaque colonies are *Enterobacter* species, while white and yellow colonies are two different species in the genus *Staphylococcus*, according to their 16S rDNA sequence data.

bacteria-feeding nematodes associated with the crown or underground plant organs when they are severely infected and decayed by bacteria. This is similar to the high density of fungus-feeding nematodes usually found in roots and rhizosphere when they are infected by fungal or oomycete pathogens (see Chapter 13). Under a compound microscope, bacterial streaming can be clearly observed from the tissue cut at the junction area between rotten and healthy crown tissue (see Fig. 3.16). The presence of bacterial fluid and a positive observation of bacterial streaming can tentatively diagnose the problem as a bacterial disease.

### Isolation and colony observation

It is diagnostically wise to perform standard plating for fungal pathogens (on PDA) and oomycete (PARP) for all crown rot cases. When fungal or oomycete pathogens are not recovered from diseased tissue, bacterial isolation can be performed. Cut a small piece of internal crown tissue at the junction between the rotten and healthy areas and proceed to the serial dilution method for bacterial isolation (see Chapter 5, Bacterial Isolation). Diluted bacterial preparation can be spread on the surface of LB or nutrient agar plates by swirling three sterilized beads. One or more types of bacterial colonies may grow on the plates (Fig. 8.33). An individual colony can be further purified by streaking on LB agar to obtain colonies developed from a single bacterial cell (see Fig. 5.13). *Enterobacter* species

appear to be creamy-coloured on the agar plate (Manter *et al.*, 2011) and opaque, smooth and off-white colour on LB agar. Isolated bacterial colonies can be further identified by their 16S rDNA sequences.

### DNA-based identification

Although there are many different methods to identify bacterial isolates, DNA-based identification is the quickest. Detailed methods are described in Chapter 5. In brief, pick a single colony and culture it in 3 ml of LB broth at 28–37°C, with shaking overnight. Pellet the bacterial cells and proceed to DNA extraction followed by PCR using a universal primer pair 16SF/16SR (see Table 5.6). Sequence the PCR amplicon directly or clone the amplicon into a T-vector to obtain its full sequence (using T7 Promoter/SP6 Upstream primers for sequencing). The 16SF/16SR primer pair generates an amplicon of approximately 1.5 kb, so obtaining a full amplicon sequence requires two rounds of Sanger sequencing using the primer walking strategy (see Fig. 5.26).

### Pathogenicity test

A pathogenicity test is used to confirm that the identified bacterial pathogen is the primary cause of the disease. For most known bacterial pathogens, this test may not be necessary, especially when many samples are being diagnosed on a daily basis.

In the case of a new strain or species, a group of hemp seedlings can be inoculated with a cell suspension prepared from a pure bacterial culture. Three inoculation methods are commonly used. The simplest one is to treat the seedling root and lower stem by dipping them in or spraying them with bacterial suspension. The second method is to wound the stem tissue by puncturing with a hypodermic needle and then pipetting a certain amount of bacterial suspension directly onto the wound. Keep the inoculation site moist during the first 48 h. The minor wound will assist bacterial infection if it is a pathogen. The third method is to inject a given amount of the bacterial suspension into the stem tissue using a syringe needle (García-González *et al.*, 2018). After inoculation, inoculated plants should be grown in a controlled environment for symptoms to develop. A pathogenic isolate should develop soft rot on the stem, while mock-inoculated plants with water should not develop any symptoms.

### Key diagnostic evidence

The crown, basal stem, or root exhibit soft rot with an accumulation of creamy bacterial fluid. The internal tissue is water-soaked or disintegrated. Fungal or oomycete pathogens are not recovered from diseased tissue (except in cases of dual or complex infections) and a creamy-coloured bacterium is consistently isolated from symptomatic plant tissue. The 16S rDNA sequence has over 99% similarity to that of an expected *Enterobacter* species.

## Stem Disease Management

The strategy for managing stem disease should focus on the avoidance, elimination, or mitigation of a given pathogen in the field, because most stem diseases are caused by soilborne pathogens. However, some pathogens, such as *Fusarium*, can be carried by cuttings and some can be transmitted by infected seeds. Therefore, the soil, cuttings and seeds should be verified to be free of the pathogens.

### Planting clean seeds and healthy cuttings

A clean seed means a pathogen-free seed. If necessary, consult the seed suppliers for disease incidence during seed production. A certified seed has usually gone through field inspections during the growing season and there is less chance for harvested seeds to harbour pathogens. Cuttings used for large-scale planting should be tested to verify the absence or presence of *Fusarium* in the tissue; and periodically testing and monitoring mother plants for *Fusarium* infections can greatly lower the chance of a *Fusarium* disease outbreak in the field.

### Knowing what is in the soil

A previous crop may build a significant amount of inocula such as fungal resting spores or sclerotia. These inocula can infect the hemp crop directly, as most soilborne pathogens can invade a broad range of plant species. Considering that many hemp cultivars may not be specifically bred for resistance against these diseases, a single or a group of pre-existing pathogens can cause high plant mortality in susceptible hemp crops (Schoener *et al.*, 2017). Therefore, select a field that has no known diseases in previous crops when planting hemp. Alternatively, soil samples systemically collected from the field can be tested for a number of specific pathogens. The land-grant university in each US state has a plant diagnostic laboratory, also known as the National Plant Diagnostic Network (NPDN) lab, which provides diagnostic services for local producers.

### Rotating crops

The best way to mitigate soilborne pathogens is to rotate hemp with other resistant or non-host crops. The key to a successful rotation is to know what species or strains of pathogen exist in the soil of a specific region and what crops or varieties are susceptible or resistant. Regional or state-specific integrated plant management (IPM) researches have generated a wealth of information on this topic and growers can consult with local university extension personnel to obtain the science-based information.

### Managing watering

Stem diseases such as charcoal rot and *Fusarium* canker may become worse under drought stress. Other diseases, such as *Pythium* crown rot, favour humid conditions. An optimum soil moisture is critical to prevent stem diseases and reduce their severity. Generally, overwatering or excessive soil moisture promote crown and root rot caused by both fungi and oomycetes as well as bacteria.

## Managing canopy and plant spacing

Crowded plants with big canopy create a favourable microclimate for pathogens to invade stem tissue. Studies on other crops have demonstrated that a wider plant spacing reduces the incidence of white mould disease. This is because extra space between plants provides adequate air movement and reduces the high moisture microclimate around the stem. Growers should consider row and plant spacing as well as the row orientation to ensure proper air circulation around each plant. The size of the canopy can be managed through the selection of cultivars and application of nitrogen fertilizers, without affecting the yield of desired product.

## Controlling weeds

Weeds can be hosts to some soilborne pathogens. Excessive weed populations can build pathogen inoculum in the soil. Volunteer plants from previous crops can propagate certain pathogens. Additionally, weeds and volunteer plants compete with hemp for water and nutrients. A clean field with complete weed control often yields healthier hemp crops than those with overly grown weeds.

## Reducing pathogen inoculum

For a field with known soilborne pathogens, a pre-planting treatment with an approved fungicide or product can lower the inoculum to a level that does not cause significant economic loss of a hemp crop. If a chemical treatment is not available, biocontrol agents such as those containing *Trichoderma* species may be applied in the field. *Trichoderma* is an effective biological agent in combating many soilborne fungal pathogens, especially *Fusarium*. As an active and live agent, *Trichoderma* can play a long-term role in disease control. In a hemp crop with severe *Fusarium* root rot, *Trichoderma* is often isolated from infected root tissue (see Chapter 9), suggesting that it is a natural biocontrol agent in the soil. *Trichoderma* species may also act as a mycoparasite to reduce the survival of *Macrophomina phaseolina*, the pathogen causing charcoal rot (Baird *et al.*, 2003).

## References

Abed-Ashtiani, F., Narmani, A. and Arzanlou, M. (2018) *Macrophomina phaseolina* associated with grapevine decline in Iran. *Phytopathologia Mediterranea* 57, 107–111. doi: 10.14601/Phytopathol_Mediterr-20691

Aegerter, B., Gordon, T. and Davis, R. (2000) Occurrence and pathogenicity of fungi associated with melon root rot and vine decline in California. *Plant Disease* 84, 224–230. doi: 10.1094/PDIS.2000.84.3.224

Andre, C.M., Hausman, J. and Guerriero, G. (2016) *Cannabis sativa*: the plant of the thousand and one molecules. *Frontiers in Plant Science* 7, 19. doi: 10.3389/fpls.2016.00019

Bains, P.S., Bennypaul, H.S., Blade, S.F. and Weeks, C. (2000) First report of hemp canker caused by *Sclerotinia sclerotiorum* in Alberta, Canada. *Plant Disease* 84, 372. doi: 10.1094/PDIS.2000.84.3.372B

Baird, R. E., Watson, C. E., and Scruggs, M. (2003) Relative longevity of *Macrophomina phaseolina* and associated mycobiota on residual soybean roots in soil. *Plant Disease* 87, 563–566. doi: 10.1094/PDIS.2003.87.5.563

Beckerman, J., Nisonson, H., Albright, N. and Creswell, T. (2017) First report of *Pythium aphanidermatum* crown and root rot of industrial hemp in the United States. *Plant Disease* 101, 1038. doi: 10.1094/PDIS-09-16-1249-PDN

Beckerman, J., Stone, J., Ruhl, G. and Creswell, T. (2018) First report of *Pythium ultimum* crown and root rot of industrial hemp in the United States. *Plant Disease* 102, 2045. doi: 10.1094/PDIS-12-17-1999-PDN

Bishop, A.L. and Davis, R.M. (1990) Internal decay of onions caused by *Enterobacter cloacae*. *Plant Disease* 74, 692–694.

Carbone, I. and Kohn, L.M. (1999) A method for designing primer sets for speciation studies in filamentous ascomycetes. *Mycologia* 91, 553–556. doi: 10.2307/3761358

Carlucci, A., Raimondo, M. L., Santos, J. and Phillips, A.J. (2012) *Plectosphaerella* species associated with root and collar rots of horticultural crops in southern Italy. *Persoonia – Molecular Phylogeny and Evolution of Fungi* 28, 34–48. doi: 10.3767/003158512X638251

Casano, S., Hernández Cotan, A., Marín Delgado, M., García-Tejero, I.F, Gómez Saavedra, O. *et al.* (2018) First report of charcoal rot caused by *Macrophomina phaseolina* on hemp (*Cannabis sativa*) varieties cultivated in southern Spain. *Plant Disease* 102, 1665. doi: 10.1094/PDIS-02-18-0208-PDN

García-González, T., Sáenz-Hidalgo, H.K., Silva-Rojas, H.V., Morales-Nieto, C., Vancheva, T. *et al.* (2018) *Enterobacter cloacae*, an emerging plant-pathogenic bacterium affecting chili pepper seedlings. *The Plant Pathology Journal* 34, 1–10. doi: 10.5423/PPJ.OA.06.2017.0128

Gilardi, G., Ortu, G., Gullino, M.L. and Garibaldi, A. (2012) Una nuova malattia della rucola selvatica causata da *Plectosphaerella cucumerina*. *Protezione delle Colture* 5, 31–33.

Kado, C.I. (2010). *Plant Bacteriology*. APS Press, St Paul, Minnesota.

Khan, A.N., Shair, F., Malik, K., Hayat, Z., Khan, M.A. *et al.* (2017) Molecular identification and genetic characterization of *Macrophomina phaseolina* strains causing pathogenicity on sunflower and chickpea. *Frontiers in Microbiology* 8, 1309. doi: 10.3389/fmicb.2017.01309

Koike, S.T., Stanghellini, H., Mauzey, S.J. and Burkhardt, A. (2019) First report of Sclerotinia crown rot caused by *Sclerotinia minor* on hemp. *Plant Disease* 103, 1771. doi: 10.1094/PDIS-01-19-0088-PDN

Li, P.L., Chai, A.L., Shi, Y.X., Xie, X.W. and Li, B.J. (2017) First report of root rot caused by *Plectosphaerella cucumerina* on cabbage in China. *Mycobiology* 45, 110–113. doi: 10.5941/MYCO.2017.45.2.110

Link, V.H. and Johnson, K.B. (2007) White mold. The Plant Health Instructor, American Phytopathological Society, St Paul, Minnesota. doi: 10.1094/PHI-I-2007-0809-01

Manter, D.K., Hunter, W.J. and Vivanco, J.M. (2011) *Enterobacter soli* sp. nov.: a lignin-degrading γ-proteobacteria isolated from soil. *Current Microbiology* 62, 1044–1049. doi: 10.1007/s00284-010-9809-9

McGehee, C.S., Apicella, P., Raudales, R., Berkowitz, G., Ma, Y. *et al.* (2019) First report of root rot and wilt caused by *Pythium myriotylum* on hemp (*Cannabis sativa*) in the United States. *Plant Disease* 103,3288. doi: 10.1094/PDIS-11-18-2028-PDN

Mersha, Z., Kering, M. and Ren, S. (2019) Southern blight of hemp caused by *Athelia rolfsii* detected in Virginia. *Plant Disease* 104, 1562. doi: 10.1094/PDIS-10-19-2178-PDN

Mullen, J. (2001) Southern blight, southern stem blight, white mold. The Plant Health Instructor, American Phytopathological Society, St Paul, Minnesota. doi: 10.1094/PHI-I-2001-0104-01

Palm, M.E., Gams, W. and Nirenberg, H.I. (1995) *Plectosporium*, a new genus for *Fusarium tabacinum*, the anamorph of *Plectosphaerella cucumerina*. *Mycologia* 87, 397–406.

Pascoe, I.G., Nancarrow, R.J. and Copes, C.J. (1984) *Fusarium tabacinum* on tomato and other hosts in Australia. *Transactions of the British Mycological Society* 82, 343–345.

Qiao, T.M., Zhang, J., Li, S.J., Han, S. and Zhu, T.H. (2016). Development of nested PCR, multiplex PCR, and loop-mediated isothermal amplification assays for rapid detection of *Cylindrocladium scoparium* on Eucalyptus. *The Plant Pathology Journal* 32, 414–422. doi: 10.5423/PPJ.OA.03.2016.0065

Schoener, J.L., Wilhelm, R., Yazbek, M. and Wang, S. (2018) First detection of *Pythium aphanidermatum* crown rot of industrial hemp in Nevada. *(Abstr.) Phytopathology* 108, S1.187. doi: 10.194/PHYTO-108-10- S1.187

Schoener, J.L., Wilhelm, R., Rawson, R., Schmitz, P. and Wang, S. (2017) Five *Fusarium* species associated with root rot and sudden death of industrial hemp in Nevada. *(Abstr.) Phytopathology* 107, S5.98. doi: 10.1094/PHYTO-107-12-S5.98

Schroeder, B.K., DuToit, L.J. and Schwarts, H.F. (2009) First report of *Enterobacter cloacae* causing onion bulb rot in the Columbia Basin of Washington State. *Plant Disease* 93, 323. doi: 10.1094/PDIS-93-3-0323A

Trabelsi, R., Sellami, H., Gharbi, Y., Krid, S., Cheffi, M. *et al.* (2017) Morphological and molecular characterization of *Fusarium* spp. associated with olive trees dieback in Tunisia. *3 Biotech* 7, 28. doi: 10.1007/s13205-016-0587-3

# 9 Diagnosing Root Diseases

A plant suspected to have a root disease may be pulled out from the ground to examine the root for any damage caused by either microorganism infections or abiotic factors. In some cases, for example a problem associated with an established tree, digging out and directly examining the entire root system is neither easy nor practical. For *Cannabis* plants, however, the root system can be easily pulled out from the pot or soil to examine any root abnormalities and evaluate its aetiological relations to above-ground symptoms. A heathy root system should contain several major roots and abundant lateral and feeding roots with a healthy-looking colour (Fig. 9.1). In contrast, an unhealthy root system may have dark lesions, blackening, or soft rot, and often lacks viable roots. Root diseases can cause above-ground symptoms such as leaf yellowing and dieback. Most root diseases are caused by soil-borne fungi, bacteria, or parasitic nematodes, among which *Fusarium* and *Pythium* are the most prevalent pathogens responsible for root diseases in *Cannabis* crops (Schoener *et al.*, 2017; Punja and Rodriguez, 2018). These pathogens may infect the root tissue exclusively, but infections often progress into the crown and stem tissue, causing root, crown and stem diseases (see Chapter 8). Furthermore, multiple pathogens in the same genus or different taxa may attack hemp roots at the same season and devastate a hemp crop. For example, up to five *Fusarium* species were found to be associated with hemp root rot and caused sudden death of a hemp crop (Schoener *et al.*, 2017). As many plant diseases, particularly soil-borne root diseases, are disease complexes driven by bi-, tri- or multi-pathogens (Wheeler *et al.*, 2019), diagnosing root diseases requires a holistic approach to find all disease-driving organisms (see Chapter 13).

## Hemp Root Rot Caused by *Fusarium* Species

Species in the *Fusarium* genus cause stem and crown rot (see Chapter 8), vascular wilt (see Chapter 10) and severe root rot on *Cannabis* crops. On marijuana plants, *Fusarium oxysporum*, *F. solani* and *F. brachygibbosum* were reported to cause root rot (Punja *et al.*, 2019), with *F. oxysporum* and *F. solani* being the major species infecting those grown hydroponically (Punja and Rodriguez, 2018). On hemp crops, several *Fusarium* species, namely *F. oxysporum*, *F. solani*, *F. redolens*, *F. tricinctum* and *F. equiseti*, were found to cause crop failure (Schoener *et al.*, 2017). The species involved in root rot may vary from case to case, but *F. oxysporum* and *F. solani* appear to be the most prevalent. Because many cultivars of *Cannabis sativa* may not have been bred for direct resistance to *Fusarium* pathogens, the incidence and severity of *Fusarium* root rot may largely depend on the composition of pathogenic species and their inoculum levels in the soil. That said, a farmland-grown hemp crop may have single-species-driven root rot or multi-species-driven root rot. In contrast, indoor-grown cannabis plants, either in substrate or hydroponically, are less prone to *Fusarium* root rot if a strict protocol to avoid these pathogens is in place. However, *Fusarium* pathogens are still introduced to cultivation facilities through contaminated nutrient solutions or pre-infected cuttings (Punja and Rodriguez, 2018). In this case, a test scheme to monitor pathogen introduction is needed.

### Field diagnosis

#### Symptoms

The initial infection may cause tiny visible necrotic lesions on the root surface and internal tissue

**Fig. 9.1.** A healthy hemp root system comprising several major roots with numerous lateral and feeding roots. Note that the colour of the roots is fresh white and lesion-free.

**Fig. 9.2.** An early symptom of hemp root rot caused by *Fusarium oxysporum*. Note the brown lesion extending from the outer layers to the inner tissue.

(Fig. 9.2). At this stage, above-ground symptoms appear subtle with leaf yellowing, edge browning, or mild curling (Fig. 9.3). However, infected plants may start to look different from other plants in terms of growth vigour and foliar colour. As the rot expands to most parts of the root system, plants exhibit significantly slower growth, stunting and even wilting, similar to symptoms caused by drought stress (Fig. 9.4). Plants are easily pulled out from the soil as the root system has partially or completely rotted with few or no lateral roots (Fig. 9.5). Pink staining and/or white mould are sometimes visible on the surface of taproots, especially when root rot is severe. The pink stain generally indicates the colonization of certain *Fusarium* species (e.g. *F. tricinctum*) that produce pink-coloured mycelia, while the white mould is the mycelia of other *Fusarium* species such as *F. oxysporum* and *F. solani*. During the late stages of root rot, *F. equiseti* overtakes *F. oxysporum* and/or *F. solani* and becomes the most prominent organism on rotted roots. Affected plants die in patches and mortality can be up to 70% (Fig. 9.4). In marijuana plants, symptoms of *Fusarium* root rot may be expressed as root browning, reduced root mass and overall stunting of plants (Punja and Rodriguez, 2018).

### Problem classification

The initial diagnosis should correctly classify the problem as a root rot disease. In the case shown in Fig. 9.2, without examining the root and further

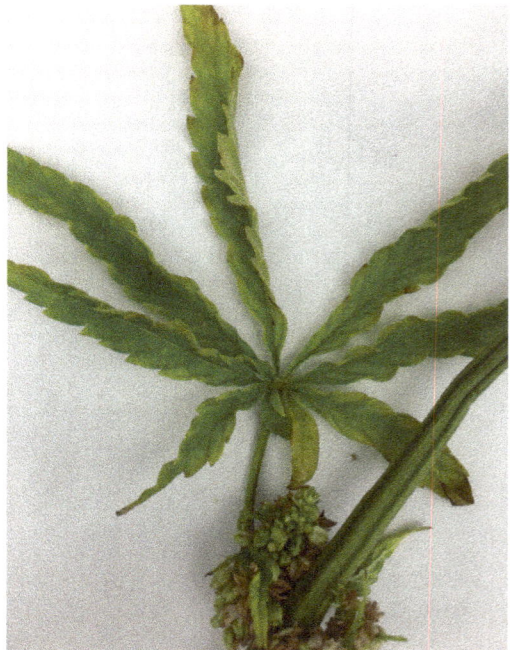

**Fig. 9.3.** Mild foliar symptoms resulting from the initial stage of root rot shown in Fig. 9.2. Note the leaf edge curling and browning as a result of insufficient water uptake by compromised root functions.

isolating the causative pathogens, the mild foliar symptom may be misdiagnosed as a nutrient or watering issue. In the case of Fig. 9.4, a mass of dead plants can be misdiagnosed as acute drought stress

**Fig. 9.4.** A patch of dead hemp plants due to root rot caused by multiple species of *Fusarium* including *F. oxysporum*, *F. solani*, *F. redolens*, *F. tricinctum* and *F. equiseti*.

**Fig. 9.5.** Severe root rot caused by multiple species of *Fusarium*. Note the healthy root on the left, the rotted roots on the right, and the pink colour stain on dead roots. Above-ground symptoms are shown in Fig. 9.4.

or other abiotic factors as there is no pathogen sign at all on the above-ground portion of plants. In both cases, pulling out the root system from the ground, closely examining the structure and texture of the roots and looking for pathogenic signs may quickly pinpoint the problem into a root rot disease. Simply pull out 10–15 symptomatic plants and the same

number of healthy plants and compare their root health (Fig. 9.6). The strong link between root rot and above-ground symptoms suggests a root problem and warrants further investigations for root-infecting pathogens. The presence of fungal signs on dead roots is also a helpful clue to fungal infection.

### Do your own diagnosis

Growers can follow the root-cause analysis process as described previously to classify the problem. *Fusarium* root rot is a soilborne disease and is generally presented as a patch or cluster of dead plants (Fig. 9.4). The pattern of affected plants should be first determined before systemically examining individual symptomatic plants. Check the foliage and stem tissue for any signs of pathogens or other types of diseases. Above-ground symptoms caused by root rot are generally uniform and similar among plants; and in most cases, infected plants are wilted and desiccated in the field. If no evidence of abiotic factors (e.g. herbicide injury or acute drought) is found, roots should be pulled out from the soil and examined for root rot. In addition, revisit previous crop disease history to see if similar diseases have occurred in the same field (*Fusarium* has a broad host range).

## Sampling

Once the problem is classified as root rot, cut several symptomatic plants from the lower stem and submit the entire root systems to a diagnostic lab. Samples should include roots with mild to severe rot and a few healthy roots. Follow the sample packing instructions described previously or provided by a diagnostic lab.

## Lab diagnosis

### Visual and microscopic examinations

Root samples can be washed under tap water to remove the soil and then placed on paper towels to absorb extra water. Compare diseased roots with presumably healthy roots to determine abnormalities such as colour and texture changes and reduction of overall root mass. Cut through the roots to reveal internal rot. Unless other pathogens such as bacteria are involved, *Fusarium* root rot is a type of dry rot without extra fluid. Under a stereo microscope, mycelia of *Fusarium* or other fungi may be observed from the root. A tiny piece of mycelia can then be picked from the root and mounted on a glass slide to observe the spores under a compound microscope. In some cases, *Fusarium* spores, especially macroconidia, are present (see Fig. 8.10). Other *Fusarium* species or even secondary non-pathogenic fungi may be observed from severely rotted roots.

## Isolation and colony observation

Hemp root rot is likely to be caused by more than one *Fusarium* species and sometimes oomycetes are also involved (see Chapter 13 and Table 13.1). Therefore, plate symptomatic root tissues on both PDA amended with streptomycin (PDA$^{+strep}$) and PARP plates and incubate the plates at 22°C in the dark. Always plate tissue pieces cut from the junction area between rotten and healthy tissue to obtain the primary pathogens. It is not uncommon that colonies with different characteristics grow on the same plate or even from the same piece of tissue (Fig. 9.7). Certain *Fusarium* species may appear more frequently in some hemp varieties (Fig. 9.8). Because of multi-species infections, representative colonies should be transferred to a fresh PDA$^{+strep}$ plate before they are overgrown and then identified individually. In the case of hemp root rot shown in Fig. 9.4, five *Fusarium* species with distinct morphology are identified from the crop (Fig. 9.9). *F. oxysporum* colonies are cottony white, generally with a dark violet or magenta pigment in the agar (seen from the back side of plate). *F. solani* colonies appear more cream-coloured with fewer aerial mycelia when compared with *F. oxysporum*, and generally produce a slight pigment or none at all. *F. redolens* colonies are flat with characteristic sparse white aerial mycelia. *F. tricinctum* is more distinct with its pink, red, or purple colour and red pigment in the agar. *F. equiseti* grows fast, has abundant white aerial mycelia and generally produces a brown

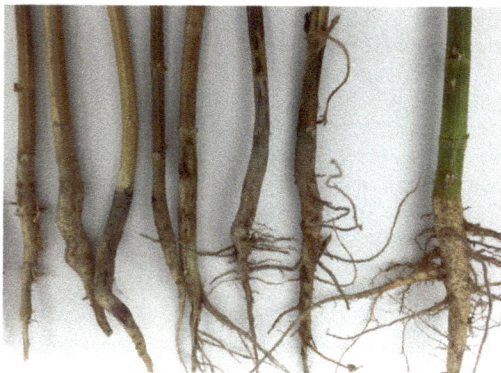

**Fig. 9.6.** Seven hemp plants exhibiting increased levels of root rot (right to left) that correspond to the severity of above-ground symptoms (not shown). Note the healthy plant on the right.

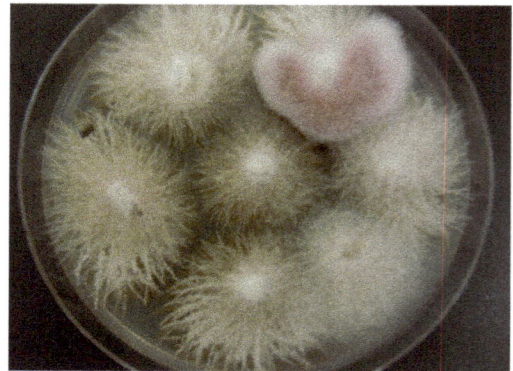

**Fig. 9.7.** Multiple species of *Fusarium* isolated from an infected hemp root. Note that both *F. redolens* and *F. tricinctum* grew out from the same tissue piece (1 o'clock position) while *F. equiseti* grew out from the piece at 5 o'clock position. All others were *F. redolens*.

**Fig. 9.8.** *Fusarium oxysporum* and *F. solani* are most commonly isolated from infected hemp roots. Note that the pieces at 6 o'clock and 11 o'clock positions were infected by both species.

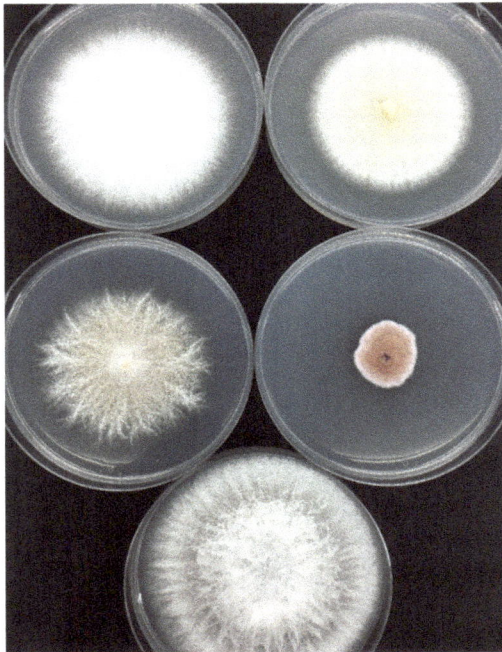

**Fig. 9.9.** Colony characters of *F. oxysporum* (upper left), *F. solani* (upper right), *F. redolens* (middle left), *F. tricinctum* (middle right) and *F. equiseti* (bottom) after 1 week of growth on PDA at 22°C. Note that *F. tricinctum* grew much more slowly than others.

pigment in the agar. However, colony morphologies and characteristics may vary among the isolates and depend on the culture media and growth age (Leslie and Summerell, 2006). Spores observed from each isolate, if produced on a plate, can be used for a preliminary identification of a species.

### DNA-based identification

While different media are used to characterize colony and spore morphology and may ultimately identify an unknown isolate to a species accurately, this approach may take a significant amount of time and resources. A diagnostic lab can use a DNA-based approach to quickly identify an isolate to a closely related species (see Chapter 5). DNA sequences obtained from the rDNA-ITS region by using ITS1/ITS4 primers are adequate to distinguish species in the *Fusarium* genus (Fig. 9.10).

### Pathogenicity test

In cases where multiple species of *Fusarium* are isolated from infected hemp roots, the pathogenicity of each species can be tested by inoculating a calculated spore suspension (e.g. $1 \times 10^6$/ml) into the rhizosphere of young hemp plants. Some species are weaker pathogens and may not be as pathogenic as *F. oxysporum* or *F. solani* when they are individually inoculated. However, these species, along with other pathogens, do drive disease development and cause the severe symptoms seen in the field.

### Key diagnostic evidence

Affected plants wilt rapidly without infectious symptoms shown on above-ground portions. The root mass is largely reduced with few or no lateral roots. Root rot is common on all affected plants. One or more *Fusarium* species are predominantly isolated from affected roots and are positively identified by their DNA sequences.

## Hemp Root Rot Caused by *Pythium* Species

*Pythium* species causes crown rot (see Chapter 8), root rot, or both. *Pythium ultimum*, *P. aphanidermatum* and *P. myriotylum* were first reported to be the pathogen causing hemp root rot in the USA (Beckerman *et al.*,

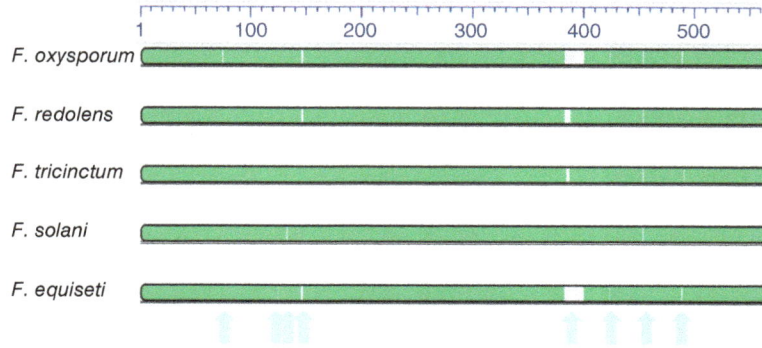

**Fig. 9.10.** Five distinct *Fusarium* isolates as shown in Fig. 9.9 were identified as *F. oxysporum* (544 bp), *F. solani* (566 bp), *F. redolens* (559 bp), *F. tricinctum* (563 bp) and *F. equiseti* (546 bp) based on their DNA sequences at the ITS region of rDNA (using ITS1/ITS4 primers). The differences (gap) among the DNA sequences of the five species are indicated by arrows.

2017, 2018; McGehee *et al.*, 2019); however, these species are well known and commonly found in agricultural fields. On marijuana plants, *P. dissotocum*, *P. myriotylum* and *P. aphanidermatum* were frequently associated with root browning and internal discoloration (Punja *et al.*, 2019). On hydroponically grown cannabis, *P. myriotylum* and *P. dissotocum* were common root rot pathogens (Punja and Rodriguez, 2018). In general, *Pythium* root rots result in plant yellowing, stunting, wilting and, in some instances, sudden death (Punja *et al.*, 2018). These symptoms are secondarily induced due to the lack of normal root functions.

### Field diagnosis

#### Symptoms

Root rot diseases often end up with above-ground symptoms such as chlorotic leaves, stunted growth, or severe wilt, but in the initial stages of root infection, above-ground symptoms may be very subtle. As the infection advances to cause essential damage to the root tissue, normal nutrient and water supplies are interrupted. Thus, the above-ground portion of the plant starts to show symptoms similar to those caused by water deficit and nutrient deficiency. When the roots are pulled out from the ground, brown to dark lesions on the roots and the loss of feeder roots are often observed. In some cases, *Pythium* species can infect the crown and lower stem in addition to the roots, and brown and water-soaked stem

lesions may be present in some plants (Beckerman *et al.*, 2017). Plants with moderately infected root systems may grow at a reduced rate and size, but those with severely infected roots may die rapidly (Fig. 9.11). Under humid conditions, aerial mycelia of *Pythium* may be visible on the surface of crown and lower stem tissue (see Fig. 8.23).

Hydroponically or aeroponically grown cannabis plants may exhibit different symptoms from those grown in the field, where a community of organisms exist and contribute to the post-infection decomposition. These indoor plants are grown in or with nutrient solution and their roots can be infected when the solution is contaminated with one or more pathogens. Infected roots look brown or darker when compared with non-infected roots (Fig. 9.12). Root tissue may look transparent under a microscope without distinct decomposition and *Pythium* oospores are often observed inside the root tissue (Fig. 9.13). *Pythium* infections lead to the loss of normal root functions and therefore affect the growth of other parts of plants.

#### Problem classification

Use the same strategy as described above for *Fusarium* root rot to define a problem as root rot disease. This is an important diagnostic process, as most encountered plant problems are open-ended. For field-grown hemp plants, a comprehensive examination of whole plants, including their root

**Fig. 9.11.** A hemp plant that died due to root rot caused by *P. aphanidermatum*. Note that there were almost no lateral roots.

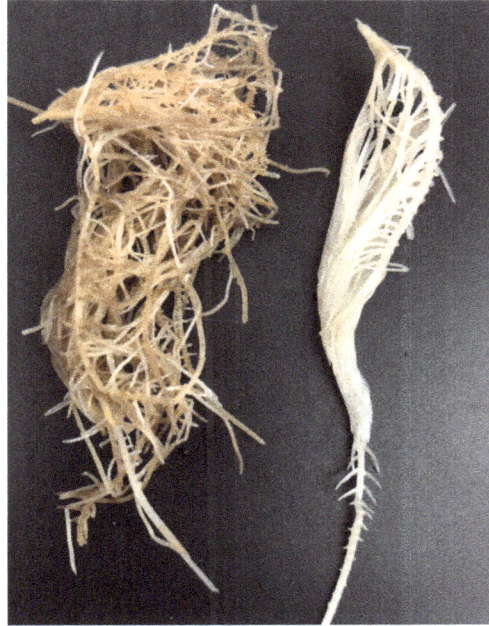

**Fig. 9.12.** Roots of aeroponically grown marijuana plants exhibited a brown colour (left) due to the infection of *P. myriotylum*. Note that uninfected roots were fresh white (right).

systems, is needed to determine whether root rot is the primary problem. Root rot of aeroponically grown plants may be easily noticed, since the entire root system is exposed. When some roots turn brown or an abnormal colour, this points to a root rot. Because symptoms caused by mild root rot often resemble those caused by a nutrient issue, a thoughtful investigation into the plant's nutrient status and potential root infection is needed.

### Do your own diagnosis

Once the problem is classified as root rot, its aetiology can be further determined. Growers can examine the roots for signs of pathogen growth. The sign of *Pythium* growth looks cottony initially and then turns to uniform grey. Under a magnifier or a portable microscope, sporangia (round or lemon-shaped) may be seen at the tip of the hyphae. This is different from the *Fusarium* root rot, where *Fusarium* species can grow on the surface of infected root or crown tissue but the colour of the pathogen signs ranges from pure white to pink

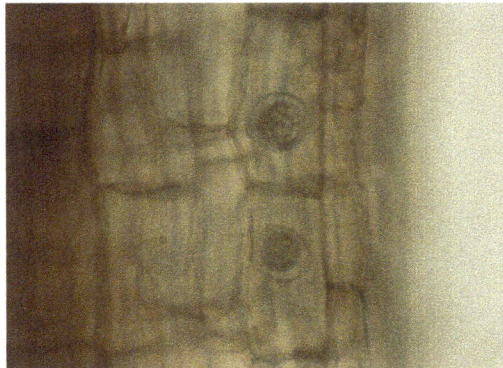

**Fig. 9.13.** Oospores of *P. myriotylum* produced inside the root tissue observed under a compound microscope. Note that the thick-walled round structure is the oospore, while the rectangular shape is the root cell.

without lemon-shaped structures. Furthermore, the roots killed by *Fusarium* look stained with colour rather than fuzzy. All these characteristics can only be used in a preliminary diagnosis in the field and an accurate diagnosis still requires tissue samples to be submitted to a plant diagnostic lab.

### Sampling

Cut several plants from the lower stems and submit entire root systems to a diagnostic lab. Samples should include roots collected from mild to severe symptomatic plants as well as apparently healthy plants.

### Lab diagnosis

#### Visual and microscopic examinations

For field-grown hemp plants, use the same procedures described above for *Fusarium* root rot. The key is to look for *Pythium* mycelia and round or lemon-shaped sporangia that may be present at the surface of infected root or crown tissue. For hydroponically or aeroponically grown plants, roots can be placed under a stereo microscope to observe the individual hyphae growing out from the root tissue (Fig. 9.14). Under a compound microscope, oospores may be observed in certain sections of root tissue (Fig. 9.13). The presence of both hyphae and oospores suggests an oomycete infection.

#### Isolation and colony observation

Hemp root rot is likely caused by more than one pathogen, including fungi and oomycetes. Therefore, it is wise to plate symptomatic root tissues on both PDA and PARP plates. Certain oomycete species are sensitive to streptomycin, so an unamended PDA plate is recommended for cases that are

suspected for *Pythium* or *Phytophthora* infection. *Pythium* grows faster than many other pathogens, so plates should be checked in 2–3 days. Transfer each distinct colony (if any) to a new CMA or PDA plate. *Pythium* colonies may appear flat and radiate on CMA but look cottony with significant aerial mycelia on PDA. Colonies of *P. aphanidermatum* (see Fig. 8.27) and *P. myriotylum* (Fig. 9.15) appear white and cottony on PDA with coenocytic hyphae (without septa or cross walls), a common characteristic for oomycete species. The hyphae are not dichotomously branched and often have significant swelling (Fig. 9.16). Sporangia are generally globose (see Fig. 8.28) and abundantly produced on a CMA. In some species, round oogonium and

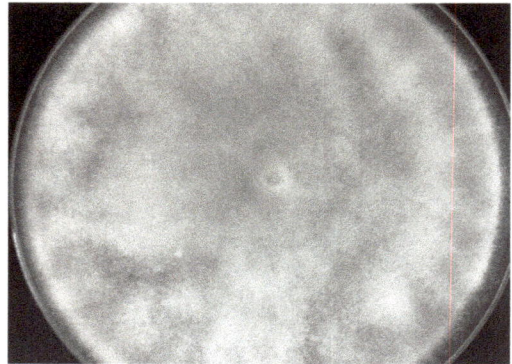

**Fig. 9.15.** A week-old *P. myriotylum* colony grown on PDA at 22°C in dark.

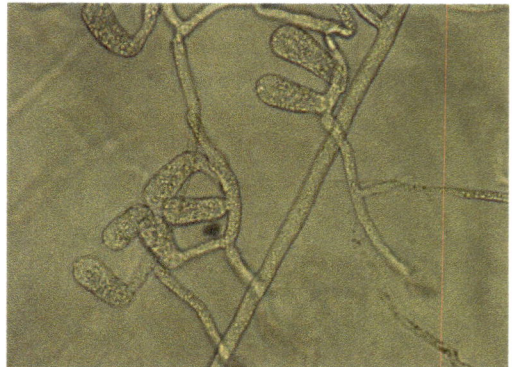

**Fig. 9.16.** Hypha swelling of *P. myriotylum* observed under a compound microscope. Note that the hyphae are not septate.

**Fig. 9.14.** Hyphae of *P. myriotylum* growing out from a root of an aeroponically grown marijuana plant.

antheridium (usually attached to the oogonium) as well as thick-walled chlamydospores are also observed. Although there are several types of spores and structures that can be used to identify species, the overlapping of morphometric data and the lack of certain structures in culture may not allow for a quick and confident identification. A DNA-based identification is often recommended.

### DNA-based identification

See section on hemp crown rot caused by *Pythium* species in Chapter 8.

### Key diagnostic evidence

Hemp plants with *Pythium* root rot should have characteristic root symptoms such as brown or dark root lesions, a lack of lateral or feeding roots, or death of the entire root system. One or more *Pythium* species are isolated from symptomatic roots but not from the healthy ones and are further confirmed by their DNA sequences.

## Hemp Root Rot Caused by *Rhizoctonia* Species

Hemp plants are susceptible to *Rhizoctonia solani*, a soilborne fungal pathogen that has a broad host range and worldwide distribution. On non-*Cannabis* crops, *R. solani* can cause root rot, collar rot, or seedling damping-off. On hemp, the pathogen causes root rot alone or with other soilborne pathogens such as *Fusarium* spp. It may also attack lower stems, causing stem rot.

### Field diagnosis

#### Symptoms

Infected roots exhibit brown lesions or rot and infection may extend to the lower portion of the main stem. Like other fungi-caused root and crown rots, *Rhizoctonia* root rot appears hardy and dry unless other pathogens are involved. Above-ground symptoms are similar to those caused by other root rot diseases.

#### Problem classification

Root rot caused by *Rhizoctonia solani* is a soil-borne disease. Use the same strategy as described in root and stem diseases above to classify a given problem as root rot disease. The determination of the specific pathogen associated with root rot may require further lab analysis. However, if there are signs of mycelial growth without the presence of spores, the cause can be narrowed down to one of the sterile fungal species.

### Do your own diagnosis

Since there are several root and crown diseases caused by a number of soilborne pathogens, diagnosing a root rot to a specific pathogen in the field may be difficult. Nevertheless, growers can rule out other diseases using the procedures described previously. *Rhizoctonia solani* produces black sclerotia but may not be present on infected roots or may be indistinguishable from those produced by other species. However, this fungus has a particular hypha morphology that can be used to separate it preliminarily from others. When a handy microscope is available, use a piece of clear tape to blot a mass of mycelia from a root and place it on a drop of water at the centre of a glass slide. Place the slide under a microscope and observe the hyphae branching and the unique positioning of the septum. *R. solani* hyphae typically branch at a 90° angle and have more than three nuclei per cell. If nuclei are difficult to see, the first septum on branched hyphae is generally very close to the branching point and the area has a slight offset (Fig. 9.17) (also see Fig. 2.7). Compared with some other sterile fungal genera, the hyphae of *Rhizoctonia* are wider. These characteristics can

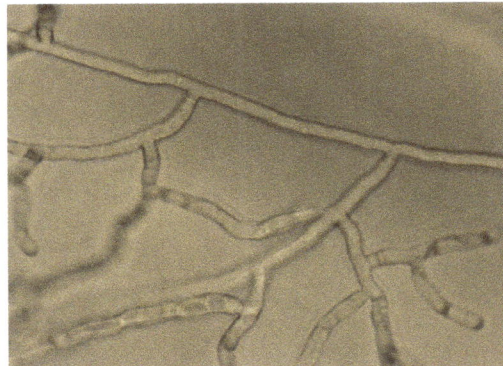

**Fig. 9.17.** Vertical branching of *Rhizoctonia solani* hyphae with characteristic offset at the base of lateral branches.

only be used for preliminary diagnosis and an accurate diagnosis requires tissue samples to be submitted to a plant diagnostic lab.

### Sampling

Cut several plants from the lower stem and submit the entire root systems to a diagnostic lab. Samples should contain roots collected from plants with mild to severe symptoms and from plants with no apparent symptoms.

### Lab diagnosis

#### Visual and microscopic examinations

Examine both the root and crown to determine how extensive the rot is and whether the rot occurs in the stem tissue. Locate the front line of the rot from which pieces of tissue can be collected for plating. Determine if the rot is dry or soft and if other pathogens are involved. Look for fungal growth on the tissue and check if the hyphae have the characteristics shown in Fig. 9.17. In samples collected during the late stages of infection, light to dark brown sclerotia may be present on infected roots. The sexual stage of *R. solani* is classified as a basidiomycete fungus, but it is rare to see its sexual spores and structures produced on infected tissue under natural conditions.

#### Isolation and colony observation

Similar to other root rot diagnosis, isolation should be considered a universal approach to catch all potential pathogens. PDA$^{+strep}$ and PARP plates can be used to isolate a wide range of suspected pathogens. *Rhizoctonia* fungi are readily isolated on PDA at 22°C. Mycelia are initially colourless but later turn light brown (Fig. 9.18). There should be no spores present on the medium, but irregular sclerotia are often produced in an old culture.

#### DNA-based identification

Use ITS1/ITS4 primers to amplify a short DNA sequence from the rDNA-ITS region and submit a purified amplicon for sequencing (see Chapter 5). Other loci can also be used in addition to the ITS region of rDNA.

### Key diagnostic evidence

Evidence includes demonstrated links between the *Rhizoctonia* species and root rot symptoms based on isolation frequency, morphology-based *Rhizoctonia* genus identification, and DNA sequence data-backed species identification.

## Hemp Root Rot Caused by *Enterobacter* Species

See the section on hemp crown rot caused by *Enterobacter* species in Chapter 8. The bacterium causes both crown and root rot. Infected roots lose their healthy colour and texture and the internal tissues decay (Fig. 9.19). The diagnosis

**Fig. 9.18.** A colony of *Rhizoctonia* species isolated from an infected root.

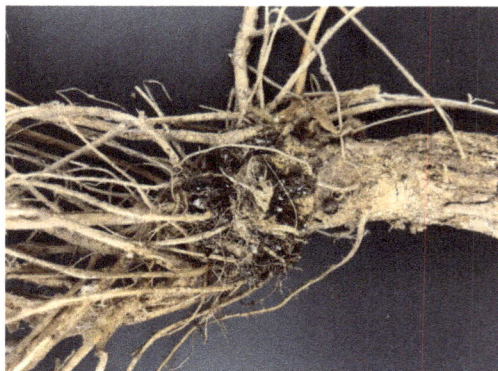

**Fig. 9.19.** Severe root rot and crown rot of a hemp plant infected by an *Enterobacter* species (see healthy hemp roots shown in Fig. 9.1).

procedures for bacterial root rot are the same as those for crown rot.

## Hemp Root Knot Caused by *Meloidogyne* Species

Root-knot nematodes (*Meloidogyne* spp.) consist of about 100 species, many of which have a broad host range. They parasitize on hundreds of plant species including fruit and vegetable crops, grasses, trees and even weeds. However, the most wide-spread and economically important species are limited to *M. incognita*, *M. javanica*, *M. arenaria*, *M. hapla*, *M. chitwoodi* and *M. graminicola* (Mitkowski and Abawi, 2003). Root-knot nematodes primarily occur in tropical to sub-tropical areas, but some species, such as *M. hapla* and *M. chitwoodi*, are major pathogens in temperate regions. Root-knot nematodes live in the soil and infect roots, causing both root and above-ground symptoms. Hemp crops are susceptible to root-knot nematodes. In South Africa, *Meloidogyne javanica* infected several hemp cultivars when inoculated with 1000 second-stage juveniles under greenhouse conditions (Pofu and Mashela, 2014). Three years later, *M. javanica* was detected on hemp cultivar (cv. 'YunMa1') in a crop field in south-west China, and this nematode caused leaf chlorosis and plant slow growth in addition to numerous root galls (Song *et al.*, 2017). In Nevada, several hemp cultivars developed typical root-knot symptoms when they grew in *Meloidogyne*-infected crop soil under greenhouse conditions. These observations, both under experimental conditions and in crop fields, suggest that root-knot nematodes are a potential threat to hemp production.

### Field diagnosis

#### Symptoms

Root-knot nematodes have several life stages, including the egg, first-stage juvenile (J1), J2, J3, J4, and female and male adults. J1 moults inside the egg and emerges from the egg to become the infectious J2. This process is often triggered by the presence of growing roots of susceptible plants. J2 moves into a young root and migrates to a site where the nematode stays and feeds permanently. Then the nematode injects oesophageal gland secretions into the site to stimulate the surrounding cells to become giant cells. The giant cells function as a feeding site or nutrient sink for the nematode to draw nutrients from other parts of root tissue. With continuous feeding, J2 (vermiform) develops to J3 and J4 with slightly enlarged body and eventually to a female adult with an almost round shape. Because of the injection of nematode oesophageal gland cell secretions into the root tissue and the increase of plant growth regulators, root cells neighbouring the feeding site divide rapidly and also enlarge, resulting in gall or knot formation (Fig. 9.20).

In addition to root knots, above-ground symptoms are visible. The most common is chronic or temporary wilt, as infected roots have a reduced ability to absorb and transport water to the rest of the plant. While wilt may occur daily, especially during the afternoon when temperatures are high, some plants may recover during the evening or in cooler weather. Unlike fungal or bacterial root rots that usually destroy the root system completely, root-knot nematodes impair root functions rather than kill the root. Therefore, above-ground symptoms resemble those caused by drought stress even though the soil has sufficient moisture. Other

**Fig. 9.20.** A hemp plant root system exhibiting numerous root knots caused by a *Meloidogyne* species.

common symptoms are plant stunting and chlorosis (Fig. 9.21) due to the reduced ability of the roots to absorb and transport nutrients from the soil. These symptoms resemble those caused by nutrient deficiency. Because root nematodes are soilborne and move slowly in the soil, symptomatic plants may appear in clusters or patches in the field.

### Problem classification

Although the disease caused by root-knot nematodes is easily diagnosed by observing the roots, the problem is often mistaken as a nutrient or water issue. A systemic examination of whole plants is critical to classifying a problem correctly as root-knot nematode infection. Nutrient deficiency generally impacts a large number of plants and the symptom may appear uniform in a field. In contrast, nematodes only affect those plants that grow in the area where nematode populations have exceeded the damage threshold. Therefore, a nematode infection often causes a small patch or a cluster of symptomatic plants. When multiple cultivars are planted in the same field, the nematode may cause more severe symptoms in certain cultivars than in others. If the root-knot disease has been found in previous crops in the same field, this record will help support classifying the problem as a root-knot nematode infection.

### Do your own diagnosis

Once the problem is determined as being root-related, a symptomatic plant can be pulled out from the ground. Shake out attached soil and examine the root system for gall-like enlargements or knots. Root-knot nematodes can cause knots on any root, small or large, and the knots may be scattered or congregated in certain areas of the root system (Fig. 9.22). The sizes of the knots vary depending on the stage of infection and the number of nematodes in each site. In some legume crops, normal root nodules may be mistaken as root knots, but one can distinguish root knots from root nodules easily by examining the morphological formation of both structures. In general, a nodule produced by the nitrogen-fixing bacterium is a structure attached to the root (Fig. 9.23), while a root knot is enlargement of a small segment of root (Fig. 9.20) (see also Fig. 2.42). Since hemp is not a legume crop, there are no nodules, or any other types of galls expected on a normal root. Therefore, any swelling or lumpy root tissue may suggest a root-knot nematode infection.

### Sampling

Root-knot nematodes live in the soil (when a host plant is absent) and on hemp roots during their infection period, so there are two ways of collecting and submitting samples to a diagnostic lab. The first method is to collect some roots or the entire root system from symptomatic plants. Use this method when knots or abnormal growths are observed on roots. Place roots in a large Ziploc bag and secure the sample in a box before submission. Root samples are more useful for directly detecting females inside the root tissue, which leads to a quick diagnosis.

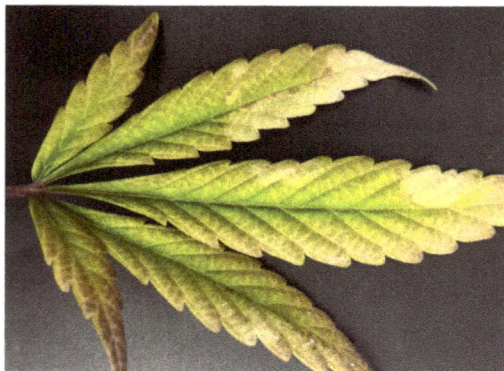

**Fig. 9.21.** Chlorosis and yellowing of a hemp leaflet due to impaired root system caused by a *Meloidogyne* species.

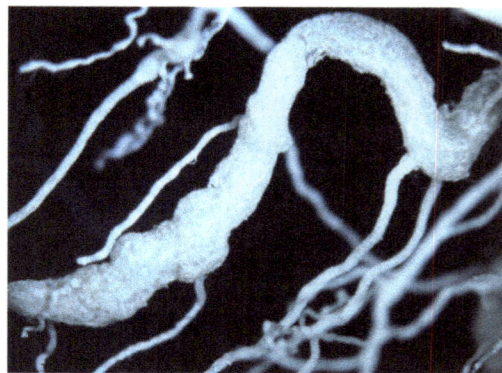

**Fig. 9.22.** Hemp roots with congregated knots in one segment (a large root at the centre) and scattered individual knots

When the root-knot nematode is suspected but root-knot symptoms are not obvious, soil around or attached to the root system (rhizosphere) can be collected from individual symptomatic plants. Simply pull a plant out from the ground and shake the soil into a plastic bag. Mix the soil thoroughly and transfer 500 ml of it into a Ziploc bag. Seal the bag to protect the soil from drying. A dried soil will kill all nematodes (except cysts of cyst nematodes) and become useless for nematode analyses. In root-knot nematode diagnosis, soil samples are used to detect second-stage juveniles and males, so they may not be as informative as root samples. Nevertheless, sampling both the root system and its rhizosphere soil is highly recommended for root-knot nematode diagnosis.

## Lab diagnosis

### Visual and microscopic examinations

When a root sample is received, place the root under running tap water to gently remove the soil and debris, then blot dry with a paper towel. Examine the root for uneven thickness or bumpy segments (Fig. 9.20). Once a suspected segment is identified (Fig. 9.24), place the root under a stereo microscope with the reflective light on. Use a dissecting needle and a handled No. 10 steel surgical blade to dissect the knot and reveal the female (see Fig. 2.36). The pearly white female can be picked out from inside the knot (Fig. 9.25). For newly developing root knots, the nematodes may still be at third or fourth stages and are not fully pear-shaped. Instead, they enlarge into a sausage shape. Matured females are pear-shaped and have a gelatinous sac into which eggs are deposited (Fig. 9.26). Eggs are often present and can be observed during the dissection of knot tissue. One female can produce more than 1000 eggs (Mitkowski and Abawi, 2003). The typical shape of eggs is ellipsoid (Fig. 9.27).

### Nematode extraction from soil

Soil samples can be processed to extract active vermiform (worm-shaped) nematodes using either a Baermann funnel or wet-sieving method as described in Chapter 5. For root-knot nematodes, the second-stage juveniles (J2s) are most likely to be present in the extracts from infected soil. The size of J2s is usually less than 500 µm in length. In some cases, males can be found from soil. All males are vermiform and much bigger than the J2s, ranging from 1100 µm to 2000 µm. In addition to their larger size, males have characteristic morphology such as a distinct lip, well-developed stylet, round tail and

**Fig. 9.23.** A Jade bean plant root system having nodules attached to the roots. Note that the nodules are produced by nitrogen-fixing *Rhizobium* bacteria.

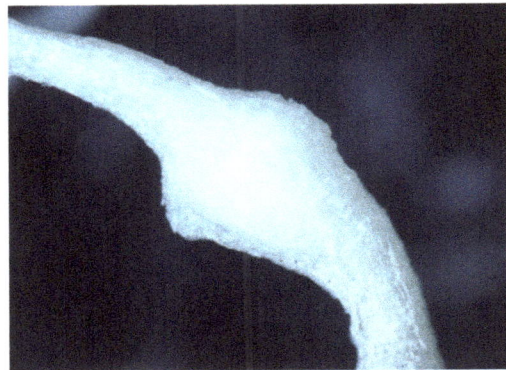

**Fig. 9.24.** An individual knot on a hemp root inside which a female of *Meloidogyne* species is parasitizing on the giant cells it has induced.

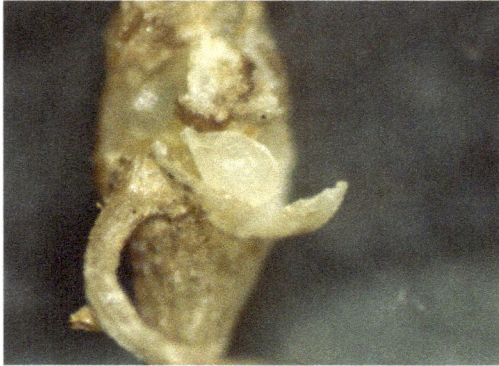

**Fig. 9.25.** A white, pear-shaped female of *Meloidogyne* species removed from inside the root knot tissue. Note the space inside the root knot where the nematode has developed from a vermiform second-stage juvenile to a globose body with a short 'neck'. The neck contains a stylet, metacorpus and oesophageal gland cells.

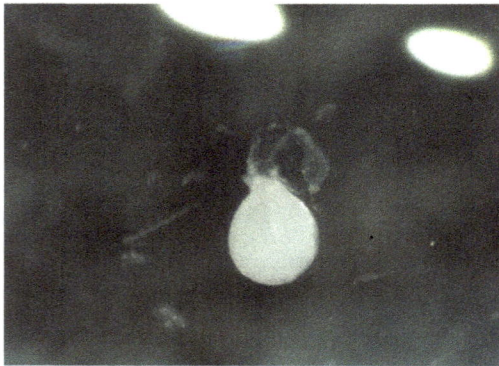

**Fig. 9.26.** A nearly mature female with a gelatinous sac. Note that the sac was dislocated during the dissection.

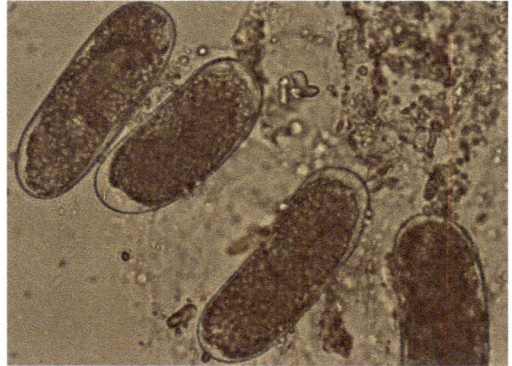

**Fig. 9.27.** Ellipsoid eggs produced by the root-knot nematode.

an obvious spicule near the tail. The spicule is for mating and a structure exclusively for males. The presence of males in a soil sample indicates root-knot nematode infection, but males in *Meloidogyne* species are often not required for reproduction. Therefore, it is not common to encounter root-knot nematode males in soil extracts.

### *Morphological identification*

It is not difficult to identify a root-knot nematode into a genus when root knots are observed and

females are found, but more procedures are required for species identification. One traditional method for species differentiation is to characterize the perineal pattern of the female nematode. Each species displays a characteristic and consistent pattern of ridges and annulations at the region surrounding the vulva and anus (the perineum).

**1.** To observe the perineal pattern, place a section of the root (containing knots) in 0.9% NaCl in a plastic Petri dish and use needles and a scalpel to dissect the females from the root under a dissecting microscope.
**2.** Transfer a live, egg-laying female to a drop of 45% lactic acid in a plastic Petri dish and then gently push the female body out from the drop so that it is held in place by surface tension.
**3.** Use a razor blade to cut off the posterior of the nematode and then gently use a dissecting needle to remove the body tissue from the posterior section.
**4.** Trim the cuticle into a square shape with the perineal pattern in the centre and then transfer it to a drop of glycerine at the centre of a glass slide.
**5.** Cover the drop with a coverslip and proceed to observation under a compound microscope.

It is recommended to prepare several perineal specimens on one slide so that they can be compared to obtain a consensus pattern. The procedures are simple, but an accurate identification requires referencing perineal pattern characters described for each known species. For the four most common *Meloidogyne* species, *M. incognita*, *M. javanica*, *M. arenaria* and *M. hapla*, the most important diagnostic characters of perineal patterns are their dorsal arch height, the presence or absence of lateral ridge and the shape of striae and tail terminus (Table 9.1).

Morphological identification to species level based on J2s is difficult. However, root-knot nematode J2s have distinct morphological characters that are different from those of other nematode genera, so they are valuable for identifying a nematode to the genus level. When nematodes are extracted from a soil sample, they are often in a community of nematodes comprising a broad range of genera and species. Root-knot nematode juveniles are usually recognized by their small size and lack of a vulva (Fig. 9.28). The oesophageal region appears light in colour and is overlapped with a denser or darker digestive system. They have a well-defined lip region, a distinct stylet with a robust basal knob (Fig. 9.29) and a distinct offset at the tail region (Mai and Mullin, 1996).

Morphological measurements can be obtained from both females and second-stage juveniles (J2s). Usually 10–50 specimens are measured to calculate the average size and size range. A drawing tube (attached to a microscope) is often needed to draw and measure a curved shape. For an *M. javanica* population infecting the hemp cultivar 'YunMa 1' (Song *et al.*, 2017), female body length (L) was 922.6 ± 133.6 µm, body width (BW) was 497.5 ± 81.6 µm, stylet 16.5 ± 1.4 µm, distance from dorsal oesophageal gland orifice to stylet base (DGO) 4.2 ± 0.6 µm and distance from vulva to anus 17.9 ± 2.1 µm. The J2s were 465.6 ± 20.0 µm long (L), 15.4 ± 1.0 µm wide (BW); their stylets were 14.8 ± 0.3 µm long and DGO 4.0 ± 0.5 µm. The J2s usually had long and tapering tails (56.7 ± 3.5 µm) but the tip was bluntly rounded instead of pointed. The tail terminus was hyaline (13.9 ± 3.4 µm). These measurements can be used as a reference for juveniles of *Meloidogyne* species. However, populations from different cultivars or geographical locations may have significantly different measurements.

## DNA-based identification

Root-knot females or juveniles can be accurately identified with a DNA-based approach by using universal primers (see Table 5.6 in Chapter 5). Alternatively, a universal primer pair designed from the 18S region of rDNA (SSUF07: 5'-AAAGATTA AGCCATGCATG-3', SSUR26: 5'-CATTCTTGG CAAATGCTTTCG-3') can be used for barcode-based nematode identification (Floyd *et al.*, 2002). Many segments of rDNA sequences were amplified by using various primers to characterize root-knot nematode species in Arkansas (Ye *et al.*, 2019) and from turfgrasses in North Carolina (Ye *et al.*, 2015), and some of those primers can be used if needed. The detailed identification process is described in Chapter 5. In brief, under a stereo microscope (transmitting light on) place a female or one to ten juveniles in a drop of 10 µl 1× TE buffer on a clear glass slide and macerate the nematode with a sterile pipette tip. Pipette the mashed nematode preparation into 50 µl 1× TE buffer in a microcentrifuge tube, store at –20°C or directly proceed to PCR. Purify PCR products followed by either direct sequencing or cloning into T-vector before sequencing. Closely related species can be found in the NCBI database by using the BLASTn search tool.

## Key diagnostic evidence

Major evidence includes abnormal root enlargement or knots, presence of root-knot females inside knot tissue and *Meloidogyne*-specific DNA sequence (from any genetic locus) obtained from the nematode specimen.

## Other Plant-Parasitic Nematodes and Root Disease Complex

As discussed in Chapter 2 (see Table 2.7), plant-parasitic nematodes contain groups that are

**Table 9.1.** Comparison of perineal patterns among four most common root-knot nematode species (Sasser and Carter, 1985).

| Species | Dorsal arch | Lateral ridge | Striae | Tail terminus whorl |
|---|---|---|---|---|
| *M. incognita* | High, square-like | Absent | Coarse | Present |
| *M. javanica* | Low, round | Present | Coarse | Present |
| *M. arenaria* | Low, round | Absent | Coarse | Absent |
| *M. hapla* | Low, round | Absent | Fine | Absent |

either migratory ectoparasitic, migratory endoparasitic, or sedentary endoparasitic. The rootknot nematode, a sedentary endoparasite, is a typical example of a nematode disease on hemp. However, there are other types of nematodes that behave differently and can cause significant root damage to hemp crops. Diagnosing those nematodes may require different approaches. In hemp crops, migratory ectoparasitic or endoparasitic nematodes are commonly found in the root zone. They use robust stylets to penetrate root tissue and draw nutrients from root cells. Migratory endoparasites cause even more damage than ectoparasites because they enter and migrate inside root tissue. Nematode feeding causes

nutritional stress on the plant. When the population of a given parasitic nematode species is high, their collective feeding may result in significant damage to the roots. Affected hemp plants exhibit symptoms ranging from mild leaf yellowing to severe stunting. Because these nematodes move very slowly in soil and are unevenly distributed in a field, an infected crop often shows up as patches of diseased plants. Further, nematode feeding or migrating creates mechanical injury to the root cells and compromises the plant's defence system against fungal, oomycete and bacterial pathogens, leading to the development of a much more severe disease or a disease complex (Back *et al.*, 2002). Even with disease-resistant cultivars, nematode invasion may break the host's resistance. Finally, some ectoparasitic nematodes are important vectors of plant viruses (Table 9.2) and they transmit viruses from plant to plant (Brown and MacFarlane, 2001). For example, *Tobacco ringspot virus* (TRSV) and *Tomato ringspot virus* (TomRSV) were reportedly isolated from hemp plants (Sevik, 2020) and these two viruses are efficiently transmitted by the North America dagger nematode, *Xiphinema americanum* (Wang and Gergerich, 1998; Wang *et al.*, 2002). Therefore, both the viruses and their nematode vectors could become problematic for hemp production. More details on root disease complex and holistic diagnosis are described in Chapter 13.

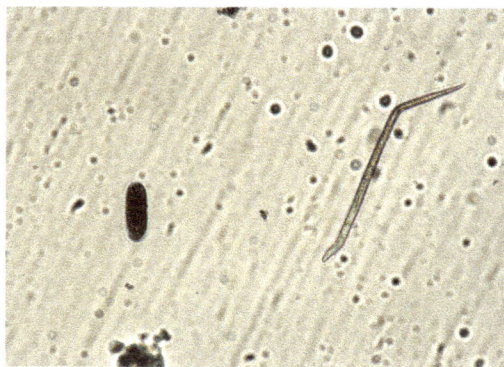

**Fig. 9.28.** A vermiform second-stage juvenile (J2) of *Meloidogyne* sp. and its size compared with an ellipsoid egg (on the left).

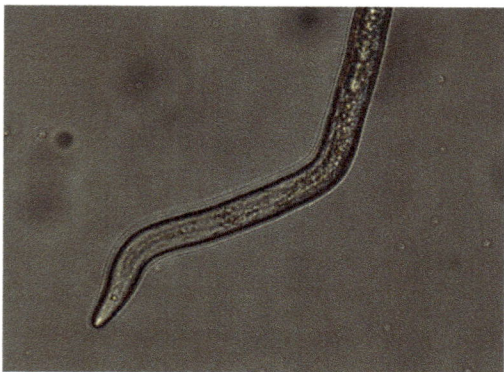

**Fig. 9.29.** The anterior portion of a second-stage juvenile of *Meloidogyne* sp.

**Table 9.2.** Transmission specificity of plant viruses by ectoparasitic nematodes.

| Nematode genera | Virus genus |
|---|---|
| *Xiphinema, Longidorus* and *Paralongidorus* | *Nepovirus* |
| *Trichodorus* and *Paratrichodorus* | *Tobravirus* |
| **Specific nematode species** | **North American nepoviruses** |
| *X. americanum, X. californicum, X. rivesi* | *Tobacco ringspot virus* |
| *X. americanum, X. bricolensis, X. californicum, X. rivesi* | *Tomato ringspot virus* |
| *X. americanum, X. californicum, X. rivesi* | *Cherry rasp leaf virus* |
| *X. index, X. italiae* | *Grapevine fan leaf virus* |
| *X. americanum* | *Peach rosette mosaic virus* |

### Field diagnosis

#### Symptoms

Infection by migratory ectoparasitic or endoparasitic nematodes alone usually does not cause visible root symptoms, especially when the nematode density is low. When a nematode population builds up, root symptoms become recognizable. For example, in an experimental condition, cucumber plants inoculated with a high population of the stubby-root nematode, *Trichodorus nanjingensis*, exhibited noticeable root tip enlargement (see Fig. 2.39). The above-ground symptoms are often noticeable when roots are substantially damaged. The typical symptoms include leaf yellowing, chlorosis, stunting and temporary wilt. In most cases, the overall symptoms resemble those of nutrient deficiency. For sedentary endoparasitic nematodes, such as root-knot and cyst nematodes, infected roots exhibit characteristic root enlargement (see Fig. 2.42 and Fig. 9.20) and attached cysts (see Fig. 2.43). When a viruliferous nematode vector is present in the soil, plants may show viral symptoms such as ringspot, mosaic, chlorosis, or distorted leaves. Plants infected by both nematodes and fungal pathogens (disease complex) may exhibit severe root rot, dieback, wilt, or death (Fig. 9.30).

#### Problem classification

Classifying a problem as a migratory nematode infection can be challenging, as the nematode-caused symptom, if any, is subtle or mild. Nevertheless, by examining a few whole plants and ruling out other common diseases, one may narrow a problem down to a nutritional or nematode issue. A soil test for both nutrients and nematodes is needed to classify a problem as a nematode infection.

#### Do your own diagnosis

Growers can perform wet sieving to extract ectoparasitic nematodes from soil or use a Baermann funnel method to extract endoparasitic nematodes from root samples (see Chapter 5). The procedures only require a 325-mesh sieve, with an optional 20-mesh sieve. Nematodes can be partially cleaned to remove large soil particles by decanting a couple of times, as nematodes float in the water longer than soil particles. The nematode preparation can be observed under a stereo microscope if equipped, or forwarded to a nematologist for identification. For those who lack a facility to extract nematodes, soil samples collected from a field can be submitted to a nematology laboratory.

#### Sampling

Most ectoparasitic plant-parasitic nematodes occur mainly in the soil around the root zone; and migratory endoparasitic nematodes occur in

**Fig. 9.30.** Cucumber roots killed by the root-knot nematode and *Pythium aphanidermatum* (a disease complex). Note that root knots and root rot were both present.

both root zone and root tissue, while sedentary nematodes (mainly females) are present inside roots. Depending on the nematode species to be tested, either the root-zone soil or roots, or both, can be sampled for nematode analysis. Because a field is more likely to have a community of nematodes species, with some doing more harm to the plant than others, it is recommended to take both soil and roots (especially symptomatic) for all nematode analyses. Nematodes are unevenly distributed in the soil and a certain pattern of soil sampling should be followed to obtain a representative sample and increase the detection rate (Fig. 9.31). For an individual plant, five to eight sub-samples can be taken from around or within the perimeter of the root system and mixed thoroughly in a bucket. Take 500 ml of the mix and place it in a plastic bag as one composite sample labelled for that specific plant. Feeder roots may be included as well as major roots if galls or other abnormalities are observed. Seal the bag to maintain the original soil moisture. Do not dry or freeze soil samples. Refrigeration of the soil sample at 4–8°C is not necessary if it is submitted within 1 week. To survey a large field for all nematode species before planting or after harvesting, a zigzag pattern is the best approach to get a smaller number of samples with maximum representation. Simply divide a field into 1-acre areas and collect about 20 sub-samples from each acre following pattern B in Fig. 9.31. Use a soil tube, auger, or shovel to collect each sub-sample and mix them thoroughly. Take 500 ml of the mix as one composite sample for nematode analysis. The total number of composite samples per field can be increased or decreased proportionally depending on the size of the field and the cost of testing. When taking soil, make sure to take soil from around the hemp roots. Superficial soil (0–5 cm) usually has fewer nematodes as it generally dries from being exposed to the sun. The deepest layer that parasitic nematodes can reach is the terminal ends of root system, but the most nematodes are present at the depth close to the middle portion of the root system, where there are the most feeder roots. Soil collected away from the root zone may have fewer nematode counts, especially for parasitic species. The timing of soil sampling can be pre-planting (to prevent), during the growing season (to diagnose), or after harvest (to evaluate), depending on the purposes.

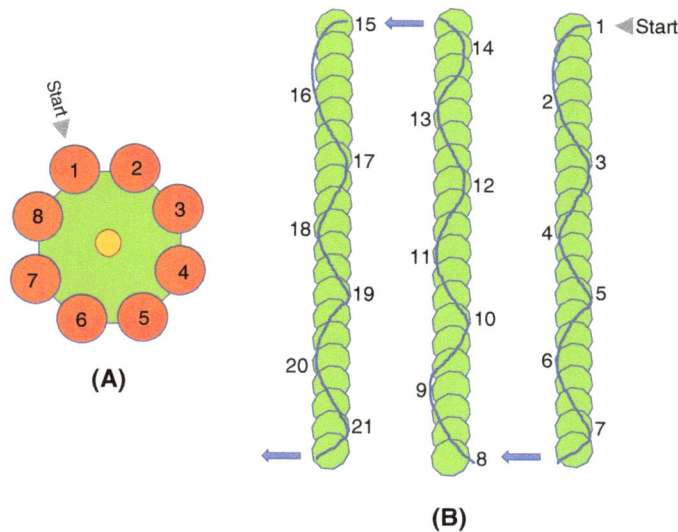

Fig. 9.31. Patterns of soil sampling for parasitic nematodes analysis. (A) The pattern of sub-sample points for an individual plant. (B) The pattern of sub-sample points for a 1-acre field. Note that the number of rows and space between sub-sample points may change from field to field.

## Lab diagnosis

### Examination for endoparasitic nematodes from roots

Root samples can be examined under a stereo microscope for any sedentary endoparasites, such as root-knot and cyst nematodes, as well as semi-endoparasites, such as citrus nematodes (*Tylenchulus* spp.) and reniform nematodes (*Rotylenchulus* spp.). Cyst nematodes can be identified by their egg-filled cysts or mature females, which are often exposed outside the root. They are often lemon-shaped or round with a very short head region. Citrus and reniform nematodes can be recognized by their posterior end swollen to a kidney shape (see Chapter 2).

### Examination of migratory endoparasitic nematodes from roots

Without obvious sedentary nematodes, root tissue can be washed clean under running tap water, chopped into small pieces, and then proceeded to Baermann funnel procedures. Migratory endoparasites such as lesion nematodes (*Pratylenchus* spp.) and burrowing nematodes (*Radopholus* spp.) are very active and easily isolated from root tissue. Fix the nematodes by adding equal amount (to nematode suspension volume) of 4% formaldehyde solution and mount nematodes on a slide to perform morphological identification under a compound microscope.

### Examination of ectoparasitic nematodes from soil

A common and quick method to extract nematodes from soil is wet sieving plus sugar centrifugation (see Chapter 5). Misting chamber and Baermann funnel procedures can also be used if there are only a few samples. However, these two methods may result in lower counts, as some nematodes are sluggish. Extracted nematodes can then be identified under a microscope.

### Morphological identification

Migratory nematodes are vermiform and structurally simple, with only a few morphological traits used for identifying a specimen to the genus level. For routine plant diagnostics, a genus level of identification is usually sufficient when it comes to disease cause and management. Species identification, when necessary, can be done with additional literature and reference materials (for example, the originally described specimens). To identify each nematode specimen encountered in a nematode extract to a genus, one may use the body size, oesophagus tract, stylet characters, the junction (with or without overlapping) between oesophageal gland and intestine, the position of vulva (middle or posterior) and tail shape. To observe these characters, pour a nematode extract, prepared from the sugar-flotation, Baermann funnel, or misting chamber extraction method, into a glass dish (with or without counting grid) and observe the nematodes under a stereo microscope with the transmitting light on. It is common to see three to five parasitic nematode genera in a sample and nematodes in each genus can be identified and counted. The following genera are considered to be common in crop soil and some have been found in hemp root-zones.

DAGGER NEMATODES (*XIPHINEMA* SPP.) The species in the *Xiphinema* genus are one of the largest parasitic nematodes by body length. The typical length is about 1.8–2.8 mm (Wang *et al.*, 1996). When relaxed, the individual nematode is in a 'C' shape. Under a stereo microscope, a long and sharp stylet is visible (Fig. 9.32). The stylet contains a spear and a long extension with basal flanges (enlargement). The total length ranges from 140 µm to 180 µm. At the base of the spear, there is a guiding ring, which is more visible under a compound microscope. Below the basal flanges, there is a slightly coiled tube (oesophageal lumen) extending into the oesophageal basal bulb. The basal bulb is clearly recognizable and often appears to be in a 'cut-off' position with the intestine (which appears as coarse refractive granules) without either dorsally or ventrally overlapping. The vulva is transverse and obvious (in female adults only). The tail shape varies by species, ranging from gradually tapered to pointed or with a significant peg. Males are not common. Juveniles are often encountered in the sample and they are similar to but smaller than female adults. Juveniles can be recognized by their lack of a vulva.

NEEDLE NEMATODES (*LONGIDORUS* SPP.) This genus is recognized by its long body, ranging from 3.3 mm to 5.1 mm, almost double the size of *Xiphinema* spp. When relaxed, the individual nematode is in an 'L' shape (Fig. 9.33). The stylet has a very long spear and extension and the total length can reach 200 µm. The conspicuous guiding ring is located at

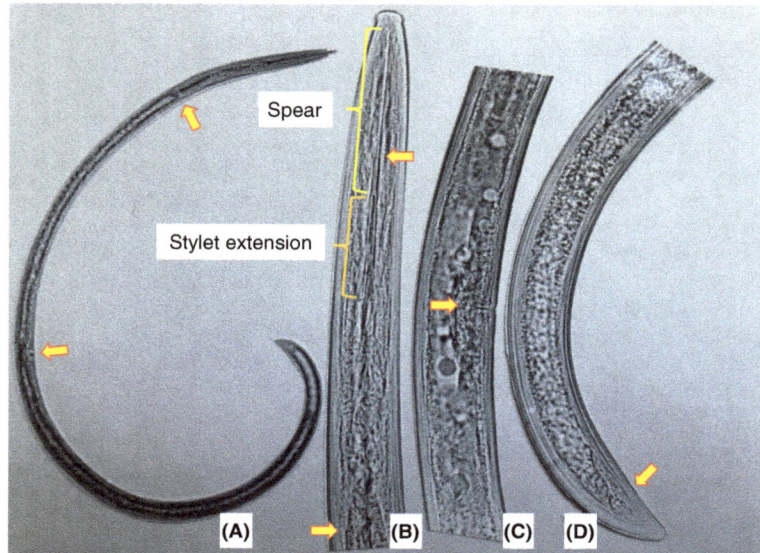

**Fig. 9.32.** Morphological characteristics of *Xiphinema thornei* Lamberti & Bleve-Zacheo, 1986. **(A)** Female adult showing a general 'C' shape, the junction of oesophageal basal bulb and intestine (arrow) and a vulva (arrow). **(B)** Anterior portion of female showing a long spear (odontostyle) and stylet extension (odontophore) in the head region. Note the guiding ring (arrow) and anterior portion of the oesophageal basal bulb. **(C)** Middle portion of female showing the transverse vulva (arrow). **(D)** Posterior portion of female showing a tapered tail and anus (arrow).

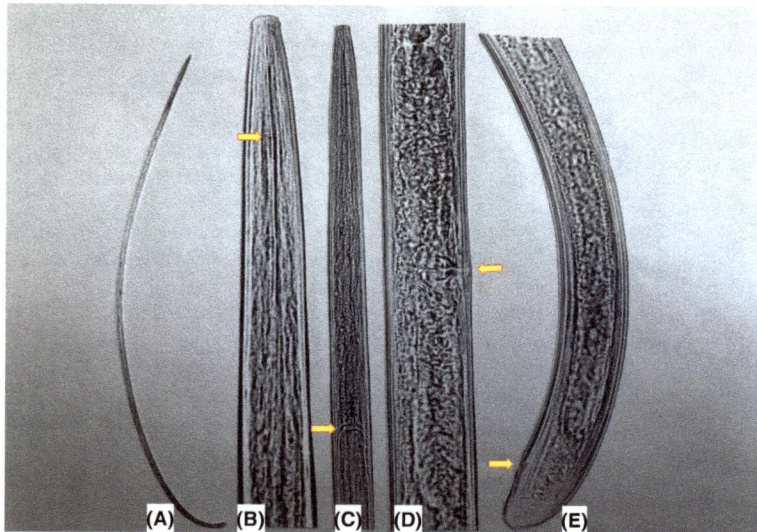

**Fig. 9.33.** Morphological characteristics of *Longidorus macromucronatus* Siddiqi, 1962. **(A)** Female adult showing typical 'L' shape when relaxed. **(B)** Anterior portion of female showing a long spear (odontostyle) and stylet extension (odontophore) in the head region. Note the guiding ring (arrow) located at the anterior portion of spear. **(C)** Junction area between the oesophageal basal bulb and intestine (arrow). **(D)** Middle portion of female showing the transverse vulva (arrow). **(E)** Posterior portion of female showing blunted tail and anus (arrow).

the upper portion of the spear, instead of at the base as in *Xiphinema* spp. At the base of the stylet extension, there are no swelling or flanges, distinguishing it from the *Xiphinema* stylet. The oesophageal tube is long and extends from the stylet base to the basal bulb. The basal bulb is distinct and elongated without overlapping with the intestine. The vulva is located at the middle portion of the body and transverse. The tail is blunted or slightly round.

**STUBBY-ROOT NEMATODES (*TRICHODORUS* SPP. AND *PARATRICHODORUS* SPP.)** *Trichodorus* spp. are much smaller than *Xiphinema* or *Longidorus*. When relaxed or heat-killed in 2% formaldehyde solution, the body becomes straight. Under a stereo microscope, the nematodes have a cylindrical shape with a slightly tapered anterior end and are chubbier when compared with other stylet-bearing nematodes (Fig. 9.34). This is because the body has a thicker cuticle layer that often inflates

upon fixation. In particular, the nematodes have a curved stylet, which is unique to *Trichodorus* and *Paratrichodorus* species. The stylet does not have a typical basal knob, as commonly seen in other parasitic nematodes. The oesophageal basal bulb is obvious and slightly overlaps with intestine. The vulva is shallow and located at the middle portion of the body. The tail is short and bluntly rounded. Males are abundant in most species, with a needle-like mating structure called spicules.

Morphologically, *Paratrichodorus* species (Fig. 9.35) are similar to *Trichodorus* species, but taxonomically they are assigned to a different genus. This separation is based on the presence of bursa in the male and more detailed structures at the spicule area. Also, almost all *Trichodorus* species have males while *Paratrichodorus* usually do not, except for a few species. When fixed, *Paratrichodorus* nematodes usually swell more than *Trichodorus* species (Fig. 9.35D). The basis used for separating these morphologically

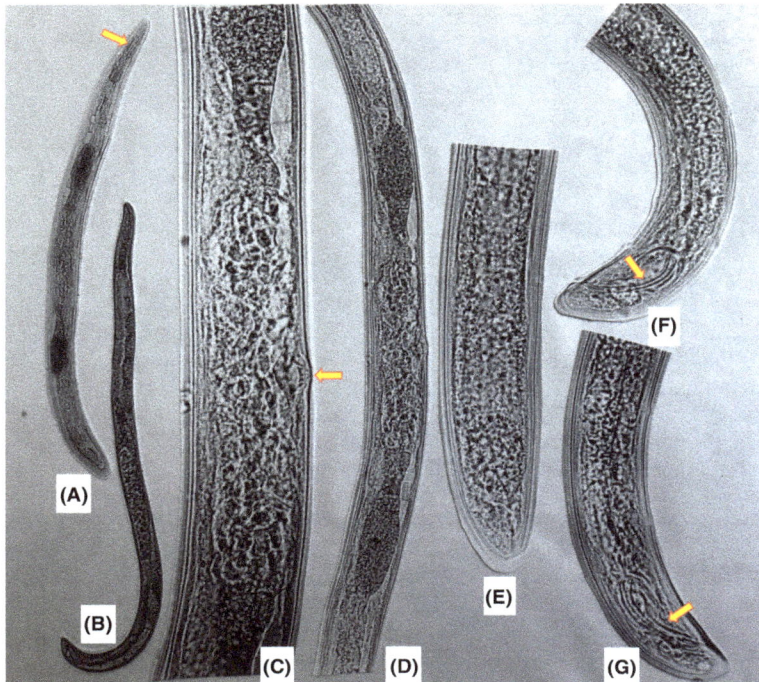

**Fig. 9.34.** Morphological characteristics of *Trichodorus nanjingensis* Liu & Cheng, 1990. **(A)** A female adult showing a general straight shape when relaxed. Note the curved stylet (arrow) at the head region. **(B)** A mature male adult. **(C)** Middle portion of female showing a shallow vulva (arrow). **(D)** A reproductive system inside the female. **(E)** Rounded tail of female. **(F, G)** Spicules (arrow) at the tail end of male.

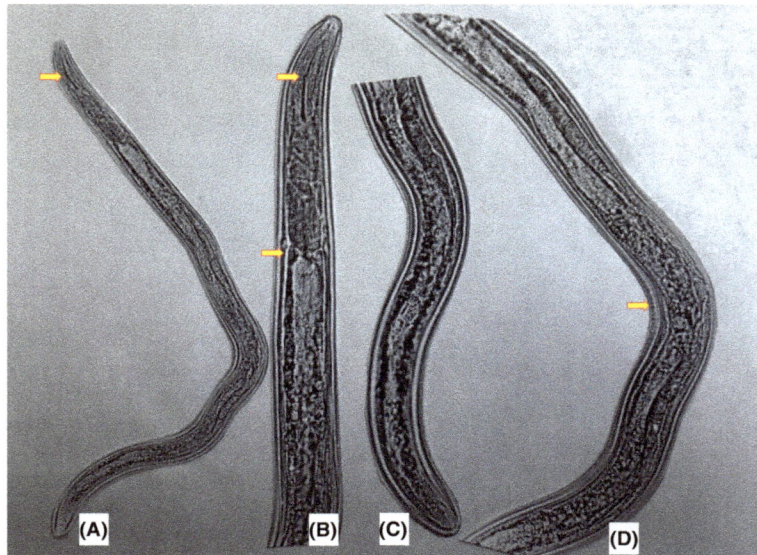

**Fig. 9.35.** Morphological characteristics of *Paratrichodorus porosus* (Allen, 1957) Siddiqi, 1974. **(A)** A female adult with a curved stylet (arrow). **(B)** Female head region showing the curved stylet (arrow) in more detail and slight overlapping between oesophageal basal bulb and intestine (arrow). **(C)** Posterior portion of female with rounded tail. **(D)** A thick layer of cuticle (arrow).

similar nematodes into two genera may be difficult to apply in a routine diagnosis, but the presence/absence of males in a nematode population may be used as one of the criteria to separate these two genera.

**STUNT NEMATODES (*TYLENCHORHYNCHUS* SPP.)** The body of a *Tylenchorhynchus* nematode is medium-sized. When relaxed, the body is in a wide 'C' shape. Under a stereo microscope, this genus of nematodes can be recognized by their short stylet (15–30 µm) but distinct basal knob, no overlapping between the oesophagus and intestine, a centrally (mid-body) located vulva and a conical to sub-cylindroid tail (Fig. 9.36). Additional characters include a slightly raised lip, a obvious to strong median oesophageal bulb and a distinct oesophageal basal bulb. These characters are visible in most female adults at the highest magnification under a stereo microscope, but better details are observed under a compound microscope. Males are common in this genus and both males and females may be present in a sample. *Tylenchorhynchus* nematodes are found from hemp root-zone soil, but their damage to hemp roots is not known. The nematode can enter the root as an endoparasite under some conditions.

**LESION NEMATODES (*PRATYLENCHUS* SPP.)** Lesion nematodes are very small (less than 1.0 mm long) when compared with other parasitic nematodes. However, they are easily recognized under a stereo microscope (with the aid of a compound microscope) by their actively moving, cylindroid body with a broad and almost square-shaped head region, a very flat lip, strong stylet basal knob, overlapping oesophageal basal bulb with intestine lateroventrally, posteriorly located vulva and bluntly rounded tail (Fig. 9.37). Males are present in most species and have conspicuous spicules and bursa.

**STEM NEMATODES (*DITYLENCHUS* SPP.)** The body of a *Ditylenchus* nematode is medium sized. When relaxed, the body becomes straight. However, under unfavourable conditions, a massive number of nematodes may coil together inside plant stem tissue (Fig. 9.38). Stem nematodes extracted from soil can be recognized under a stereo microscope by their plain lip region, weak but visible stylet, large median oesophageal bulb, distinct oesophageal basal bulb without overlapping with the intestine, posteriorly located vulva and elongated or tapered

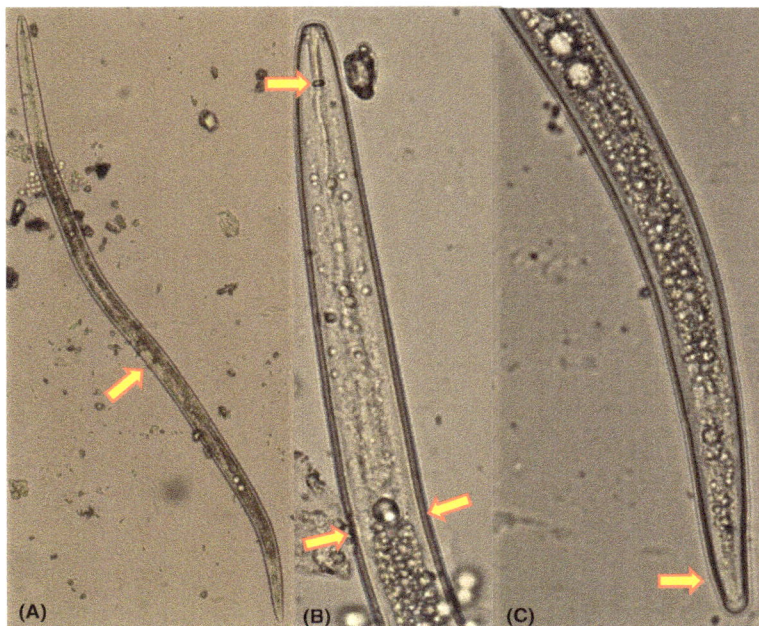

**Fig. 9.36.** Morphological characteristics of *Tylenchorhynchus* sp. extracted from a hemp root zone. **(A)** A female adult with a centrally located vulva (arrow). **(B)** Female head region showing a distinct stylet basal knob (arrow) and non-overlapping between the oesophageal basal bulb and intestine (arrows). **(C)** Posterior portion of female with sub-cylindroid tail (arrow). Note that tail length is about 2–3 times the tail width.

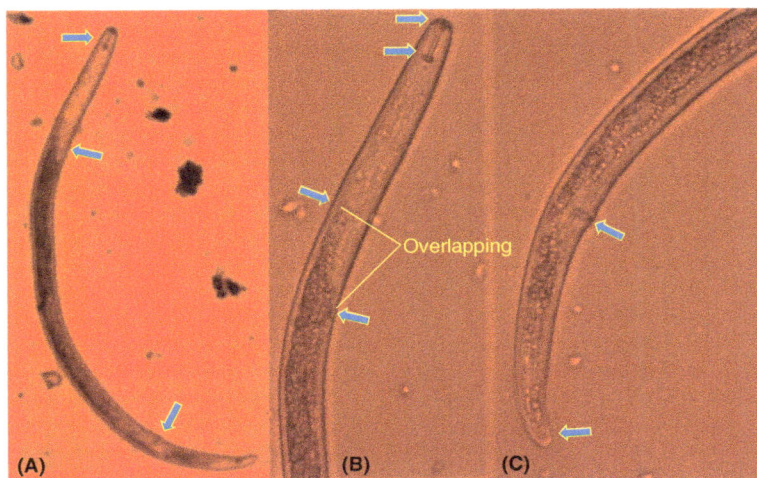

**Fig. 9.37.** Morphological characteristics of *Pratylenchus* sp. **(A)** A female adult at a lower magnification showing a strong stylet, overlapping oesophageal bulb (with intestine) and posteriorly located vulva (arrows). **(B)** Female head region at a higher magnification showing flattened lip, strong stylet basal knob and overlapping region between the oesophageal basal bulb and intestine (arrows). **(C)** Posterior portion of female with bluntly rounded tail and distinct vulva (arrows).

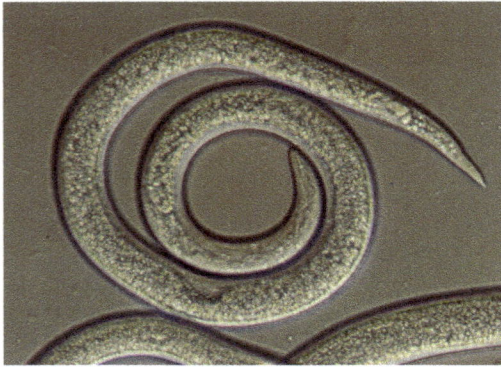

**Fig. 9.38.** A stem nematode (*Ditylenchus dipsaci*) coiled inside stem tissue. Note the small but conspicuous stylet with distinct knob and a long oesophageal lumen.

tail with a pointed terminus (Fig. 9.39). Males are abundant, with conspicuous spicules and bursa.

### DNA-based identification

All nematodes can be identified with a DNA-based approach by using universal primers (see Chapter 5, Table 5.6). Alternatively, 18S (TTGATTACG TCCCTGCCCTTT) and 26S (TTTCACTCGCC GTTACTAAGG) primers can be used to amplify a 1.5 kb DNA sequence from nematodes (Vrain *et al.*, 1992). This pair of primers is relatively universal, as it also amplifies sequences efficiently from dagger nematodes (Vrain *et al.*, 1992) and stubby-root nematodes (Lopez-Nicora *et al.*, 2014). There are many 'species-specific' primers designed for nematode species identification. However, these specific primers may not fit well into the identification process for an unknown species. Therefore, try to use a primer pair that can amplify a targeted region from a broad range of genera or species. To extract genomic DNA from a migratory nematode, place a female or male in a drop of 10 μl 1× TE buffer on a clear glass slide and macerate the nematode with a pipette tip. Pipette the mashed nematode preparation into 50 μl 1× TE buffer in a microcentrifuge tube, store at –20°C or directly proceed to PCR. Purify PCR products followed by either direct sequencing or cloning into T-vector before sequencing. Closely related species can be found in the NCBI database by using the BLASTn search tool (see Chapter 5).

### Key diagnostic evidence

A high population of stylet-bearing nematodes is present in hemp root-zone soil, which suggests that the root system is supporting nematode reproduction. Moderate to severe root and above-ground symptoms are observed and associated with the high density of a given nematode species in the soil. The nematode species is confirmed by its morphological characters and DNA-sequences.

## Root Disease Management

Nematodes and root pathogens live in soil. Some pathogens are very difficult, and also expensive, to eradicate. Unlike foliar diseases, which can be treated as soon as noticed with an approved product, a root disease may be difficult or too late to treat when it is detected during the growing season. That being said, a strategy for root disease management should focus on the prevention and periodical monitoring.

### Preventing pathogen introduction

Common ways of introducing a root pathogen into a field are through soil, growth media, tractors or machinery, infected cuttings or seeds and contaminated shoes. Contaminated irrigation water and fertilizer solutions can be another pathway for a pathogen to enter into hydroponic or aeroponic facilities. A field selected for hemp production should be secured without heavy traffic between the hemp field and non-hemp fields. Disposable shoe-coverings may be used when entering the field. Machinery should be thoroughly cleaned and sanitized before being used in the hemp field. Seeds and cuttings obtained from external sources should be verified to be free from *Fusarium* and other soil- or seed-borne pathogens.

### Testing and monitoring soil health

Soil with the optimal physical and chemical properties is critical for a healthy crop, which is why a soil test for nutrients, acidity and other characteristics has become a routine practice. Similarly, a soil test for potential pathogens and nematodes is equally important for preventing root diseases. Therefore, soil samples should be collected and submitted to a plant diagnostic lab for nematode and pathogen

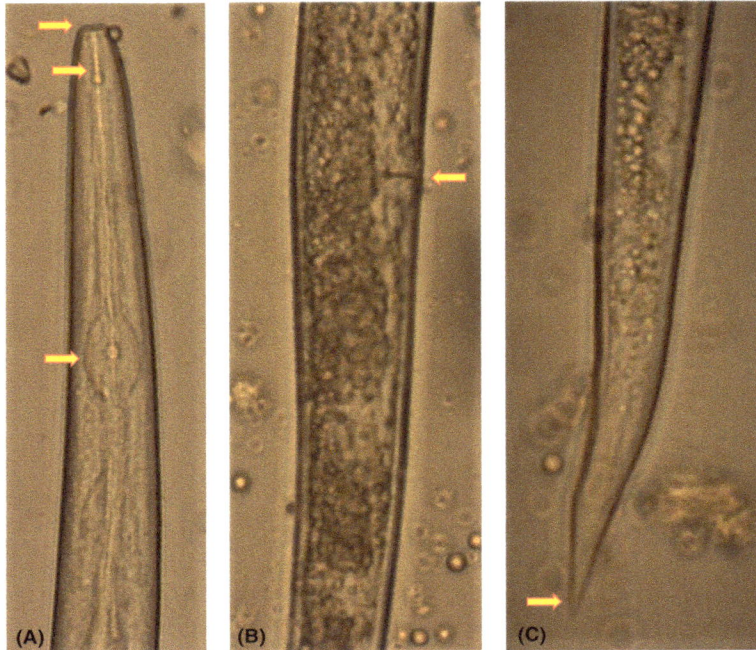

**Fig. 9.39.** Morphological characteristics of *Ditylenchus dipsaci*. **(A)** Female head region showing plain lip, a small stylet and a large median oesophageal bulb (arrows). **(B)** Posteriorly located vulva (arrow). **(C)** Tapered tail with pointed terminus (arrow).

counts (see Fig. 9.31 for sampling pattern). *Fusarium* species and endoparasitic nematodes are high-risk pathogens, as they cause serious diseases in hemp. A field found to contain a high level of pathogen inoculum or nematodes should not be selected for hemp production.

### Mitigating environmental stress

In arid and semi-arid areas, drought stress and pathogen infections may occur simultaneously in the field. When plants are drought-stressed, some weak or opportunistic soil-inhabiting pathogens can attack roots, causing severe root rot. In barley, chronic drought stress promoted *Fusarium* crown rot by extending the initial infection phase and enhancing the proliferation and spread of the pathogen (Liu and Liu, 2016). In chickpea, drought stress increased the incidence of root rot caused by either *Rhizoctonia bataticola* or *Fusarium solani* (Sinha *et al.*, 2019). In the case of hemp root rot caused by multiple *Fusarium* species, the crop was initially under drought stress that triggered root rot

caused by multiple species, including those considered as weak pathogens (Schoener *et al.*, 2017). On the other hand, overwatering or saturated soil also stresses plants by reducing oxygen in the soil. Excessive soil moisture reduces the root vigour and worsens *Pythium* or bacterial root rot.

### Amending soil with biocontrol agents

*Trichoderma* species are an effective biological control agent against *Fusarium* and other soilborne pathogens and can be applied in the soil to manage root rot diseases. *Trichoderma* species occur naturally in soil and they are sometimes isolated from hemp roots, especially when the roots are infected with *Fusarium* or other pathogens. In Petri-dish experiments, *Trichoderma* are highly active in parasitizing certain fungal pathogens isolated from diseased hemp roots (Fig. 9.40). *Trichoderma* not only help suppress the disease naturally, but also have other benefits, such as inducing host resistance, increasing crop tolerance to drought stress and, to some extent, degrading hazardous chemical

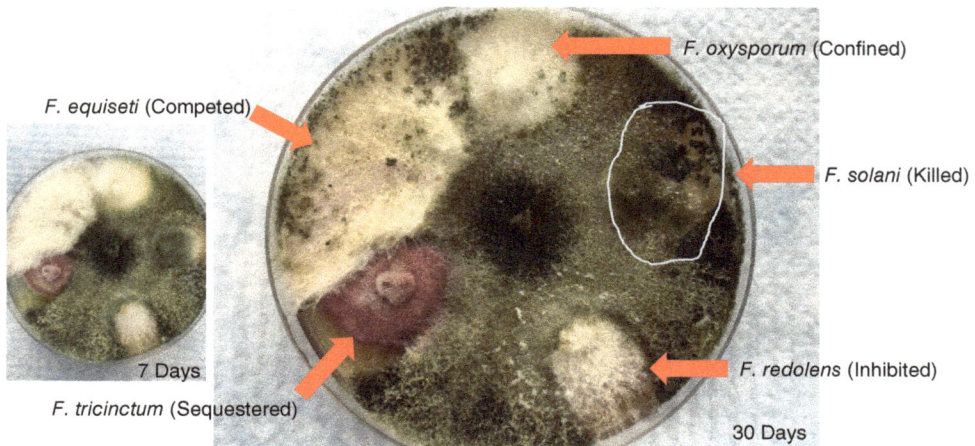

**Fig. 9.40.** *Trichoderma virens* strain CsR-34 killed or suppressed the growth of major *Fusarium* species isolated from hemp plants having a severe root rot disease. Note that *T. virens* was originally at centre of PDA plate and outgrew all *Fusarium* species within 7 days (left) and completely destroyed *F. solani* colony (white circle) within 30 days (right).

compounds. Applying *Trichoderma* products can directly introduce these beneficial organisms into the soil and lower disease severity. Hypothetically, *Trichoderma* strains isolated from hemp roots may more readily colonize hemp roots with a greater efficacy in disease control.

### Controlling nematodes

There are some organic measures used to lower nematode density in soil, such as tillage, solarization and rotation with resistant or non-host plants. Crop rotation is the most effective and inexpensive method. Biological control of nematodes has been explored but is generally ineffective. Nematicides are often used in nematode control in some high-value crops. However, chemical-based nematicides or fumigants are generally toxic. For hemp production, it is better to choose a field that has zero to low density of harmful nematodes in the soil, so that nematicides can be avoided.

### Fallowing fields and rotating crops

Fields infected with certain pathogens can be fallowed, if practical, to lower pathogen inocula before planting a hemp crop. Alternatively, a known resistant cultivar or crop can be planted in a field for 1–2 years before a hemp crop is returned. A rotation scheme is generally effective for root disease control, but it requires annual testing to monitor the status of pathogens and nematodes so that an appropriate crop can be planted. Certain hemp varieties appear to be more tolerant to some root diseases than others and different varieties may be rotated if fallowing or rotating with a non-hemp crop is not an option.

### Treating soil with fungicides

Application of a fungicide is the last resort for root disease control, but its usage is ruled by the product label. There may not be many options available when it comes to chemical treatment for a field when a hemp crop is growing.

### References

Back, M.A., Haydock, P.P.J. and Jenkinson, P. (2002) Disease complexes involving plant parasitic nematodes and soilborne pathogens. *Plant Pathology* 51, 683–697. doi: 10.1046/j.1365-3059.2002.00785.x

Beckerman, J., Nisonson, H., Albright, N. and Creswell, T. (2017) First report of *Pythium aphanidermatum* crown and root rot of industrial hemp in the United States. *Plant Disease* 101, 1038. doi: 10.1094/PDIS-09-16-1249-PDN

Beckerman, J., Stone, J., Ruhl, G. and Creswell, T. (2018) First report of *Pythium ultimum* crown and root rot of industrial hemp in the United States. *Plant Disease* 102, 2045. doi: 10.1094/PDIS-12-17-1999-PDN

Brown, D.J.F. and MacFarlane, S.A. (2001) 'Worms' that transmit viruses. *Biologist* 48, 35–40.

Floyd, R., Abebe, E., Papert, A. and Blaxter, M. (2002) Molecular barcodes for soil nematode identification. *Molecular Ecology* 11, 839–850. doi: 10.1046/j.1365-294x.2002.01485.x

Leslie, J.F. and Summerell, B.A. (2006) *The Fusarium Laboratory Manual*, 1st edn. Blackwell Publishing, Ames, Iowa. doi: 10.1002/9780470278376

Liu, X. and Liu, C. (2016) Effects of drought-stress on *Fusarium* crown rot development in barley. *PLoS ONE* 11, e0167304. doi: 10.1371/journal.pone.0167304

Lopez-Nicora, H.D., Mekete, T., Sekora, N. and Niblack, T. L. (2014) First report of the stubby-root nematode (*Paratrichodorus allius*) from a corn field in Ohio. *Plant Disease* 98, 1164. doi: 10.1094/PDIS-11-13-1180-PDN

Mai, W.F. and Mullin, P.G. (1996) *Plant-parasitic Nematodes A Pictorial Key to Genera*. Cornell University Press, Ithaca, New York.

McGehee, C.S., Apicella, P., Raudales, R., Berkowitz, G., Ma, Y. et al. (2019) First report of root rot and wilt caused by *Pythium myriotylum* on hemp (*Cannabis sativa*) in the United States. *Plant Disease* 103, 3288. doi: 10.1094/PDIS-11-18-2028-PDN

Mitkowski, N.A. and Abawi, G.S. (2003) Root-knot nematodes. The Plant Health Instructor, American Phytopathological Society, St Paul, Minnesota. doi: 10.1094/PHI-I-2003-0917-01

Pofu, K.M. and Mashela, P.W. (2014) Density-dependent growth patterns of *Meloidogyne javanica* on hemp cultivars: establishing nematode-sampling timeframes in host-status trials. *American Journal of Experimental Agriculture* 4, 639–650. doi: 10.9734/AJEA/2014/4853

Punja, Z.K. and Rodriguez, G. (2018) *Fusarium* and *Pythium* species infecting roots of hydroponically grown marijuana (*Cannabis sativa* L.) plants. *Canadian Journal of Plant Pathology* 40, 498–513. doi: 10.1080/07060661.2018.1535466

Punja, Z.K., Collyer, D., Scott, C., Lung, S., Holmes, J. et al. (2019) Pathogens and molds affecting production and quality of *Cannabis sativa* L. *Frontiers in Plant Science* 10, 1120. doi: 10.3389/fpls.2019.01120

Punja, Z.K., Scott, C. and Chen, S. (2018) Root and crown rot pathogens causing wilt symptoms on field-grown marijuana (*Cannabis sativa* L.) plants. *Canadian Journal of Plant Pathology* 40, 528–541. doi: 10.1080/07060661.2018.1535470

Sasser, J.N. and Carter, C.C. (1985) *An Advanced Treatise on Meloidogyne. Volume 1. Biology and Control*. North Carolina State University Graphics, Raleigh, North Carolina.

Schoener, J.L., Wilhelm, R., Rawson, R., Schmitz, P. and Wang, S. (2017) Five *Fusarium* species associated with root rot and sudden death of industrial hemp in Nevada. *(Abstr.) Phytopathology* 107, S5.98. doi: 10.1094/PHYTO-107-12-S5.98

Sevik, M.A. (2020) Kenevir (*Cannabis sativa* L.) Bitkilerinde Görülen Virüs Kaynaklı Hastalıklar. *Turkish Journal of Agricultural Research* 7, 111–119. doi: 10.19159/tutad.663715

Sinha, R., Irulappan, V., Mohan-Raju, B., Suganthi, A. and Senthil-Kumar, M. (2019) Impact of drought stress on simultaneously occurring pathogen infection in field-grown chickpea. *Scientific Reports* 9, 5577. doi: 10.1038/s41598-019-41463-z

Song, Z.Q., Cheng, F.X., Zhang, D.Y., Liu, Y. and Chen, X.W. (2017) First report of *Meloidogyne javanica* infecting hemp (*Cannabis sativa*) in China. *Plant Disease* 101, 842. doi: 10.1094/PDIS-10-16-1537-PDN

Vrain, T.C., Wakarchuk, D., Lévesque, A.C. and Hamilton, R. (1992) Intraspecific rDNA restriction fragment length polymorphism in the *Xiphinema americanum* group. *Fundamental and Applied Nematology* 15, 563–573.

Wang, S. and Gergerich, R.C. (1998) Immunofluorescent localization of tobacco ringspot virus in the vector nematode *Xiphinema americanum*. *Phytopathology* 88, 885–889. doi: 10.1094/PHYTO.2002.92.6.646

Wang, S., Chiu, W. F., Yu, C., Li, C. and Robbins, R.T. (1996) The occurrence and geographical distribution of longidorid and trichodorid nematodes associated with vineyards and orchards in China. *Russian Journal of Nematology* 4, 145–153.

Wang, S., Gergerich, R.C., Wickizer, S.L. and Kim, K.S. (2002) Localization of transmissible and nontransmissible viruses in the vector nematode *Xiphinema americanum*. *Phytopathology* 92, 646–653. doi: 10.1094/PHYTO.2002.92.6.646

Wheeler, D.L., Scott, J., Dung, J.K.S. and Johnson, D. A. (2019) Evidence of a trans-kingdom plant disease complex between a fungus and plant-parasitic nematodes. *PLoS ONE* 14, e0211508. doi: 10.1371/journal.pone.0211508

Ye, W., Robbins, R.T. and Kirkpatrick, T. (2019) Molecular characterization of root-knot nematodes (*Meloidogyne* spp.) from Arkansas, USA. *Scientific Reports* 9, 15680. doi: 10.1038/s41598-019-52118-4

Ye, W., Zeng, Y. and Kerns, J. (2015) Molecular characterization and diagnosis of root-knot nematodes (*Meloidogyne* spp.) from turfgrasses in North Carolina, USA. *PLoS ONE* 24, 1–16. doi: 10.1371/journal.pone.0143556

# 10 Diagnosing Systemic Diseases

A systemic disease is defined as one where a pathogen infects a plant systemically or affects the plant as a whole. There are certain types of diseases in hemp and cannabis crops that are viewed more as systemic rather than localized, based on both infection behaviour and symptom development. For example, systemic wilt caused by *Fusarium* species affects the plant as a whole by destroying the vascular system. Many viruses and viroids can move from cell to cell after initial infection and then spread systemically after reaching the vascular system (Rodrigo *et al.*, 2014). Fastidious bacteria such as phytoplasmas cause systemic symptoms even though they are mainly restricted to phloem tissue (Doi *et al.*, 1967; Hogenhout *et al.*, 2008). Plants infected by these pathogens may show various symptoms without organ boundaries, such as systemic wilt, malformation, chlorosis, witches' broom and abnormal growth. Viruses and viroids are difficult to cure partly because they live inside plant cells and hijack the host's nucleic acid replication systems to reproduce many copies of themselves. Effective treatment largely depends on the successful identification of chemicals that can effectively target the viruses or the replication system these viruses use. Similarly, diseases caused by phytoplasmas and other fastidious bacteria are also difficult to cure. Both viral and phytoplasma diseases are transmitted by arthropods and are categorized as vector-borne diseases. Therefore, controlling vectors is a major strategy for mitigating disease spread in the field. Both viruses and fastidious bacteria are obligate pathogens and their diagnoses commonly rely on direct detection of pathogen DNA or RNA from plant tissues via molecular methods.

## Vascular Wilt Caused by *Fusarium* Species

As discussed in previous chapters, species in the *Fusarium* genus cause root rot and stem canker diseases in *Cannabis* plants. They also cause vascular wilt: they infect and move systemically inside vascular tissue, damage the xylem vessels and interrupt the vessels' normal function of translocating water from the roots throughout the plant. Common species causing hemp and cannabis wilt are *Fusarium oxysporum* and *F. solani*.

### Field diagnosis

#### Symptoms

The initial symptoms of vascular wilt show up on newly grown leaves as chlorosis or yellowing and on older leaves as brown lesions along the edge. As the infection progresses systemically, plants start to exhibit temporary or permanent wilt even with adequate soil moisture. Infected plants grow poorly and are not responsive to fertilization. Stems mostly appear normal on their surfaces but may have a necrotic lesion at the base (Fig. 10.1). Internal vascular tissue and pith become dark brown and necrotic (Fig. 10.2), which is the most diagnostic symptom for *Fusarium* wilt. *Fusarium* pathogens can be passed from mother plants to cuttings.

#### Problem classification

When plants exhibit wilt and poor growth, other potential causes such as chronic or acute drought stress and nutrient deficiency should be evaluated and ruled out. If adequate watering and fertilization do not improve growth, the problem can be narrowed down to potential vascular disease. Examine stems of symptomatic plants for internal vascular browning. Any discoloration may indicate a fungus-caused vascular disease.

#### Do your own diagnosis

For a dicotyledon plant, a cross-section of the stem typically exhibits layers of (from outer to inner)

DOI: 10.1079/9781789246070.0010

epidermis, hypodermis, cortex, phloem, xylem and pith. The phloem and xylem form an integrated structure called a vascular bundle. Certain portions of the stem, especially in young plants, may lack the pith tissue and look hollow at the centre. If a plant is infected by *Fusarium*, one can see the brownish

and damaged vascular bundle from cross- or longitudinal sections of the lower and middle portions of the stem. A cross-section of stem with a hollow typically shows a brown circle with a hole at the centre (Fig. 10.3) and a longitudinal section exhibits brownish streaking inside the stem (Fig. 10.4). In an older woody stem, infection can be observed as a light brown colour along the pith and vascular tissue (Fig. 10.5), while the inside of a healthy stem

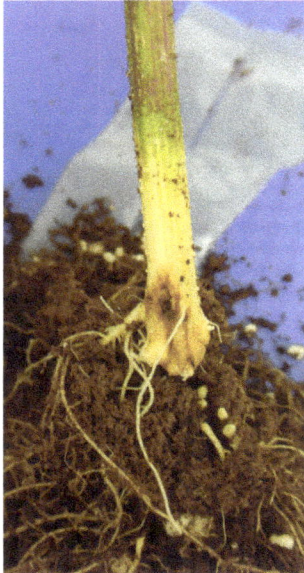

**Fig. 10.1.** A brown lesion at the base of a cannabis stem caused by *F. oxysporum*. Note that the root system was healthy.

**Fig. 10.3.** A cross-section of a cannabis stem infected with *Fusarium oxysporum*. Note that the xylem was dark brown and the phloem was lightly discoloured.

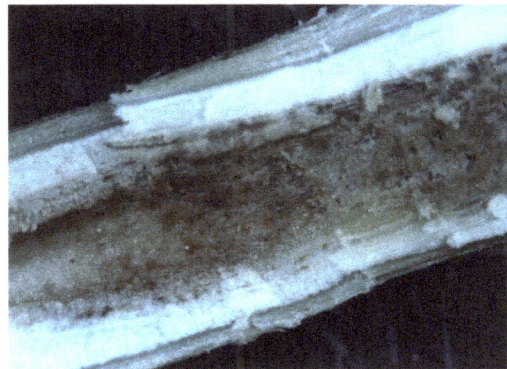

**Fig. 10.2.** A healthy cannabis stem (left) and one infected by *F. oxysporum* (right). Note the dark brown colour of pith and surrounding tissue. The infection appeared to start from the stem base and move upwards.

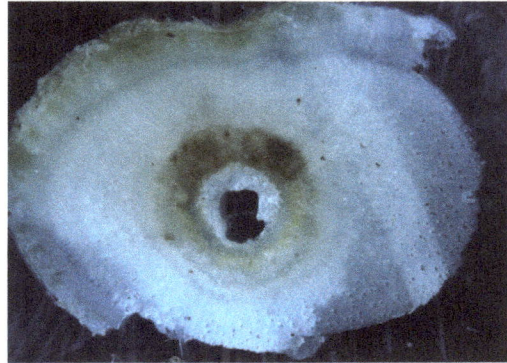

**Fig. 10.4.** A longitudinal cut of a cannabis stem (infected by *Fusarium oxysporum*) showing distinct brown streaking inside the stem. Note that the outer layers of stem, including hypodermis and cortex, remain intact (normal colour) and that *Fusarium*-caused vascular wilt differs from *Fusarium*-caused stem canker or rot symptomatically, as the former is xylem and phloem specific while the latter affects all layers of the stem (see Chapter 8).

should be fresh white without discoloration. Vascular discoloration is not a diagnostic characteristic solely for *Fusarium* infection, but the presence of this symptom is very likely to indicate *Fusarium* vascular wilt, especially for *Cannabis sativa* plants. To confirm *Fusarium* wilt, a pre-poured PDA plate can be used to isolate *Fusarium* from discoloured vascular tissue and plates can then be submitted to a diagnostic lab for species identification.

### Sampling

Once the problem has been determined to be vascular wilt, cut a stem showing typical vascular streaking from symptomatic plants into smaller pieces. Place the stem pieces in a Ziploc bag as one sample. Repeat the same procedures to collect stem samples from an additional one or two symptomatic plants. Do not add extra moisture or paper towels, and do not freeze the samples. Submit samples as they are to a diagnostic lab.

### Lab diagnosis

### Visual and microscopic examinations

Visually examine a stem sample and cut through the stem to reveal the internal texture and structure. Depending on the severity of the disease, vascular discoloration may vary from dark brown with rot to lightly coloured. Examine the epidermis,

**Fig. 10.5.** A typical discoloration seen in a longitudinal cut of a hemp stem infected by *Fusarium oxysporum*. The top is a healthy-looking stem and the bottom is an infected stem.

hypodermis and cortex layers for any signs of decay and determine if the disease belongs to *Fusarium* stem rot or vascular wilt. As shown in Fig. 10.4, *Fusarium* wilt is a disease mainly affecting vascular tissue. Under a stereo microscope, examine the vascular bundle's texture and rule out other types of vascular diseases or even arthropods. In the case of *Fusarium* wilt, the vascular tissue may look darker, necrotic and often decayed, but remains solid. If the tissue is decaying, place a tiny amount of tissue on a glass slide and observe it under a compound microscope to see if any fungal mycelia or spores are present, which may provide additional information to the diagnosis.

### Isolation and colony observation

Isolating *Fusarium* species from vascular tissue is the best way to prove *Fusarium*-caused vascular wilt. It is likely that either *F. oxysporum* or *F. solani* is recovered from infected tissue, but it is not uncommon that a plant is infected by both species (Schoener *et al.*, 2017). Refer to the hemp root rot section in Chapter 9 for isolation procedures and colony characteristics of *Fusarium* species. In brief, plate symptomatic vascular tissues on PDA amended with streptomycin (PDA$^{+strep}$) and incubate the plates at 22°C in the dark. *F. oxysporum* and *F. solani* are the most common species associated with vascular wilt in *Cannabis* plants. In 3–5 days, a white and cotton-like colony growing out from the tissue is an indication of *F. oxysporum* infection, while an off-white colony suggests *F. solani* infection (Fig. 10.6). Other species of *Fusarium* may be

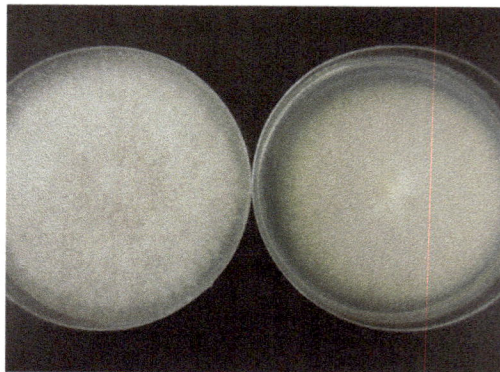

**Fig. 10.6.** Colonies (on PDA) of *Fusarium oxysporum* (left) and *F. solani* (right) isolated from cannabis vascular tissue.

isolated but their roles in vascular wilt require further investigations. Microconidia are abundant in the culture for both species (see Fig. 2.10).

### DNA-based identification

*F. oxysporum* and *F. solani* can be identified on culture media based on their morphological characteristics. However, it may be difficult and time-consuming to distinguish between these two species based on the colony morphology alone. Their identity can be easily determined using a DNA-based approach (see Chapter 5). DNA sequences obtained from the rDNA-ITS region by using ITS1/ITS4 primers are quite different between *F. oxysporum* and *F. solani* (Fig. 10.7) and a quick BLAST search can assign an isolate to a closely related species. When sequencing is not available, running an agarose gel (2%) can quickly distinguishes these two species, with a slightly larger PCR amplicon for *F. solani* isolates versus a smaller amplicon for *F. oxysporum* isolates.

### Pathogenicity test

In general, isolation and identification of either *F. oxysporum* or *F. solani* from hemp or cannabis plants exhibiting typical vascular wilt symptoms conclude *Fusarium* wilt diagnosis. However, a pathogenicity test may be performed on the same cultivar from which the disease is diagnosed.

1. In a greenhouse or growth chamber, prepare a set of young hemp plants each individually grown in a pot with three to five leaflets for an inoculation test.
2. At the same time, transfer an isolate to be tested to several fresh PDA plates and incubate the plates at 22°C for 1–2 weeks or until fully grown over.
3. Scrape mycelia and spores from three fully grown over plates into 100 ml sterile deionized water and adjust the spore concentration to approximately 3–4 $\times 10^6$/ml by using a haemocytometer (see below).
4. Once a spore suspension is prepared with sufficient quantity and a desired concentration, punch two holes in the soil around the root system of a test plant with a 5 ml pipette tip and then add 5 ml of the spore suspension into each hole.
5. Fill the holes with adjacent soil after inoculation. Inoculated plants can be assessed for symptom development within 2–4 weeks.

The same *Fusarium* species should be isolated and identified from the inoculated plants after symptom development to confirm the pathogenicity of an isolate.

A protocol for the use of a haemocytometer in calculating spore concentration in a suspension is as follows.

1. Pipette the spore suspension up and down five to seven times with a 5 ml pipette to obtain a uniform

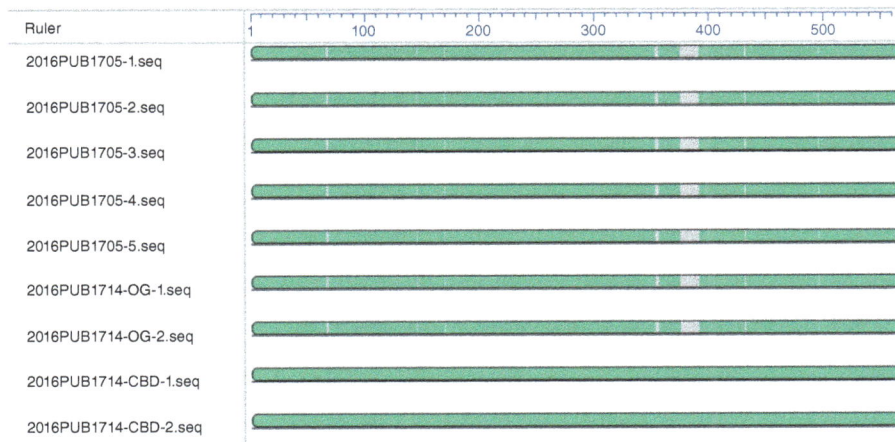

**Fig. 10.7.** Differences in DNA sequences of the ITS region of rDNA (using ITS1/ITS4 primers) between *Fusarium oxysporum* and *F. solani* isolated from *Cannabis sativa* plants. Note that the first seven isolates were identified as *F. oxysporum* (544 bp) and the last two isolates were *F. solani* (569 bp), and that there are seven gaps (grey bars) in *F. oxysporum* isolates when compared with sequences of *F. solani* isolates.

suspension. In case of a large quantity (e.g. over 200 ml), a gentle swirling can resuspend spores evenly.

2. Place the cover slip (provided with the haemocytometer) over the haemocytometer's chambers.

3. Pipette 10 μl of spore suspension and gently load into the chamber on one side (through capillary action) and load another 10 μl to another chamber on the other side. The amount required to fill each chamber completely is provided by the manufacturer.

4. Under a compound microscope, adjust the focus to locate the grid lines of the haemocytometer and the spores within.

5. Count the number of spores in the four large squares (each containing 16 small squares) at the four corners of the counting chambers.

6. Count the total spores in four large squares and write down the number.

7. Divide the total number by 4 and multiply by $10^4$. This is the concentration of spores per millilitre.

8. Multiply by the total volume of the cell suspension to obtain the total number of spores.

9. Clean the haemocytometer and cover slip with a diluted bleach solution or 70% alcohol and then rinse in deionized water. Air dry and store until next use.

10. Note that trypan blue is used to stain unviable cells. If it has to be used for calculating viable spores, only unstained spores are considered alive. The dilution factor (usually diluted by adding an equal amount of trypan blue solution) will be considered in the final concentration calculation.

### Key diagnostic evidence

Vascular bundle-specific discoloration or decay and predominant isolation of *Fusarium* species from symptomatic vascular tissue warrant a diagnosis of *Fusarium* wilt. All isolates obtained from diseased plants are identified to be *Fusarium* species through either morphological and/or DNA sequence data. Pathogenicity data are supportive in cases where *Fusarium* species other than *F. oxysporum* or *F. solani* are found to be associated with vascular tissue.

### Witches' Broom Caused by Phytoplasmas

The phytoplasmas, formally named as 'Candidatus Phytoplasma' (IRPCM Phytoplasma/Spiroplasma Working Team, 2004), are a group of wall-less prokaryotes that live in plant phloem and insects. Unlike other bacteria that can be cultured, characterized, or deposited in publicly accessible culture collections, phytoplasmas cannot be cultured, purified and deposited *in vitro*. Therefore, these organisms are mostly classified by their DNA sequence data from certain conserved genes and phylogenetic analyses. Because of the lack of organism characteristics, phytoplasmas are placed into the novel genus 'Candidatus (Ca.) Phytoplasma', created uniquely to accommodate various phytoplasmas associated with many host plants. Under this new genus, several sub-taxa at the species level are given to distinguish phytoplasmas that have less than 97·5% similarity in their 16S rRNA gene sequences. Based on this nomenclature, a phytoplasma organism is named as 'Candidatus (Ca.) Phytoplasma' followed by a species name. For example, phytoplasmas causing clover proliferation are named as 'Ca. Phytoplasma trifolii'. Phytoplasmas are also classified by phylogenetic group based on 16S rRNA gene sequences (Lee *et al.*, 2000). For example, the clover proliferation phytoplasma is classified as group 16SrVI. Both classification systems are used to describe a specific phytoplasma species (Table 10.1).

The phytoplasmas listed in Table 10.1 cause a variety of symptoms in plants, including clustering of small branches (witches' broom), phyllody (developing floral organs to leafy structure), greenish pigmenting in non-green flowers or shoots (virescence) and yellowing and stunting of plants. These symptoms are induced by phytoplasmas through their interference with plant development (Hogenhout *et al.*, 2008) and are often expressed systemically. Phytoplasmas are vectored by phloem-feeder insects in the order Hemiptera, particularly leafhoppers, planthoppers and psyllids (Weintraub and Beanland, 2006). They can survive in the guts and salivary glands and traverse them to reach the saliva. When a phytoplasma-carrying vector feeds on a plant, it delivers phytoplasmas into the phloem. Therefore, the host range of a phytoplasma largely depends on the plant feeding range of its vector, which is generally broad. One of the most effective practices to manage phytoplasma diseases is to periodically monitor and control insect vectors and practise timely removal of infected plants from fields.

Hemp witches' broom has been reported from China (Zhao *et al.*, 2007), India (Raj *et al.*, 2008),

**Table 10.1.** Phylogenetic groups and species names of common phytoplasmas.

| Phylogenetic group | Species name | Typical common name | Reference |
|---|---|---|---|
| Aster yellows group (16SrI) | 'Ca. Phytoplasma asteris' | Aster yellow phytoplasma | Lee et al., 2004a |
| Peanut witches'-broom group (16SrII) | 'Ca. Phytoplasma aurantifolia' | Peanut witches' broom phytoplasma | Zreik et al., 1995 |
| X-disease group (16SrIII) | 'Ca. Phytoplasma pruni' | Western X-disease mycoplasma-like organism | IRPCM Phytoplasma / Spiroplasma Working Team, 2004 |
| Coconut lethal yellowing group (16SrIV) | 'Ca. Phytoplasma palmae' | Coconut lethal yellowing phytoplasma | IRPCM Phytoplasma / Spiroplasma Working Team, 2004 |
| Elm yellows group (16SrV) | 'Ca. Phytoplasma ziziphi' | Ziziphus jujube witches'-broom phytoplasma | Jung et al., 2003a |
|  | 'Ca. Phytoplasma vitis' | Flavescence dorée phytoplasma | IRPCM Phytoplasma / Spiroplasma Working Team, 2004 |
|  | 'Ca. Phytoplasma ulmi' | Elm yellows phytoplasma | Lee et al., 2004b |
| Clover proliferation group (16SrVI) | 'Ca. Phytoplasma trifolii' | Clover proliferation mycoplasma | Hiruki and Wang, 2004 |
| Ash yellows group (16SrVII) | 'Ca. Phytoplasma fraxini' | Ash yellows phytoplasma | Griffiths et al., 1999 |
| Loofah witches'-broom group (16SrVIII) | 'Ca. Phytoplasma luffae' | Loofah witches'-broom phytoplasma | IRPCM Phytoplasma / Spiroplasma Working Team, 2004 |
| Pigeon pea witches'-broom group (16SrIX) | 'Ca. Phytoplasma phoenicium' | Almond witches'-broom phytoplasma | Verdin et al., 2003 |
| Apple proliferation group (16SrX) | 'Ca. Phytoplasma mali' | Apple proliferation phytoplasma | Seemüller & Schneider, 2004 |
|  | 'Ca. Phytoplasma pyri' | Pear decline phytoplasma | Seemüller & Schneider, 2004 |
|  | 'Ca. Phytoplasma prunorum' | European stone fruit yellows phytoplasma | Seemüller & Schneider, 2004 |
| Rice yellow dwarf group (16SrXI) | 'Ca. Phytoplasma oryzae' | None | Jung et al., 2003b |
| Stolbur group (16SrXII) | 'Ca. Phytoplasma australiense' | Australian grapevine yellows phytoplasma | Davis et al., 1997 |
|  | 'Ca. Phytoplasma solani' | Stolbur phytoplasma | IRPCM Phytoplasma / Spiroplasma Working Team, 2004 |
| BGWL group (16SrXIV) | 'Ca. Phytoplasma cynodontis' | Bermuda grass white leaf phytoplasma | Marcone et al., 2004 |
| 'Ca. Phytoplasma brasiliense' group (16SrXV) | 'Ca. Phytoplasma brasiliense' | Hibiscus witches'-broom phytoplasma | Montano et al., 2001 |

Iran (Schain et al., 2011) and the USA (Feng et al., 2019; Wang, 2019), where 'Ca. Phytoplasma ulmi', 'Ca. Phytoplasma asteris', 'Ca. Phytoplasma solani' and 'Ca. Phytoplasma trifolii' were detected, respectively. However, these detections are based solely on the presence of a small segment of genomic DNA similar to the reported phytoplasma DNA sequences and their aetiological role is not confirmed experimentally. In Nevada, several hemp crops had a low percentage of plants exhibiting witches' broom, and 'Ca. Phytoplasma trifolii' was detected in some but not all. That being said, a witches' broom-like symptom in hemp may not necessarily be caused by a phytoplasma. Additionally, a plant that is tested positive for phytoplasmas may also have symptoms caused by other pathogens, such as viruses. For example, leaf curling, mottling and chlorosis are often associated with witches' broom in some hemp crops in Nevada, and *Beet curly top virus* is more frequently detected than phytoplasmas (see next section of this chapter).

## Field diagnosis

### Symptoms

The typical symptoms of phytoplasma infection in hemp are the development of clusters of highly proliferating branches with significantly shortened internodes and smaller leaves on the affected branches (Fig. 10.8). Leaves in a cluster are crowded, sometimes yellowish, and tiny (Fig. 10.9). Leaves outside the cluster turn to yellow and sometimes are deformed and twisted. Early infected plants may be severely stunted when compared with a normal plant (Fig. 10.10). Young plants may be killed during the early season. Those infected during the middle season generally survive but still exhibit mild to severe symptoms in late season. Some plants may show clusters of witches' broom on certain branches, while the rest of the plant looks normal (Fig. 10.11), which suggests that only a portion of the plant is infected. Symptomatic plants generally have no value to harvest.

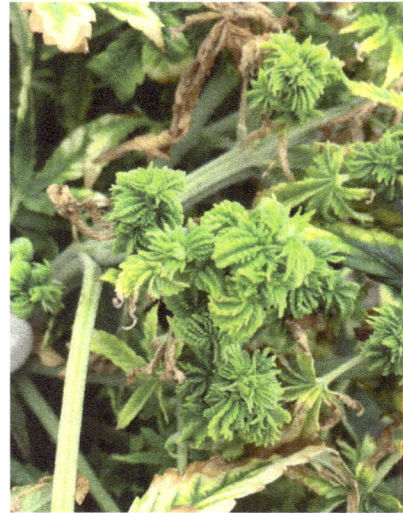

**Fig. 10.9.** A closer look of each cluster with underdeveloped new leaves. Note that older leaves turned yellow to brown.

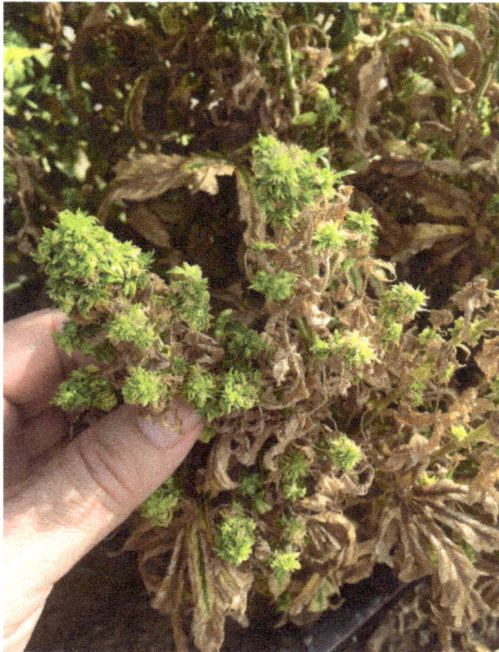

**Fig. 10.8.** A hemp plant exhibiting severe witches' broom symptom with numerous clusters of new growth on a single branch. Note that older leaves were killed.

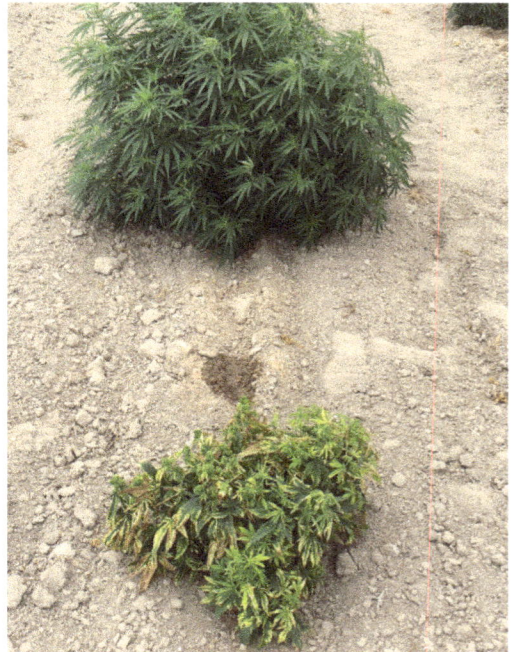

**Fig. 10.10.** An infected plant was severely stunted (bottom) compared with a healthy plant (top).

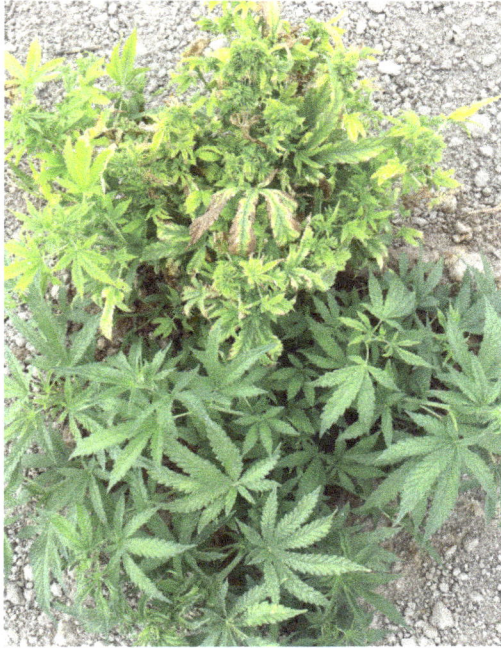

Fig. 10.11. A portion of hemp plant exhibiting witches' broom symptom. Note the significant difference in leaf morphology between the symptomatic portion and the apparently healthy portion.

### Problem classification

Abnormal growth is the key character for phytoplasma-induced diseases. Witches' broom and stunting with yellow foliage are the most notable signs for phytoplasma infection. Infected plants may also have leaf curling, chlorosis, mosaic, or leaf browning (Fig. 10.12). These symptoms are more likely to be caused by a virus or viroid. Since both phytoplasmas (in phloem) and viruses (within plant cells) are obligate pathogens, a hemp plant can be infected by both. More commonly, some plants are infected by one or the other, developing complex symptoms in the field. Such a case should be classified into phytoplasma plus viral diseases. Other abiotic factors that can cause broom-like symptoms are herbicides or products containing plant hormones. These factors can be ruled out if no such product has ever been used in a crop.

Fig. 10.12. A hemp plant that was tested positive for 'Ca. Phytoplasma trifolii' also exhibited mosaic (green islands) and chlorosis (left) in addition to witches' broom (right).

### Do your own diagnosis

Phytoplasma-caused witches' broom can be easily identified in the field when some plants are showing abnormal growth or clustering of small leaves on young shoots. This abnormality can appear during any stage of production. Another key diagnostic symptom is foliage yellowing, twisting and stunting. All these symptoms signal a systemic infection in plants. However, if the majority of plants become symptomatic quickly during a short time window, the initial diagnosis should be focused on the potential cause being from the application of chemical-based products that may contain or be contaminated with a herbicide or growth regulator. If symptomatic plants are sporadically distributed in a field and insect vectors are present during the growing season, the problem is likely related to a phytoplasma infection. Note that the symptoms of phytoplasma diseases are unique, as they are mostly expressed as altering plant growth rather than killing specific plant tissues or organs. Infected plants should not show root or stem rot unless there are other diseases involved. Secondly, co-infection by some viruses is likely in hemp and some plants may show both witches' broom and viral symptoms such as leaf curl, mosaic, chlorosis and yellowing. To determine which symptom is the most prevalent, select a number of plants in the field and record the type of symptoms observed in each plant (Fig. 10.13). This symptom polling process greatly helps determine if a phytoplasma, a virus, or both are contributing to the disease,

**Fig. 10.13.** An example of symptom polling to determine the prevalence of each symptom type caused by phytoplasma and/or viruses in hemp. Use the pictures as a guide to classify a symptom correctly.

which will guide subsequent sampling processes for targeted testing.

### Sampling

As mentioned above, phytoplasmas are mainly restricted to phloem tissue, especially in the leaf petioles and midveins. For a lab to detect a phytoplasma, a sample should comprise symptomatic leaflets. Do not remove petioles when collecting leaf tissue. Stem pieces are not necessary unless there is a pathological issue with the stem. In most cases, a small branch with typical witches' broom is an ideal sample for phytoplasma testing. Multiple samples (five to ten) from different plants at different locations in a field and a few (two to three) samples collected from asymptomatic plants are a good pool of samples for a lab to determine the prevalence of the phytoplasma infection in a field and whether it is associated with the symptoms observed. Make sure each sample corresponds to one plant and do not mix plant materials from different plants. Place each sample in a clean Ziploc bag, seal and label the bags, place them in a sturdy box and submit to a diagnostic lab.

### Lab diagnosis

#### Visual and microscopic examinations

A plant sample with witches' broom or abnormal growth should be first examined visually and then under a stereo microscope to rule out any mites or tiny insects that may cause similar symptoms. Pay attention to the curled leaves inside the cluster, as some arthropods may be present and feeding on leaf tissue. For example, eriophyid mites are microscopic and often go undetected if a sample is only

examined visually. Under a stereo microscope, a witches' broom symptom is often seen as numerous tiny leaves packed together at the end of new growth (see Fig. 3.19). In some cases when witches' broom is not so typical, infected branches may have much shortened internodes and smaller than normal leaves (Fig. 10.14).

#### Dienes' staining and light microscopy

A quick method a lab can use to detect phytoplasma infection is Dienes' staining (Musetti, 2013). This method is effective only in cases where the phytoplasma concentration inside the phloem tissue is high. However, it is very useful to localize *in situ* the phytoplasmas in the tissue and helps assess the distribution of phytoplasmas in symptomatic plants.

**1.** Take the midveins or midribs and petioles from infected shoots.
**2.** Cut them into thin sections using a razor blade under a stereo microscope, or use a sectioning machine to obtain semi-thin sections for staining. Tissue can also be resin-embedded to obtain ultrathin sections. Freezing microtome sections can be used to get fresh thin sections for staining without resin-embedding.
**3.** Prepare Dienes' stain stock solution by adding 2.5 g of methylene blue, 1.25 g of Azure II, 10 g of maltose and 0.25 g of sodium carbonate ($Na_2CO_3$) into 100 ml distilled water followed by filtering through a filter paper.
**4.** Dilute the stock solution to 0.2% (v/v) as a working solution for staining.
**5.** Immerse sections in Dienes' stain working solution at room temperature for 10 min, wash them in distilled water, mount them in a drop of water

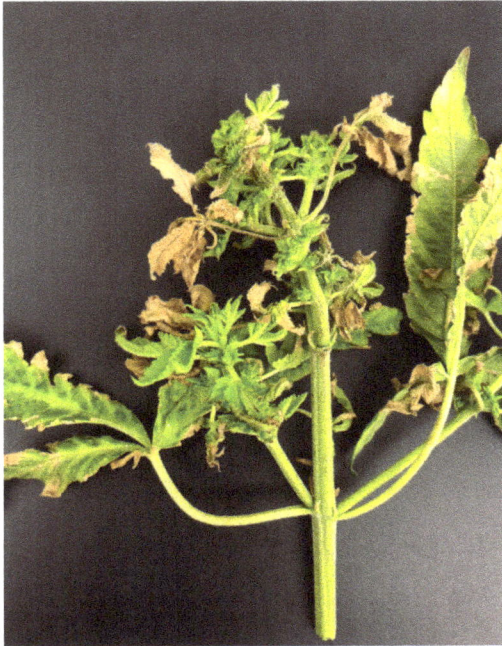

**Fig. 10.14.** Hemp shoots that were tested positive for 'Ca. Phytoplasma trifolii' had shortened internodes and reduced leaf sizes.

at the centre of a glass slide and observe under a compound microscope.

If the phloem sieve tubes are infected by phytoplasmas, they should be stained with patches of dark blue, while healthy tissue is not stained. Xylem and the cortex tissue may be stained light blue. Positive staining only indicates the presence of mollicutes, such as phytoplasmas and spiroplasmas, but it does not tell which group or species of phytoplasma is stained.

### DAPI staining and fluorescence microscopy

The chemical 4′,6-diamidino-2-phenylindole (DAPI) is a blue-fluorescent DNA stain that binds preferentially to AT regions of dsDNA. Since phloem sieve tubes do not have a nucleus, a positive staining (blue colour) by DAPI suggests the presence of DNA or nuclei and this can be used to detect the presence of phytoplasmas in either hand-cut or freezing microtome sections. This method was successfully used to localize and visualize phytoplasma infection in a phloem tissue under a

fluorescence microscope (Andrade and Arismendi, 2013). The procedure is quick and easy. In brief, wash sections in PBS solution several times as needed, add a sufficient amount of 300 nM DAPI stain solution to cover the sections and incubate for 1–5 min without exposure to light (wrapped or covered with a piece of aluminium foil). Remove the stain solution and then wash the sections two to three times in PBS. Mount the sections on a slide and observe under a fluorescent microscope. Note that DAPI, when it binds to DNA, is maximally excited by UV light at 358 nm and emits maximally in the blue range at 461 nm (Karg and Golic, 2018). The right length of UV light should be used.

### Real-time quantitative PCR (qPCR)

Phytoplasmas can be broadly detected by using one of the established qPCR methods validated by the USDA-APHIS-PPQ Beltsville Laboratory (see section on Real-Time Quantitative PCR in Chapter 5). This method can screen samples for various phytoplasma infections. In brief, collect petioles and midribs from an infected plant sample and extract total DNA from these tissue pieces (see Chapter 5). Note that petioles and midribs are the most appropriate tissue for DNA extraction, as these phloem-containing tissues have the highest copies of phytoplasma cells. The total DNA obtained should contain both plant DNA and phytoplasma DNA (if infected) and can be proceeded to qPCR screening. Once a sample has tested positive, it can be proceeded to DNA-based identification described in the next section.

### DNA-based identification

Even though rapid staining and universal qPCR test are available to screen plants for a phytoplasma infection, an accurate identification of phytoplasmas to the species level still relies on its rRNA gene sequence. As shown in Table 10.1, phytoplasmas are classified into a number of phylogenetic groups based on their 16S rRNA gene sequences (Lee *et al.*, 2000); therefore, DNA sequence data are the most useful variables for classifying a phytoplasma to a species or a group. To obtain a segment of phytoplasma DNA sequence, run conventional PCRs using three universal primer pairs (see Table 5.6). The first round of PCR should be run using primers P1/P7 and the second round using R16F2n/R16R2 (nested PCR). If P1/P7 generates a strong band (amplicon), no nested PCR is needed. Use the nested PCR only when P1/P7 generates a very weak band. If both primer pairs fail to amplify a phytoplasma DNA, the P1/Tint primer pair can be used. These two sets of PCRs generally amplify phytoplasma DNA efficiently from total DNA prepared from plant tissue, and the amplicon can be directly proceeded to PCR product purification, cloning and sequencing. Note that amplicons generated by these primers are larger than 1 kb and direct sequencing of a purified PCR product only read partial 5'- and 3'- sequences. To obtain a full amplicon sequence, clone the amplicon into a plasmid vector and sequence twice using the primer-walking strategy (see Chapter 5).

### Pathogenicity test

A phytoplasma detected from a hemp plant requires further biological tests to determine if it is the cause of the disease. Since phytoplasmas are non-culturable organisms, traditional isolation and inoculation methods do not apply. However, the system of using an insect vector to transmit phytoplasmas from a symptomatic to a healthy plant can be used to determine the pathogenicity (Namba, 2019). In this method, the same phytoplasma must be detected from the donor plant, insect vector and inoculated plant (Fig. 10.15). Other methods include the use of direct grafting (Aldaghi *et al.*, 2007) and dodder (*Cuscuta* spp.) transmission (Pribylova and Spak, 2013). No matter which method is employed, a phytoplasma should be detected in both donor and recipient plants (after transmission) by PCR assays and the transmission should induce phytoplasma-like symptoms on recipient plants.

### Key diagnostic evidence

The detection of phytoplasma DNA from an infected plant may conclude a diagnosis of phytoplasma disease. However, additional evidence such as phytoplasma-specific symptoms and transmission tests help determine the aetiological role of the phytoplasma.

### Leaf Curl Caused by *Beet Curly Top Virus* (BCTV)

*Beet curly top virus* (BCTV) is a member of the genus *Curtovirus* in the family *Geminiviridae* (Hull, 2014). Unlike the majority of plant viruses that have genomes consisting of single-stranded RNA (ssRNA), BCTV has a circular single-stranded DNA (ssDNA) genome. The particles of BCTV are called geminate particles (22 × 38 nm), meaning that each virus contains two joined icosahedra structures (Zhang *et al.*, 2001). The twinned or geminate particles are characteristic for geminiviruses. BCTV is a well-known plant virus that affects many crops, including sugar beet, tomato and pepper, and it is mainly transmitted by the beet leafhopper (*Circulifer tenellus*), an important pest in

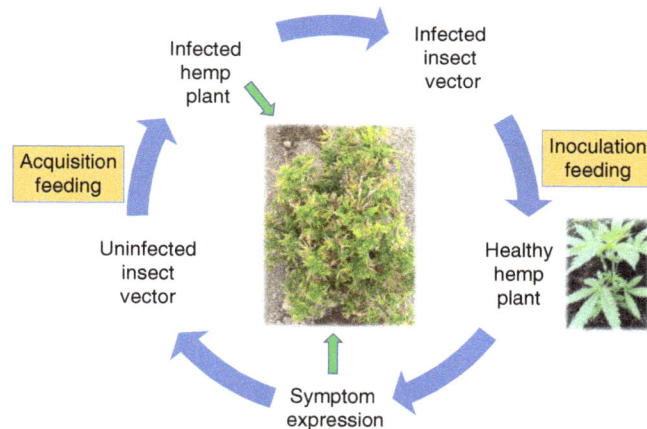

**Fig. 10.15.** A vector-based transmission system to confirm the pathogenicity of a detected phytoplasma on hemp plants. Note that recipient plants must be healthy and test negative for phytoplasmas prior to transmission test.

western USA. In natural beet leafhopper populations, a significant percentage of leafhoppers carry BCTV (Rondon *et al.*, 2016), and therefore this insect plays a critical role in spreading the virus from one crop to another. Hemp, as a new crop, has been severely affected by this 'old' virus in Colorado (Giladi *et al.*, 2020), Nevada and California, where some hemp crops exhibited leaf curl, twisting and other symptoms.

## Field diagnosis

### Symptoms

A hemp plant infected with BCTV may exhibit a variety of symptoms ranging from mild to severe. The typical symptom is leaf rolling and, in almost all cases, leaves roll upwards instead of downwards (Fig. 10.16). Infected plants may exhibit a fading green or partially yellow colour in some leaves (Fig. 10.17). Older leaves may roll upwards slightly and the back side of leaves appears silver coloured (Fig. 10.18). Some leaves appear thicker than normal, with a wavy or malformed upper surface (Fig. 10.19). Newly grown leaves are generally narrowed and curled with slightly faded green colour (Fig. 10. 20). Severely infected plants are stunted with curled and distorted foliage (Fig. 10.21). Symptoms occur throughout the vegetable and flowering stages, but they are often noticeable during the middle of season. However, a plant infected with BCTV may not show obvious symptoms, depending on the growth stage at which the virus infects the plant, or the plant may recover from the symptoms, even though it still contains the virus. This characteristic of BCTV infection was observed on tobacco plants (Benda and Bennett, 1964). In some of our clinical samples, some plants appearing asymptomatic contained an amplifiable level of BCTV DNA. The development of or recovery from viral symptoms in plants is influenced by the environment conditions, host varieties and virus strains. Hemp leaf curl is prevalent mainly in outdoor hemp fields, because beet leafhoppers, the vector of BCTV, are common pests in agricultural fields. However, the insects may not always be seen in the field when plants start to show viral symptoms, as they may have already migrated out from the field after transmitting the virus to hemp plants.

## Problem classification

Leaf curling is sometimes considered a result of a temporary water deficit during hotter days. It is

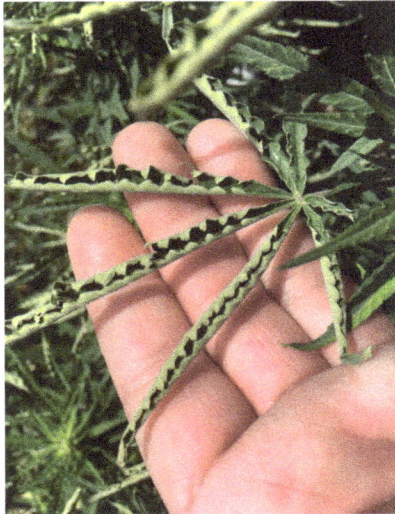

**Fig. 10.16.** A BCTV-infected hemp leaflet with narrowed and completely rolled leaves.

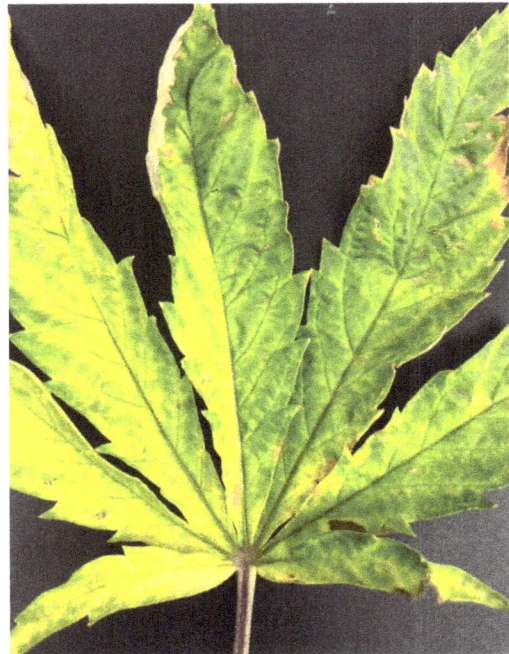

**Fig. 10.17.** A BCTV-infected hemp leaflet exhibiting faded green, chlorosis and yellow colours.

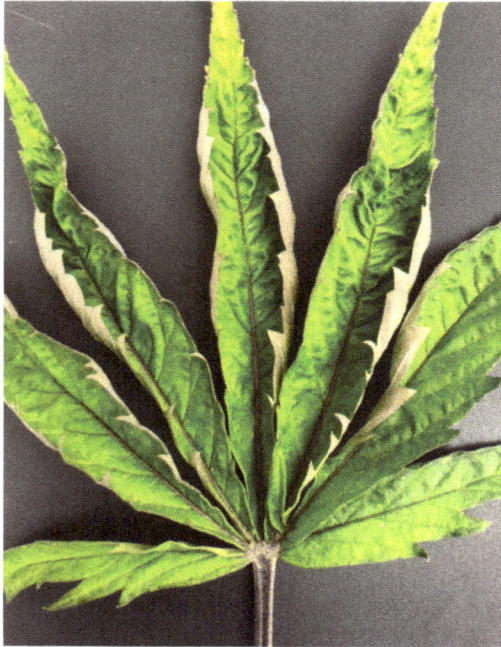

**Fig. 10.18.** A BCTV-infected hemp leaflet slightly curled inwards.

**Fig. 10.19.** A BCTV-infected hemp leaflet appearing abnormally darker in colour and thicker.

one of the drought avoidance mechanisms for plants to prevent water from evaporating during drought stress. Before assuming BCTV-related diseases, drought-induced temporary leaf roll should be assessed and ruled out. If continued irrigation does not reverse the symptom, an infectious agent such as BCTV can then be suspected. In addition to BCTV, other viruses or viroids may cause similar symptoms or co-infect with BCTV, causing more complex disease symptoms in the field. In this case, a symptom polling (Fig. 10.13) can be conducted to determine whether one or more causative agents are associated with the plants.

### Do your own diagnosis

Leaf rolling is easily recognized in the field, especially when plants are also stunted. BCTV-infected hemp plants generally have four major characteristic symptoms: (i) upward-rolled leave; (ii) narrowed and twisted new leaves; (iii) a yellow to pale-green colour; and (iv) stunting. Not all plants have these four symptoms during the same period, but these symptoms can

be seen among plants of a crop. One simple way to diagnose leaf roll disease in the field is to rub a curled leaf using one's fingers to hear if it produces a crispy sound. This technique works well for identifying potato plants infected by *Potato leafroll virus*. A BCTV-induced and curled leaf is usually stiff, coarse and thick, which creates more audible sounds than a normal leaf. Another hint is that beet leafhopper may be found on some plants. Because BCTV is carried by beet leafhopper naturally (Rondon *et al.*, 2016), the coexistence of leaf curl and the insect vector suggests a BCTV-induced disease. Many weed species are naturally infected with BCTV and they serve as virus reservoirs as well as food plants for the beet leafhopper. In the western USA, weed species naturally infected with BCTV include water smartweed, swamp smartweed, common knotweed, lady's thumb, wire grass, fogweed, spear orache, red orache, sowbane, Mexican tea, Russian thistle, rough pigweed, tumbleweed, charlock, shepherd's purse, Spanish clover, dwarf mallow, cheeseweed, prickly sow-thistle and cotton-batting plant (Severin, 1934). These plants, when infected with BCTV, exhibit the following characteristic symptoms: (i) leaf curling, rolling, cupping,

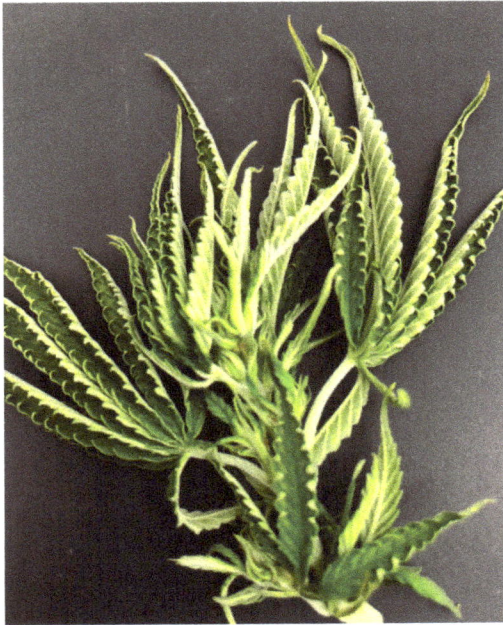

**Fig. 10.20.** A BCTV-infected hemp shoot having narrowed and curled leaves.

**Fig. 10.21.** A stunted hemp plant exhibiting leaf curl and distortion due to the BCTV infection.

balling, twisting, chlorosis, mottling, puckering; (ii) distorted or transparent leaf veins and protuberances on the lower surface of the leaves; (iii) abnormal development of secondary shoots with underdeveloped leaves arising from the axils; (iv) shortened internodes; and (v) stunting. Many of these symptoms are also expressed in hemp plants, some of which are similar to those caused by phytoplasmas. The presence of any of these plants that show viral symptoms in a hemp cultivation area provides an epidemiological clue to hemp leaf curl disease.

### Sampling

Once the problem is determined to be a viral disease, leaf samples can be taken for lab analysis. Use the symptom polling system shown in Fig. 10.13 to record the symptom types and collect leaves with each type of symptom. Place three to five leaflets collected from each plant into a Ziploc bag as one sample and repeat the process to obtain a total of five to ten samples from a field. Label each sample with a unique identification number and note the typical symptom type. A few healthy or apparently asymptomatic plants should also be sampled and submitted. The asymptomatic samples not only help determine if the virus is latent in some asymptomatic plants, but also help explain whether the virus detected is associated with the symptoms observed. A lab testing result based on both symptomatic and asymptomatic samples will provide a more accurate diagnosis in terms of the aetiological role of a detected virus. Leafhoppers, if seen in the field, can also be collected and submitted to a laboratory for BCTV analysis. The results can determine if and roughly what percentage of leafhoppers are viruliferous (virus-carrying).

### Lab diagnosis

#### Symptom observation and sub-sampling

Leaf samples received from clients need to be examined again to rule out any abiotic factors causing leaf cupping, chlorosis and distortion and to determine whether the symptoms are virus-related before proceeding to virus testing. Small pieces of leaf tissue can be re-sampled from symptomatic areas of leaves. In general, chlorotic or yellow tissue may contain a high titre of viruses, while the dark green islands (DGIs), a common symptom of virus infection in plants, are usually free of viruses (Moore *et al.*, 2001).

#### ELISA

ELISA should be considered as the first method to be used for a bulk of samples. This method can test or screen several hundred samples within 2 days (see Chapter 5). Follow the specific testing protocol provided by the manufacturer from which ELISA reagents (antibodies) are ordered.

### PCR and DNA-based identification

BCTV contains a single-stranded circular DNA (ssDNA), which makes it easy to detect the virus directly by PCR without reverse transcription. There are several primers designed for BCTV detection (Table 10.2) and more primers can be picked up from the BCTV isolate BCTV-Can complete genome (2931 bp, accession no. MK803280) (Giladi *et al.*, 2020). As the specific curly top virus species is unknown before testing, universal paired primers such as BGv377/BGc1509 are recommended for use at first to screen for all types of curly viruses. BGv377/BGc1509 have successfully and consistently detected BCTV from symptomatic hemp plants in Nevada and all positive PCR amplicons (approximately 1.1 bp) were confirmed to be BCTV DNA after sequencing. With this primer pair, a general 35 cycles of PCR can be run at the parameters of denaturing at 94°C for 1 min, annealing at 58°C for 1 min and extending at 72°C for 1 min with final extension at 72°C for 10 min. A shaped single band should be seen on agarose gel (Fig. 10.22). In the case of weak amplification, the original DNA can be diluted 100 times before PCR to dilute the PCR inhibitors. The 1.1 kb amplicon generated by BGv377 and BGc1509 can be purified for direct sequencing or cloning into T-vector before sequencing (see Chapter 5). Some diagnostic primers generate multiple bands. In this case, the primers may not be specific to the targeted viruses or non-specific amplifications have occurred in PCR; and a specific amplicon (if known by size) can be extracted from the gel and then sequenced. Note that a positive detection should be based on the DNA sequence instead of a positive amplification; and the amplified DNA fragment should be sequenced to confirm that the PCR product is indeed amplified from the genome of the suspected virus in a plant sample.

### Virus purification, characterization and electron microscopy (EM)

The purification of virus particles and characterization of their physical and molecular properties are seldom performed in a routine diagnostic setting unless a novel virus or disease is speculated, but a simple negative staining may be performed to see the virus particles (see Chapter 5). Since BCTV has unique geminate particles (22 × 38 nm), which are easily differentiated from rod- or linear-shaped virus particles under an electron microscope, a direct observation of geminate particles in plant sap can quickly diagnose a BCTV infection. If the virus titre is low in a leaf sap preparation, a partially purified virus preparation can be used for a negative staining.

**Table 10.2.** Primer pairs used in the curly top virus detections.

| Oligo name | Sequence (5' to 3') | Amplicon size | Forward/ Reverse | Pair with | Detection target | Reference |
|---|---|---|---|---|---|---|
| BMCTVv2195 | CTAAAAGGCCGCGCAG | 1.2 kb | Forward | BMCTVc514 | General[a] | Chen *et al.*, 2010 |
| BMCTVc514 | CCTCAGTAGCTTCTTCACTTCC | | Reverse | | General[a] | |
| BCTVv2557 | GCTTGGTCAAGAGAAGT | 965 bp | Forward | BGc396 | BCTV | |
| BSCTVv2688 | GCTGGTACTTCGATGTTG | 720 bp | Forward | BGc396 | BMCTV | |
| BMCTVv2825 | TGATCGAGGCATGGTT | 506 bp | Forward | BGc396 | BSCTV | |
| BGc396 | CAACTGGTCGATACTGCTAG | | Reverse | | | |
| BGv377 | CTAGCAGTATCGACCAGTTG | 1.1 kb | Forward | BGc1509 | Universal[b] | Soto and Gilbertson, 2003; Chen and Gilbertson, 2008 |
| BGc1509 | GACATTGACTGGAGACCGTT | | Reverse | | Universal[b] | |
| Forward-337 | ATGGGACCTTTCAGAGTGGA | Unspecified | Forward | Reverse-1,278 | BCTV | Giladi *et al.*, 2020 |
| Reverse-1,278 | TGTATGCCACATTGTTTGGC | | Reverse | | | |

[a]This primer pair may be considered as universal primers. [b]This primer pair can detect viral DNA from plant tissue infected with BCTV, beet mild curly top (BMCTV), or beet severe curly top (BSCTV).

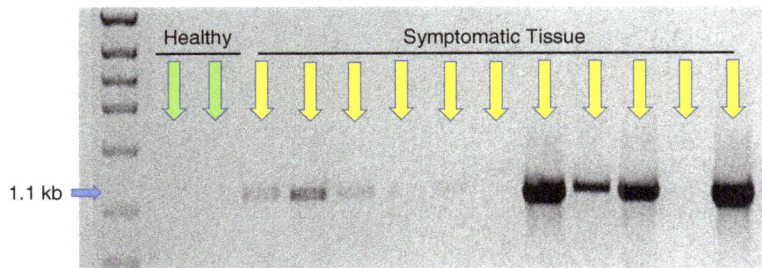

**Fig. 10.22.** Positive detection of *Beet curly top virus* from hemp plants using BGv377/BGc1509 primers. Note that some symptomatic plant tissue samples contained low copies of BCTV DNA, as shown by their weak bands (1.1 kb) in the gel.

### Key diagnostic evidence

The positive detection of BCTV DNA from an infected plant and typical curly top symptoms may conclude a diagnosis of curly top disease in hemp. In a routine diagnosis, a pathogenicity test may not be necessary, as this virus is widely known to cause diseases in a diverse range of crops. However, a vector-based transmission system as illustrated in Fig. 10.15 may be performed to demonstrate virus pathogenicity in hemp.

### Plant Stunting Caused by *Hop Latent Viroid*

Several reports have suggested that *Hop latent viroid* (HpLVd) is associated with a stunting disease in *Cannabis sativa*. In one report, the symptoms of plants infected with HpLVd were described as stunting, leaf malformation, chlorosis, brittle stems and yield reduction (Warren *et al.*, 2019). In another report, infected plants were found to have an outwardly horizontal plant structure, brittle stems and reduction of flower mass and trichomes (Bektaş *et al.*, 2019). HpLVd was described by Puchta *et al.* (1988) as a 256-nucleotide circular RNA. Because this viroid does not seem to produce any visible disease symptoms, it was named *Hop latent viroid*. The word 'latent' means that the viroid is present in a host plant but does not induce any visible abnormalities. Unlike viruses that have a defined particle structure, HpLVd only has a secondary rod-shaped RNA structure that is infectious and can be mechanically inoculated onto a hop plant. HpLVd has been found in hops worldwide, but it was detected only recently from *Cannabis sativa* through next-generation sequencing. Pathogenicity tests by inoculating infectious RNA

to healthy *C. sativa* plants confirmed that this viroid is not latent and did cause stunting, leaf malformation and chlorosis (Warren *et al.*, 2019).

The symptoms observed from plants with positive detection of HpLVd are not unique enough for visual diagnosis, as other viruses can cause similar symptoms. In certain cases, mild phytoplasma infection can result in stunting and malformation. Therefore, the diagnosis of a virus-like disease requires a holistic approach. In *Cannabis* plants, especially hemp crops, BCTV and phytoplasmas may contribute more to the symptoms observed in the field, as they are widespread and vectored by insects. An accurate diagnosis of a disease for true causative agents may require a panel of testing to target several common pathogens that may co-infect plants.

Specific detection of HpLVd can be performed by reverse transcriptase PCR (RT-PCR) developed by Hataya *et al.* (1992). First, extract RNA from symptomatic plant tissue and convert the RNA to first strand cDNA in 20 µl of RT reaction by adding reverse transcriptase and HpLVd-specific primer HLVd-1M (5'-TAGTTTCCAACTCCG GCTGG-3') as well as other components, following the manufacturer's instructions. Incubate the mixture at 37°C for 1 h. Next, use 10 µl of RT reaction mixture to set up a regular PCR reaction by adding primers HLVd-1M and HLVd-1P (5'-GGATACAACTCTTGAGCGCC-3') as well as other components provided with the *Taq* DNA polymerase kit. Run 30–35 cycles and then run a gel (3%) to examine amplification. A fragment of 250 bp is expected in a sample positive for HpLVd. Alternatively, HpLVd can be detected using HpLVd-specific RT-PCR primers HLVdF (5'-ATACA ACTCTTGAGCGCCGA-3') and HLVdR (5'-CCA CCGGGTAGTTTCCAACT-3') (Warren *et al.*,

2019). The PCR amplicon can be further purified and sequenced. A simple BLAST search can determine if the RNA sequence matches the published HpLVd genome.

## Leaf Chlorosis Caused by *Lettuce Chlorosis Virus*

*Lettuce chlorosis virus* (LCV), a member of the *Crinivirus* genus in the Closteroviridae family, was detected from *Cannabis sativa* plants in Israel (Hadad *et al.*, 2019). LCV caused yellowing, chlorotic and necrotic foliage. Interveinal chlorosis, leaf thickness and brittleness were also observed in infected plants. The virus can be transmitted from infected *Cannabis* plants to healthy ones by the sweet potato whitefly (*Bemisia tabaci*) Middle Eastern Asia Minor1 (MEAM1) biotype in a semi-persistent manner. This can be a major means of virus transmission in fields. In addition, shoots from infected plants, especially mother plants used for propagation, may serve as a primary source of infection for indoor cultivation. LCV does not spread via *Cannabis* seeds. Weeds are hosts of LCV and serve as natural reservoirs of the virus.

A diagnosis of LCV based solely on symptoms is difficult, as many viral diseases appear similar. In the case of co-infection with several viruses, plants may exhibit more complex symptoms. Therefore, it is important to use the strategy described for *Beet curly top virus* and phytoplasma to narrow down the cause to a specific virus or determine if one or more diseases exist. Refer to Fig. 10.13 and assess the percentage of each symptom type observed in a crop. In general, witches' broom is more likely to be caused by a phytoplasma infection, leaf curl by *Beet curly top virus*, foliar chlorosis by *Lettuce chlorosis virus*, leaf mosaic by cannabis cryptic virus and stunting by *Hop latent viroid*, although there are significant overlaps in symptoms caused by these pathogens.

*Lettuce chlorosis virus* has a bipartite ssRNA genome and an RT-PCR test can quickly detect the presence or absence of this virus in plant tissue. Three primer sets designed from RNA1 and RNA2 (Hadad *et al.*, 2019) can be used for RT-PCR detection: forward primer RNA1-F-7170 (5'-TCACAG CCGAGATCAACAGA-3') and complementary primer RNA1-R-8433 (5'-GTTACCAGCCTTGA GTCAATCA-3'); RNA2-F-6090 (5'-TCATCTTCA GGCCAAACACGG-3') and RNA2-R-7094 (5'-TC CACCTAATCCGATTCCAC-3'); and RNA2-F-7628 (5'-GCAGGTCATGACGTCAGATTT-3') and RNA2-R-8189 (5'-TGAACAATCACTACAGGT TTGG-3').

## Other Viral Diseases

A recent review indicated that several other viruses could induce hemp diseases (Sevik, 2020). However, the majority of them are cited from unverified literature. These viruses include hemp streak virus, hemp mosaic virus, *Alfalfa mosaic virus*, *Cucumber mosaic virus*, *Arabis mosaic virus*, *Tobacco mosaic virus*, *Tobacco ringspot virus*, *Tobacco streak virus*, *Tomato ringspot virus*, euonymus ringspot virus, elm mosaic virus and *Foxtail mosaic virus*. None of these have yet been reported from *Cannabis sativa* with solid data on their association with *Cannabis* plants and aetiological roles in the symptom's development. Some, such as hemp streak virus and hemp mosaic virus, have not been described as valid viruses or named officially. Hemp streak virus is speculated to be a pathogen causing interveinal chlorosis and leaf margin wrinkling in *Cannabis* plants. An investigation into the existence of such a virus and the relationship to the symptoms resulted in a sole detection of cannabis cryptic virus (CanCV) instead of hemp streak virus; and even with the presence of CanCV in plant tissue, this virus is not responsible for the disease symptoms concerned (Righetti *et al.*, 2018). That said, a virus that is claimed to be a causative agent in *Cannabis* plants must be a valid virus that has been characterized and has proven aetiological roles in a disease from which it is isolated or detected.

## Systemic Disease Management

Systemic diseases are mostly caused by viruses, viroids and phloem-inhabiting bacteria. *Fusarium* species as vascular pathogens cause systemic wilt in plants. Treating these diseases using chemicals is difficult, as most of these pathogens are obligate and live inside plant tissue or cells. Furthermore, many viruses and phytoplasmas are transmitted by insects. Therefore, the strategy of systemic disease control should focus on inoculum prevention and vector control.

### Monitoring mother plant health

*Fusarium* wilt and certain viral diseases such as *Lettuce chlorosis virus* are efficiently transmitted by shoots or cuttings. Screening propagative plants by symptoms alone is generally not sufficient, as pathogens can infect plants without external symptoms being visible in the early stages of disease development. Therefore, a strict disease monitoring programme that routinely tests mother plants for targeted pathogens should be in place. A culture method is sufficient for *Fusarium* pathogens and PCR (for DNA viruses) and RT-PCR (RNA viruses) are quick enough to determine the presence of certain viruses or viroids. Refer to Chapter 4 on how to build your own lab to meet the specific diagnostic needs.

### Managing insect vectors

Leafhoppers spread phytoplasmas and the *Beet curly top virus* among hemp plants or from other crops to hemp crops. Monitor leafhoppers in the field early in season and control their populations using approved products. These leafhoppers may unnoticedly feed on seedlings or young plants and deliver pathogens into the plants. When plants start to show disease symptoms, the leafhoppers may migrate out of the field. Systemic diseases can occur anytime during the season and vector control should be in force all the time to mitigate the spread of disease.

### Controlling weeds

Weeds are reservoirs of viruses and certain bacteria and also serve as food to insect vectors. When a field is planted with hemp, all weeds in the field and surrounding areas should be removed or killed. This is especially important for *Beet curly top virus* management.

### Selecting a right field location

It is always a good strategy to plant a hemp crop away from tomato or other crops that share the same vector-borne diseases. Since most viruses and vectors have a broad host range, a reasonable distance between a hemp crop from other common crops can minimize the cross-over of those diseases. This should be considered from both pathogen and vector perspectives.

### Using resistant varieties

Certain varieties are more prone to certain viral diseases than others. Unless there are resistant lines or cultivars developed against systemic diseases, growers can compare the disease incidence among cultivars to determine the one that may have a better disease tolerance. A plant diagnostic lab that receives hemp samples may have data to show which varieties are less commonly found with a specific disease and those data may suggest that certain varieties are more tolerant than others.

### Avoiding contact transmission

Certain viruses can be transmitted from plant to plant via contact. Employees working on cannabis or hemp plants should wash their hands with soap before handling growing plants. Spacing between plants should be adequate to avoid rubbing between leaves. Pruning or cutting tools should be thoroughly disinfected by dipping in a disinfectant solution between use on each plant. Tool disinfection is especially critical to prevent the spread of *Fusarium* wilt.

### Reducing abiotic stresses

Systemic diseases can be aggravated under environmental and nutritional stresses. These factors can contribute to some symptoms, such as leaf chlorosis (either uniform or interveinal) or overall stunting. Healthy growing conditions support plant vigour and help plants overcome infections. This is especially important for certain viral diseases in which plants can recover from symptoms under certain environmental conditions.

### References

Aldaghi, M., Massart, S., Roussel, S., Steyer, S., Lateur, M. *et al.* (2007) Comparison of different techniques for inoculation of 'Candidatus Phytoplasma mali' on apple and periwinkle in biological indexing procedure. *Communications in Agricultural and Applied Biological Sciences* 72, 779–784.

Andrade, N.M. and Arismendi, N.L. (2013) DAPI staining and fluorescence microscopy techniques for phytoplasmas. In: Dickinson, M. and Hodgetts, J. (eds) *Phytoplasma. Methods and Protocols.* Methods in Molecular Biology, Vol 938. Humana Press, Totowa, New Jersey, pp. 115–121. doi: 10.1007/978-1-62703-089-2_10

Bektaş, A., Hardwick, K.M., Waterman, K. and Kristof, J. (2019) Occurrence of hop latent viroid in *Cannabis sativa* with symptoms of *Cannabis* stunting disease in California. *Plant Disease* 103, 2699. doi: 10.1094/PDIS-03-19-0459-PDN

Benda, G.T.A. and Bennett, C.W. (1964) Effect of curly top virus on tobacco seedlings: infection without obvious symptoms. *Virology* 24, 97–101. doi: 10.1016/0042-6822(64)90152-7

Chen, L.-F. and Gilbertson, R.L. (2008) Beet mild curly top virus. In: Rao, G.P. *et al.* (eds), *Characterization, Diagnosis & Management of Plant Viruses*. Studium Press LLC, Houston, Texas, pp. 195–215.

Chen, L.F., Brannigan, K., Clark, R. and Gilbertson, R.L. (2010) Characterization of curtoviruses associated with curly top disease of tomato in California and monitoring for these viruses in beet leafhoppers. *Plant Disease* 94, 99–108. doi: 10.1094/PDIS-94-1-0099

Davis, R.E., Dally, E.L., Gundersen, D E., Lee, I.-M and Habili, N. (1997) 'Candidatus Phytoplasma australiense,' a new phytoplasma taxon associated with Australian grapevine yellows. *International Journal of Systematic Bacteriology* 47, 262–269. doi: 10.1099/00207713-47-2-262

Doi, Y., Teranaka, M., Yora, K. and Asuyama, H. (1967) Mycoplasma- or PLT group-like microorganisms found in the phloem elements of plants infected with mulberry dwarf, potato witches' broom, aster yellows, or paulownia witches' broom. *Japanese Journal of Phytopathology* 33, 259–266. doi: 10.3186/jjphytopath.33.259

Feng, X., Kyotani, M., Dubrovsky, S. and Fabritius, A.L. (2019) First report of 'Candidatus Phytoplasma trifolii' associated with witches' broom disease in *Cannabis sativa* in Nevada, USA. *Plant Disease* 7, 103. doi: 10.1094/PDIS-01-19-0098-PDN

Giladi, Y., Hadad, L., Luria, N., Cranshaw, W., Lachman, O. *et al.* (2020) First report of beet curly top virus infecting *Cannabis sativa* in western Colorado. *Plant Disease* 104, 999. doi: 10.1094/PDIS-08-19-1656-PDN

Griffiths, H.M., Sinclair, W.A., Smart, C.D. and Davis R.E. (1999) The phytoplasma associated with ash yellows and lilac witches'-broom: 'Candidatus Phytoplasma fraxini.' *International Journal of Systematic Bacteriology* 49, 1605–1614. doi: 10.1099/00207713-49-4-1605

Hadad, L., Luria, N., Smith, E., Sela, N., Lachman, O. *et al.* (2019) Lettuce chlorosis virus disease: a new threat to cannabis production. *Viruses* 11, 802. doi: 10.3390/v11090802

Hataya, T., Hikage, K., Suda, N., Nagata, T., Li, S. *et al.* (1992) Detection of hop latent viroid (HLVd) using reverse transcription and polymerase chain reaction (RT-PCR). *Annals of the Phytopathological Society of Japan* 58, 677–684.

Hiruki, C. and Wang, K. (2004) Clover proliferation phytoplasma: 'Candidatus Phytoplasma trifolii.' *International Journal of Systematic and Evolutionary Microbiology* 54, 1349–1353. doi: 10.1099/ijs.0.02842-0

Hogenhout, S.A., Oshima, K., Ammar, E., Kakizawa, S., Kingdom, H.N. *et al.* (2008) Phytoplasmas: bacteria that manipulate plants and insects. *Molecular Plant Pathology* 9, 403–423. doi: 10.1111/j.1364-3703.2008.00472.x

Hull, R. (2014) *Plant Virology*, 5th edn. Academic Press, London.

IRPCM Phytoplasma/Spiroplasma Working Team-- Phytoplasma Taxonomy Group (2004) 'Candidatus Phytoplasma,' a taxon for the wall-less, non-helical prokaryotes that colonize plant phloem and insects. *International Journal of Systematic and Evolutionary Microbiology* 54, 1243–1255. doi: 10.1099/ijs.0.02854-0

Jung, H.-Y., Sawayanagi, T., Kakizawa, S., Nishigawa, H., Wei, W. *et al.* (2003a) 'Candidatus Phytoplasma ziziphi', a novel phytoplasma taxon associated with jujube witches'-broom disease. *International Journal of Systematic and Evolutionary Microbiology* 53, 1037–1041. doi: 10.1099/ijs.0.02393-0

Jung, H.Y., Sawayanagi, T., Wongkaew, P., Kakizawa, S., Nishigawa, H. *et al.* (2003b) 'Candidatus Phytoplasma oryzae', a novel phytoplasma taxon associated with rice yellow dwarf disease. *International Journal of Systematic and Evolutionary Microbiology* 53, 1925–1929. doi: 10.1099/ijs.0.02531-0

Karg, T.J. and Golic, K.G. (2018) Photoconversion of DAPI and Hoechst dyes to green and red-emitting forms after exposure to UV excitation. *Chromosoma* 127, 235–245. doi: 10.1007/s00412-017-0654-5

Lee, I.-M., Davis, R.E. and Gundersen-Rindal, D.E. (2000) Phytoplasma: phytopathogenic mollicutes. *Annual Review of Microbiology* 54, 221–255. doi: 10.1146/annurev.micro.54.1.221

Lee, I.-M., Gundersen-Rindal, D.E., Davis, R.E., Bottner, K.D., Marcone, C. *et al.* (2004a) 'Candidatus Phytoplasma asteris', a novel phytoplasma taxon associated with aster yellows and related diseases. *International Journal of Systematic and Evolutionary Microbiology* 54, 1037–1048. doi: 10.1099/ijs.0.02843-0

Lee, I.-M., Martini, M., Marcone, C. and Zhu, S.F. (2004b) Classification of phytoplasma strains in the elm yellows group (16SrV) and proposal of 'Candidatus Phytoplasma ulmi' for the phytoplasma associated with elm yellows. *International Journal of Systematic and Evolutionary Microbiology* 54, 337–347. doi: 10.1099/ijs.0.02697-0

Marcone, C., Schneider, B. and Seemüller, E. (2004) 'Candidatus Phytoplasma cynodontis', the phytoplasma associated with Bermuda grass white leaf disease. *International Journal of Systematic and Evolutionary Microbiology* 54, 1077–1082. doi: 10.1099/ijs.0.02837-0

Montano, H.G., Davis, R.E., Dally, E.L., Hogenhout, S., Pimentel, J.P. *et al.* (2001) 'Candidatus Phytoplasma brasiliense', a new phytoplasma taxon associated with hibiscus witches'-broom disease. *International Journal of Systematic and Evolutionary Microbiology* 51, 1109–1118. doi: 10.1099/00207713-51-3-1109

Moore, C.J., Sutherland, P.W., Forster, R.L., Gardner, R.C. and MacDiarmid, R.M. (2001) Dark green islands in plant virus infection are the result of post-transcriptional gene silencing. *Molecular Plant-Microbe Interactions* 14, 939–946. doi: 10.1094/MPMI.2001.14.8.939

Musetti, R. (2013) Dienes' staining and light microscopy for phytoplasma visualization. In: Dickinson, M. and Hodgetts, J. (eds) *Phytoplasma. Methods and Protocols.* Methods in Molecular Biology, Vol. 938. Humana Press, Totowa, New Jersey, pp. 109–113. doi: 10.1007/978-1-62703-089-2_9

Namba, S. (2019) Molecular and biological properties of phytoplasmas. *Proceedings of the Japan Academy. Series B, Physical and Biological Sciences* 95, 401–418. doi: 10.2183/pjab.95.028

Přibylová, J. and Špak, J. (2013) Dodder transmission of phytoplasmas. In: Dickinson, M. and Hodgetts, J. (eds) *Phytoplasma. Methods and Protocols.* Methods in Molecular Biology, Vol. 938. Humana Press, Totowa, New Jersey, pp. 41–46. doi: 10.1007/978-1-62703-089-2_4

Puchta, H., Ramm, K. and Sänger, H. L. (1988) The molecular structure of hop latent viroid (HLV), a new viroid occurring worldwide in hops. *Nucleic Acids Research* 16, 4197–4216. doi: 10.1093/nar/16.10.4197

Raj, S.K., Snehi, S.K., Khan, M.S. and Kumar, S. (2008) 'Candidatus Phytoplasma asteris' (group 16SrI) associated with a witches'-broom disease of *Cannabis sativa* in India. *Plant Pathology* 57, 1173–1173. doi: 10.1111/j.1365-3059.2008.01920.x

Righetti, L., Paris, R., Ratti, C., Calassanzio, M., Onofri, C. et al. (2018). Not the one, but the only one: about *Cannabis cryptic virus* in plants showing 'hemp streak' disease symptoms. *European Journal of Plant Pathology* 150, 575–588. doi: 10.1007/s10658-017-1301-y

Rodrigo, G., Zwart, M.P. and Elena, S.F. (2014) Onset of virus systemic infection in plants is determined by speed of cell-to-cell movement and number of primary infection foci. *Journal of the Royal Society, Interface* 11, 20140555. doi: 10.1098/rsif.2014.0555

Rondon, S.I., Roster, M.S., Hamlin, L.L., Green, K.J., Karasev, A.V. et al. (2016) Characterization of beet curly top virus strains circulating in beet leafhoppers (Hemiptera: Cicadellidae) in northeastern Oregon. *Plant Disease* 100, 1586–1590. doi: 10.1094/PDIS-10-15-1189-RE

Schain, F.V., Bahar, M. and Zirak, L. (2011) Characterization of Stolbur (16srxii) group phytoplasmas associated with *Cannabis sativa* witches'-broom disease in Iran. *Plant Pathology Journal* 10, 161–167. doi: 10.3923/ppj.2011.161.167

Schoener, J.L., Wilhelm, R., Rawson, R., Schmitz, P. and Wang, S. (2017) First detection for *Fusarium oxysporum* and *F. solani* causing wilt of medical marijuana plants in Nevada. (Abstr.) *Phytopathology* 107, S5.98. doi: 10.1094/PHYTO-107-12-S5.98

Seemüller, E. and Schneider, B. (2004) 'Candidatus Phytoplasma mali', 'Candidatus Phytoplasma pyri' and 'Candidatus Phytoplasma prunorum', the causal agents of apple proliferation, pear decline and European stone fruit yellows, respectively. *International Journal of Systematic and Evolutionary Microbiology* 54, 1217–1226. doi: 10.1099/ijs.0.02823-0

Severin, H. (1934) Weed host range and overwintering of curly-top virus. *Hilgardia* 8, 261–280. doi: 10.3733/hilg.v08n08p261

Sevik, M.A. (2020) Kenevir (*Cannabis sativa* L.) Bitkilerinde Görülen Virüs Kaynaklı Hastalıklar. *Turkish Journal of Agricultural Research* 7, 111–119. doi: 10.19159/tutad.663715

Soto, M.J. and Gilbertson, R.L. (2003) Distribution and rate of movement of the curtovirus beet mild curly top virus (family Geminiviridae) in the beet leafhopper. *Phytopathology* 93, 478–484. doi: 10.1094/PHYTO.2003.93.4.478

Verdin, E., Salar, P., Danet, J.-L., Choueiri, E., Jreijiri, F. et al. (2003) 'Candidatus Phytoplasma phoenicium', a novel phytoplasma associated with an emerging lethal disease of almond trees in Lebanon and Iran. *International Journal of Systematic and Evolutionary Microbiology* 53, 833–838. doi: 10.1099/ijs.0.02453-0

Wang, S. (2019) Common marijuana and hemp diseases. *Cannabis Business Times,* April 2019, 29–38.

Warren, J.G., Mercado, J. and Grace, D. (2019) Occurrence of hop latent viroid causing disease in *Cannabis sativa* in California. *Plant Disease* 103, 2699. doi: 10.1094/PDIS-03-19-0530-PDN

Weintraub, P.G. and Beanland, L. (2006) Insect vectors of phytoplasmas. *Annual Review of Entomology* 51, 91–111. doi: 10.1146/annurev.ento.51.110104.151039

Zhang, W., Olson, N.H., Baker, T.S., Faulkner, L., Agbandje-McKenna, M. et al. (2001) Structure of the maize streak virus geminate particle. *Virology* 279, 471–477. doi: 10.1006/viro.2000.0739

Zhao, Y., Sun, Q., Davis, R.E., Lee, I.M. and Liu, Q. (2007) First report of witches'-broom disease in a *cannabis* spp. in China and its association with a phytoplasma of elm yellows group (16srv). *Plant Disease* 91, 227. doi: 10.1094/PDIS-91-2-0227C

Zreik, L., Carle, P., Bové, J.M. and Garnier, M. (1995) Characterization of the mycoplasmalike organism associated with witches'-broom disease of lime and proposition of a 'Candidatus' taxon for the organism, 'Candidatus Phytoplasma aurantifolia'. *International Journal of Systematic Bacteriology* 45, 449–453. doi: 10.1099/00207713-45-3-449

# 11 Diagnosing Non-infectious Diseases

Non-infectious diseases are not transmissible from plant to plant, because they are not caused by pathogens such as fungi, bacteria, viruses and nematodes. For this reason, they are often called abiotic diseases or disorders. Abiotic disorders are caused by adverse environmental conditions, hazardous chemicals, imbalanced nutrient levels, or genetic factors (Fig. 11.1). Unlike an infectious disease that is developed by interactions among the host, pathogen and environment (see Fig. 2.4), a non-infectious disease is a direct outcome of the interaction between the host plant and the environment. Diagnosing non-infectious diseases may be difficult, as many causative factors are not present in or detectable from plant tissue or in the surrounding environment at the time of diagnosis. Furthermore, many of these diseases may resemble those caused by pathogens. Even worse, secondary microorganisms may invade plant tissue predisposed by abiotic diseases, which may complicate the diagnosis. However, an accurate diagnosis of abiotic diseases is achievable through the observation of symptom characteristics and the assessment of potential links to cultural practices and weather conditions. There are some tips often used to determine if a disease is abiotic during a field diagnosis (Table 11.1), but specific tests for nutrients, chemicals, or pathogens are still required to validate the initial assessment.

## Environmental Injuries

### Drought stress

Environmental factors such as air temperature, soil moisture, light and soil pH play critical roles in plant growth. Indoor cannabis plants are generally grown healthily under controlled environments, but outdoor hemp crops are subject to variable and sometimes unpredictable environmental conditions.

A dry and hot climate often contributes to acute or chronic drought stress. Plants under chronic drought stress commonly exhibit leaf scorch (Fig. 11.2), have less green foliage and are underdeveloped. Leaf scorch is characterized by browning of the leaf tip and edge and is often seen first in older leaves. Leaf scorch can be caused by the bacterium *Xylella fastidiosa* in other plant species (Almeida and Nunney, 2015), but it has as yet not been reported in *Cannabis sativa*. Drought-caused leaf scorch may be expressed throughout the foliage, while bacterial leaf scorch may be restricted to certain portions of the plant that contains the bacterium (see Fig. 2.24)

### Sub-zero temperatures

Hemp plants may get injured when air temperatures drop to below zero (freeze damage) or when frost is formed due to low air temperature and high humidity (frost damage). This is because sub-zero temperatures damage plant cells by forming ice crystals in the apoplast (Pearce, 2001), a space outside the plasma membrane for cellular materials to diffuse freely. The initially formed ice crystals may continue to grow through a process known as ice recrystallization. This growth eventually damages plasma membranes, resulting in the collapse of cell membrane structures, dehydration and loss of cell volume. Ice crystals may also form on the surface of the plant and then enter into the plant tissue through stomata and hydathodes, which may trigger ice crystal growth inside the intracellular space. Some plant species are freeze-tolerant because of their expression of ice-binding proteins that can adsorb to ice crystals and modify their growth (Bredow and Walker, 2017). *C. sativa* is sensitive to sub-zero temperatures even though some varieties may be frost tolerant. In Nevada, an overnight drop of temperature to below zero may cause foliage colour change, dehydration and collapse

DOI: 10.1079/9781789246070.0011

(Fig. 11.3). Stems can also be damaged, with symptoms shown as superficial lesions (Fig. 11.4). In most cases, frost or freeze damage occurs in the fall, when plants are almost ready to be harvested. Crops planted at higher elevations, such as high desert regions, may get low-temperature injuries more frequently, due to the significant diurnal temperature variations.

### Non-optimal lighting

Indoor-grown hemp and cannabis plants are largely affected by light quality, intensity and photoperiod.

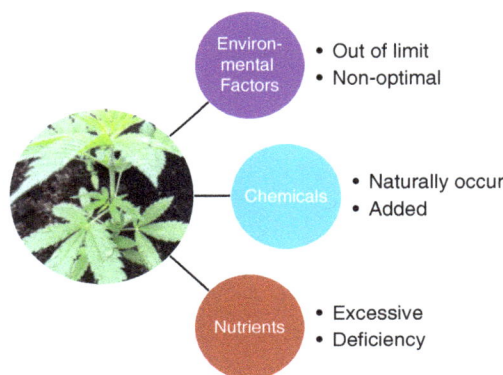

**Fig. 11.1.** Common factors that may cause abiotic diseases in hemp plants.

Light not only affects plant growth and morphology but also affects the content of cannabinoids, such as tetrahydrocannabinnol (THC) and cannabidiol (CBD). In a study conducted by Magagnini *et al.* (2018), *Cannabis* plant morphology could be manipulated by light spectrum as they found that plants under high-pressure sodium (HPS) treatment were taller and had a higher flower dry weight than those under AP673L (LED) and NS1 (LED) treatments. Additionally, plants with NS1 treatment had the highest cannabigerol (CBG) content and those with NS1 or AP673L treatments had higher CBD and THC concentrations when compared with HPS treatment. Photoperiod is another factor that can change plant morphology significantly. In one case, cannabis plants were found to grow an enlarged stem after they were transited to an incorrect photoperiod (Fig. 11.5).

### Alkaline soil

The soil of many agricultural fields has pH levels ranging from 5.5 to 8.0 and for most crops the optimum pH is between 6.0 and 7.5. Alkaline soil is generally clay with a pH level over 8.5 and its physical and chemical properties are not considered ideal for the production of certain crops. In the arid west of the USA, some hemp crops planted in unamended alkaline soils have been observed to have poor growth and are often prone to fungal diseases (Fig. 11.6). Besides a high pH

**Table 11.1.** The key differences between infectious diseases and non-infectious diseases.

|  | Infectious diseases | Non-infectious diseases |
| --- | --- | --- |
| Symptom appearance | Gradual, progressive | Sudden (in hours, overnight, or a few days) |
| Symptom distribution on individual plants | Random, centralized, or connected | Fairly uniform, widely distributed |
| Symptoms on plant organs | Generally specific to one organ, e.g. leaf or stem | Multiple or all parts of plants affected at the same time |
| Symptoms on different varieties | May be variety-specific | Generally appear on all varieties planted |
| Symptoms on different plant species | Usually one species affected | Generally multiple species in the same location affected |
| Locations of affected plants | Random, sporadic | Field edge, corner, or special sites |
| Patterns of affected plants in a field | Scattered, patched, or clustered | Straight lines, circles, squares, etc. |
| Percentage of affected plants | Usually 1–10% in early stage | Very high percentage |
| Range of symptoms | Generally broader | Generally narrow |
| Junction between healthy and symptomatic tissue | Usually vague with a progression zone between dead and healthy tissue | Distinct without a progression zone |
| Pathogen signs | Commonly present | Absent |

**Fig. 11.2.** Hemp plants exhibiting uniform leaf scorch due to chronic drought stress.

**Fig. 11.4.** Superficial brown lesions on hemp stems caused by sub-zero temperatures. Note that leaves turned brown or light purple.

**Fig. 11.3.** A hemp plant grown in a high desert region injured by overnight sub-zero temperatures.

**Fig. 11.5.** Enlarged stem (right) induced by an incorrect photoperiod compared with a normal stem (left).

that hinders plant growth, other factors associated with alkaline soil also contribute to the overall reduction of growth vigour and plant mass. *C. sativa* can grow in substrate with a pH ranging from 4.0 to 7.0, but more optimally at pH levels of 5.5–6.5 (Whipker *et al.*, 2019). In field conditions, soil pH may be difficult to adjust, but a field with a soil pH range from 6.0 to 7.0 can be selected for a hemp crop. Higher soil pH can make micronutrients such as iron unavailable for the plant to uptake, even if they are adequate in the soil. At the other extreme, a lower soil pH, for example lower than 5.0, may lead to increased micronutrient uptake and cause iron and/or manganese toxicity. Diagnosing a plant problem associated with alkaline soil should focus on nutrient-related symptoms, because nutrient availability to plants is largely affected by soil pH levels. Most nutrients have highest absorption rates at pH levels of 6.2–7.3 and alkaline soil decreases nutrient absorption significantly. A hemp field with a significant number of plants showing retarded or delayed growth may be assessed for its suitability for hemp production. A comprehensive soil analysis is often needed to determine the field eligibility.

## Phytotoxicity

Hemp plants can be injured by air pollution, herbicides, excessive salts, fertilizer overdose, pesticides, plant growth regulators, or other chemical-based products. Some of these may naturally exist in soils or environments, while others are introduced into plants through human activities, such as chemical applications for controlling weeds or pests. Although approved chemicals are important in managing crop health in modern agriculture, there are many cases of phytotoxicities caused by overdose use, misuse, product drift or runoff, or residue accumulation in the soil or on the plant. Chemical injuries can be expressed as a broad range of symptoms, including poor germination, death of seedlings or growing tissues, distorted or needle-like leaves, leaf spots and scorch, and delayed plant development. Among many products used in agricultural practices, herbicides are one of the most common causes of plant injuries and affected plants usually exhibit twisted, cupping, or narrow leaves on newly grown shoots. Excessive salts or high salinity in soil can prevent water uptake by the roots, causing drought-like symptoms such as leaf scorch or browning along the edges. Salts can be accumulated in leaf tissue, causing necrosis. Other chemicals can cause leaf spots along the leaf veins (Fig. 11.7), uniform discoloration (Fig. 11.8), or dieback. Diagnosing phytotoxicity may be difficult, but there are some indications that often point to a chemical injury. For example, there should be no signs of plant pathogens observed in plant tissue affected or killed by chemicals. If leaf spots are observed, there should be no transition between dead and healthy areas and the spots may appear uniform in colour throughout the plants. Further, chemical injury generally appears in a relatively short period of time after exposure and it can result from a direct mistreatment or a treatment for adjacent crops. Nevertheless, certain symptoms caused by herbicides may look similar to those caused by viruses, insects, or mites. In this case, a set of soil or plant tissue samples should be taken and submitted to a chemistry laboratory for an analysis. It is always helpful to advise the lab of what products

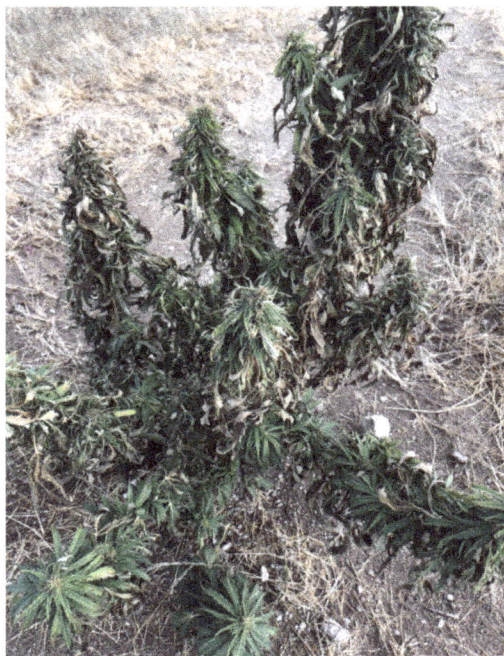

**Fig. 11.6.** A hemp plant grown poorly in a field with alkaline soil.

**Fig. 11.7.** Leaflets from an indoor-grown cannabis plant exhibiting necrotic lesions along the leaf veins due to a chemical injury.

have been used so that the lab can specifically test the samples for the suspected ingredients.

## Nutrient Deficiency

Deficiency in one or more essential nutrients may occur in hemp crops when soil fertility is poor. Indoor-cultivated cannabis plants are less likely to have nutrient issues, as they are grown under the optimal fertilization regime to maximize the product yield. Plant nutrients are classified into three major groups of elements. The first is macronutrients, containing nitrogen, phosphorus, potassium, calcium, magnesium and sulfur; the second is micronutrients, which include boron, chlorine, copper, iron, manganese, molybdenum, nickel and zinc; and the third is beneficial elements such as aluminium, cobalt, lanthanides, selenium, silicon, sodium and vanadium (Barker and Pilbeam, 2015). Both macronutrients and micronutrients are essential elements for many plant species. When a plant is not taking up sufficient nutrients from soil or foliar applications, it will show one or more symptoms to reflect its nutrient deficiency. The symptoms range from leaf yellowing and aberrant development to severe stunting and yield loss (Table 11.2) and the expression of such symptoms may depend on the plant species. *C. sativa* plants deficient in certain nutrients may exhibit leaf chlorosis in older leaves (Fig. 11.9) or leaf tip necrosis (Fig. 11.10). These symptoms are not unique to a specific nutrient deficiency and they should be carefully assessed for potential causes, including infectious diseases and other abiotic factors. In general, a diagnosis of nutrient deficiency based on symptoms alone is difficult and unreliable. Therefore, a tissue or soil test is often needed to determine whether the symptoms result from deficiencies in certain nutrient elements.

## Chimera

Chimeras are heterogenomic organisms that contain cells of more than one genotype and they occur in the vast majority of multicellular organisms (Frank and Chitwood, 2016). Phenotypically, chimeras are generally considered a rare phenomenon,

**Fig. 11.8.** A leaflet of an indoor-grown cannabis plant exhibiting one-sided and uniform reduction of green colour after a spray of an unspecified product.

**Table 11.2.** Common plant symptoms associated with the deficiency of essential macronutrients and micronutrients.

| Nutrient element | Key deficiency symptoms |
| --- | --- |
| Nitrogen | Yellow foliage due to reduced synthesis of chlorophyll and proteins |
| Phosphorus | Dark green or purpling of leaves and stems due to the accumulation of photosynthates and anthocyanin |
| Potassium | Suppressed plant growth and development, leaf margin scorching, first appearing on old leaves |
| Calcium | Aberrant development and reduced yields, chlorosis and necrosis of youngest leaves or shoot apices |
| Magnesium | Interveinal chlorosis on older leaves |
| Sulfur | Interveinal chlorosis, purpling, cupping, or whitening, depending on plant species |
| Boron | Inhibition of apical growth, underdevelopment of leaves, production of lateral shoots, short internodes, reduction of root growth and poor flower production |
| Chlorine | Leaf bronzing and chlorosis, leaf curling and wilting, and root growth inhibition |
| Copper | Lack of green colour in leaves |
| Iron | Chlorosis of young leaves, or of all leaves when iron supply is exhausted |
| Manganese | Interveinal chlorosis on young leaves and necrotic spots or streaks |
| Molybdenum | Suppression in leaf size and irregularities of leaf blade formation (whiptail), chlorosis and stunted growth |
| Nickel | Chlorosis of young leaves, reduced leaf size, altered leaf blades and necrosis of leaf tips |
| Zinc | Interveinal chlorosis, smaller leaves and short internodes |

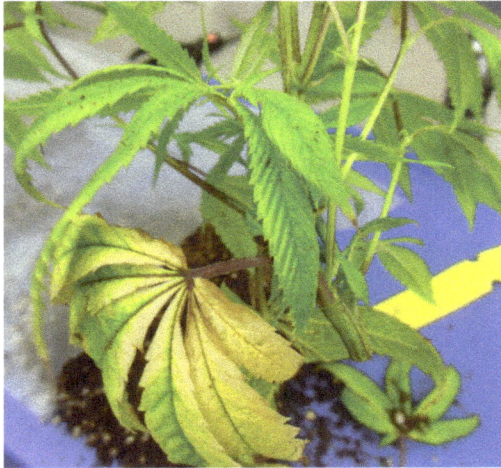

**Fig. 11.9.** A cannabis plant shoot exhibiting chlorosis and yellowing symptoms on an older leaf, due to nutrient deficiency.

**Fig. 11.11.** A hemp plant shoot showing variegated mosaic. Note that this plant grew normally with healthy foliage except this shoot.

**Fig. 11.10.** A leaflet from a cannabis plant exhibiting leaf tip necrosis and mild interveinal chlorosis, due to nutrient deficiency.

but genotypically, many organisms are actually heterogenomic within an individual because of somatic mutations. A hemp crop may have a few chimeric plants showing different colours of leaves within a plant (Fig. 11.11). This symptom is usually described as variegated mosaic and is limited to certain plants. Chimeric plants are not desirable and can be rogued as off-type plants, especially when a crop is produced for seed. During a field diagnosis, chimeric symptoms may be mistaken as

a viral disease, but the latter generally causes more symptoms and is more prevalent in the field.

## Non-infectious Disease Management

Cultural practices are the key to preventing and mitigating non-infectious diseases. In arid and semi-arid regions, drought stress should be minimized by practising efficient irrigation. In clay and alkaline soil, irrigation needs to be optimized to avoid excessive water in soil due to poor drainage. The timing of planting and harvesting should reflect the best time window for hemp to grow in a region. Planting too early risks cold damage to seedlings and planting too late risks freeze damage to flower buds before harvest. Avoid or limit the use of pesticides, herbicides and other chemicals unless such a use is justified scientifically and approved regulatorily. Overuse or unnecessary use of commercial products to prevent pests or diseases can cause phytotoxicity. That said, it is important to read the product label and determine if the product can be used to prevent or treat a disease or pest. In most cases, a product is used only when a pest is identified and when a disease is diagnosed. Managing soil fertility and amending soil to an optimal pH level are critical for a healthy *Cannabis* crop. It is recommended to have the soil tested before planting a hemp crop. The soil test should cover major plant-available nutrients, pH and other chemical and physical characteristics. Fields with high concentrations of salts or at toxic levels of any element should be avoided for planting. During the growing season, plant tissue may be taken to assess potential nutrient deficiencies. The procedures for sampling plant tissue and soil should be based on the instructions of a soil and plant nutrition testing laboratory to which samples are to be submitted.

## References

Almeida, R.P.P. and Nunney, L. (2015) How do plant diseases caused by *Xylella fastidiosa* emerge? *Plant Disease* 99, 1457–1467. doi: 10.1094/PDIS-02-15-0159-FE

Barker, A.V. and Pilbeam, D.J. (2015) *Handbook of Plant Nutrition*, 2nd edn. CRC Press, Boca Raton, Florida.

Bredow, M. and Walker, V.K. (2017) Ice-binding proteins in plants. *Frontiers in Plant Science* 8, 2153. doi: 10.3389/fpls.2017.02153

Frank, M.H. and Chitwood, D.H. (2016) Plant chimeras: the good, the bad, and the 'Bizzaria'. *Developmental Biology* 419, 41–53. doi: 10.1016/j.ydbio.2016.07.003

Magagnini, G., Grassi, G. and Kotiranta, S. (2018) The effect of light spectrum on the morphology and cannabinoid content of *Cannabis sativa* L. *Medical Cannabis and Cannabinoids* 1, 19–27. doi: 10.1159/000489030

Pearce, R.S. (2001) Plant freezing and damage. *Annals of Botany* 87, 417–424. doi: 10.1006/anbo.2000.1352

Whipker, B., Smith, J.T., Cockson, P. and Landis, H. (2019) New research results: optimal pH for *Cannabis*. *Cannabis Business* Times, April 2019.

# 12 Diagnosing Insect and Mite Damage

Arthropods, including both insects and mites, are invertebrate animals classified in the phylum Arthropoda or Euarthropoda (Ortega-Hernández, 2016). They have an exoskeleton, a tough cuticle layer for body protection, a segmented body typically consisting of the head, thorax and abdomen, and appendages. In some arthropods, the head and thorax are combined to become a cephalothorax. There are over 800,000 named arthropod species and possibly many more that have not even been discovered. Insects and mites combined represent the majority of species found in the phylum Arthropoda and are morphologically diverse. Taxonomically, insects are classified in the subphylum Uniramia, while mites are classified in the subphylum Chelicerata (Fig. 12.1). The majority of harmful species are in the classes Insecta and Arachnida.

Insects and mites have their own unique life cycles. For most insects, each individual develops from an egg to an adult through complete or incomplete metamorphosis, a process that changes body shape dramatically. The life cycle of complete metamorphosis typically has four life stages: egg, larva (caterpillar), pupa and adult (Fig. 12.2). An egg, laid by a female adult, hatches into a larva that is usually worm-like and quite different morphologically from the adult-to-be. A larva can eat plant tissue, grow larger, moult several times to shed its exoskeletons and then transform into a pupa. A pupa is a non-feeding life stage in which some insects transform from the larva to adult. After the transformation is complete, the adult breaks the pupal case and exits. The adult is a mature stage and can reproduce. Other insects go through incomplete metamorphosis. They only have three life stages: egg, nymph and adult (Fig. 12.3). An egg hatches into a nymph, morphically similar to an adult. The nymph eats plant tissue, grows, moults and becomes an adult. Mites have four life stages: egg, larva with six legs, nymph with eight legs and adult (Fig. 12.4). The length of a life cycle varies greatly among species, and the shortest can be only a few weeks. On infested plants, multiple stages may be observed but usually not all of them are present at the same time. This creates some difficulties for morphological identification, as certain stages, especially the adult, have more morphological characters used in identification.

Insects and mites are not considered pathogens even though they cause significant damage to plants. They use host plants as food and shelter to support their growth and reproduction. Insects do have limited selection on food plants (a monophagous insect restricts itself to a single host species), but many have broad host ranges. Insects and mites have different feeding behaviours, including chewing, piercing-sucking, leaf mining, stem boring, or tissue galling. These feeding behaviours may cause different damage symptoms on plants. For example, insects such as aphids, whiteflies, mealybugs, soft scales and thrips have piercing-sucking mouthparts and they cause symptoms ranging from patches of silver colour and leaf distortion to the stunted growth of buds. Insects with chewing mouthparts are generally less selective in their food and can eat a portion of leaves or flowers directly, making the damage easily noticeable (Fig. 12.5). Insects like leaf-miner larvae live and feed between the leaf surfaces and create internal tunnels but show obvious tracks on the leaf surface. The damage symptoms vary greatly among different groups of insects or mites, and symptoms may not be specific to an insect genus or even family.

Field diagnosis of insect damage is largely based on the presence of a harmful pest species and its significant level of infestation. Most insect pests are visible to the naked eye, but some are microscopic. With a hand lens, almost all stages of insect, if present, can be seen on the plant surface or sometimes inside plant tissue. The coexistence of active insects and specific damage to plant tissue often leads to a

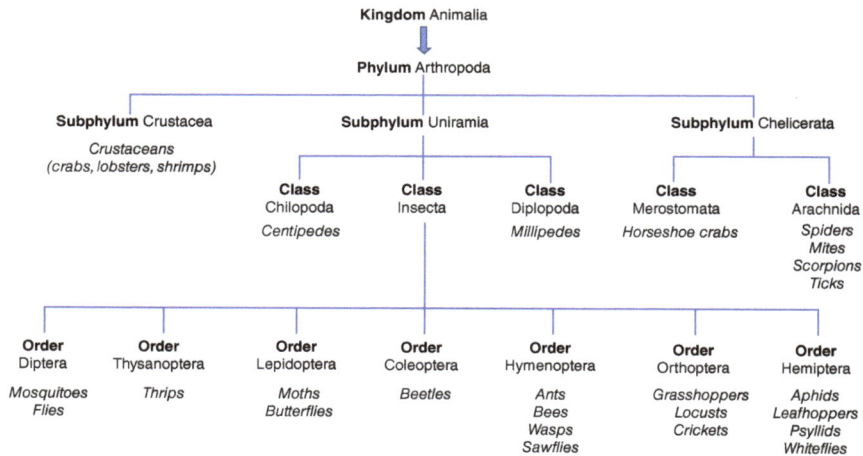

**Fig. 12.1.** Taxonomic positions of insects and mites in the Animal Kingdom and representative organisms in each taxonomic rank.

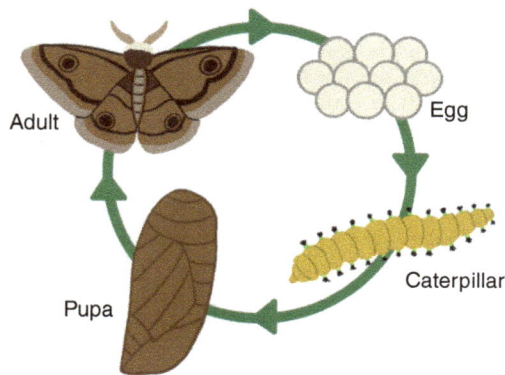

**Fig. 12.2.** An example of insect life cycle showing a complete metamorphosis.

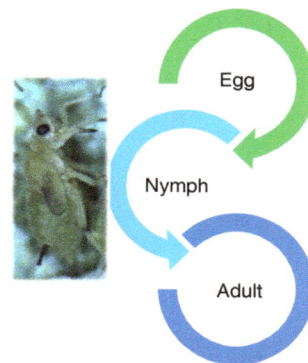

**Fig. 12.3.** A life cycle consisting of three stages with a gradual metamorphosis.

diagnostic conclusion. Sometimes, however, damage symptoms exist without insects in place. For example, the insect larva, the actual damaging stage, may no longer be associated with food plants later on. In this case, insect leftovers may play a significant role in diagnosis. Many insects leave some remains, including moulted skins, eggs, honeydew, black and hardened faecal deposits, etc., from their development while feeding on plants. In other cases, patterns and characteristics of damage are very typical of insect infestation even without the presence of active insects and/or their leftovers. In this situation, it is not prudent to rule out an insect problem just based

on the absence of an insect pest in plants or samples at the time of examination. Many pest-damage symptoms are diagnosable even without the presence of the pest. For example, pear leaf blister mite causes brownish leaf spot with a tiny hole in the centre of the spot (Fig. 12.6). The presence of holes suggests arthropod damage rather than a fungal leaf spot disease, as fungal infections do not uniformly create entry or exit holes to the inside of leaf tissue. Finally, the severity of damage is proportionally related to the level of pest population in the field and the population level should be considered in insect damage diagnosis. Not all harmful insects found on plants

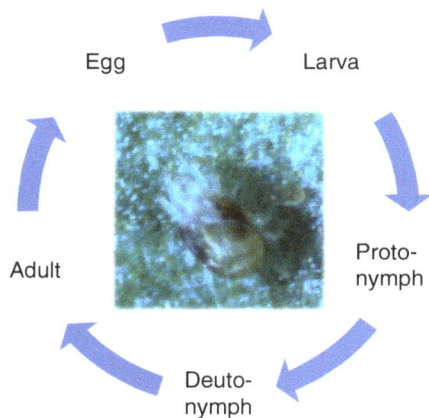

Egg — Larva

Adult — Proto-nymph

Deuto-nymph

**Fig. 12.4.** A spider mite life cycle with four stages.

**Fig. 12.5.** Damage of hackberry leaves caused by an insect with a chewing mouthpart.

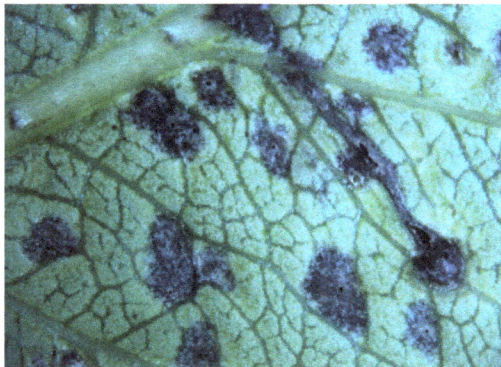

**Fig. 12.6.** A pear leaf exhibiting multiple brown spots with a tiny hole at the centre of each spot.

are responsible for the symptoms. Only those that occur with a significant high number per plant are considered potential pests. Unlike macroscopic insects, mites are microscopic, which makes field diagnosis difficult. Symptoms caused by mites can be similar to diseases or abiotic disorders. Therefore, mite damage diagnosis mainly relies on laboratory procedures such as microscopic examination.

There are a number of insect pests observed from hemp crops (Cranshaw *et al.*, 2019; Anonymous, 2020). Insects that chew on hemp leaves include beet webworm, beet armyworm, cotton square borer, earwigs, flea beetles, grasshoppers, spotted cucumber beetle, painted lady, saltmarsh caterpillar, yellow woollybear, zebra caterpillar, variegated cutworm, yellowstriped armyworm and tree crickets. Insects that suck fluids from leaves include cannabis aphid, rice root aphid, various leafhoppers, lygus bug, stink bug and thrips. Insects that infest hemp stems and make tunnels include Eurasian hemp borer and European corn borer. There are also several insect species found to be associated with buds and flowers. Not all insects found on hemp plants are harmful. Some are beneficial as natural biocontrol agents or pollinators. The majority of these insects are visible and some can be easily identified by common names. However, an accurate identification of every encountered species requires insect taxonomic expertise and insect specimens should be submitted to a trained entomologist for identification.

This chapter is intended to illustrate the process of diagnosing insect and mite damage with an emphasis on microscopic insects and mites, as well as the hidden symptoms caused by them. As discussed in Chapter 3, diagnosing an insect problem is different from insect identification. A hemp problem should be diagnosed as a type of insect damage first before any taxonomic identification of a suspected insect is performed. That said, growers can make a preliminary diagnosis based on insect population trends in the field and levels of plant damage observed, collect insect specimens of all available stages and submit them to a plant clinic for identification.

## Hemp Russet Mites (*Aculops cannibicola*)

*Aculops* is a genus of mites belonging to the family Eriophyidae. Eriophyidae mites may look similar: a microscopic size (usually under 200 μm long and approximately 65 μm wide), cylindrical or sausage shaped but tapered sharply at the posterior. Under a dissecting microscope, the mites may look like tiny 'worms' with a pale white to yellow colour (Fig. 12.7). Hemp russet mite (*Aculops cannibicola*) infests hemp leaves and stems and its damage behaviour is similar to that of the tomato russet mite (*Aculops lycopersici*) (Fig. 12.8).

## Symptoms

Russet mites feed on hemp leaf surfaces with their piercing-sucking mouthparts and remove cell contents from leaves and sometimes stems if infested. At a low population, the plant may show subtle abnormalities when compared with unaffected plants and the symptoms are usually expressed as slight leaf curl and thickness of leaf tissue (Fig. 12.9). At high populations, young leaves may distort with edges curling upwards, inside which there is a high density of russet mites (Fig. 12.10). Some leaves

may have sporadic dark green patches (not green island caused by viruses) and thickness as the rest of leaves have a faded green colour due to mite feeding (Fig. 12.11). Young shoots are less developed, as the mite infestation depletes many of the

**Fig. 12.7.** A russet mite feeds on a hemp leaf.

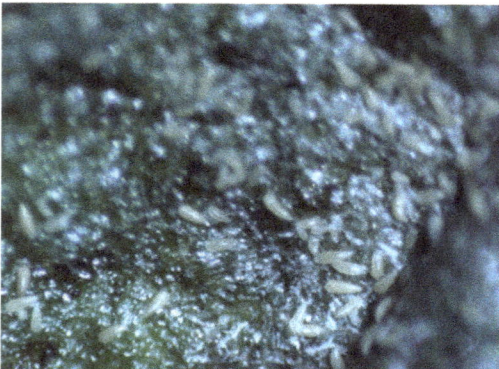

**Fig. 12.9.** A cannabis seedling infested with hemp russet mite at a low population. Note that the plant has a healthy stem and root system.

**Fig. 12.8.** A crowd of tomato russet mites (*Aculops lycopersici*) feeding on the upper surface of a tomato leaf.

**Fig. 12.10.** A hemp leaf twisted and curled, caused by a high population of hemp russet mite.

nutrients needed for growth (Fig. 12.12). At the late stage of infestation, leaves may become bronze or russet coloured. Spotting may appear but without distinct margins. Mites may move from well-damaged areas to other healthy areas of the plant, but they move slowly. Severely damaged portions of foliage may wilt or die.

### Diagnosis

Russet mites are very small and it requires at least a 20× hand lens to see them. Because of this, russet

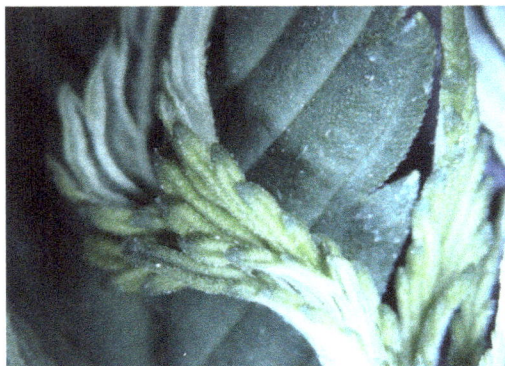

**Fig. 12.11.** A variegated dark green and faded green on hemp leaves infested with russet mites. Note that russet mites were mostly observed at the faded green areas.

mites, along with other eriophyid mites, are rarely noticed until plants start to show symptoms. Additionally, most symptoms caused by russet mites may look similar to other issues such as virus infection, nutrient deficiency, or damage caused by other fluid-sucking arthropods, all of which complicate the diagnosis. The first step to diagnose russet mite damage is to place symptomatic leaves under a stereo microscope (with reflected light on) to observe the leaf surface. Turn the magnification to the highest level, as russet mites may not be spotted under lower magnifications at first glance, especially when mite populations are low. Examine most areas of both sides of leaves to determine whether or not russet mites are associated with leaf tissue. When a russet mite is seen, it is usually sausage-shaped, translucent to pale or yellow, and moving slowly on the leaf surface (Fig. 12.13). Note that mites can be present on both upper and lower sides of leaves, but the majority of them feed at the lower side. To see more detailed morphology of a russet mite, a short piece of clear tape can be gently applied to the leaf surface where a number of mites are present (see blotting in Chapter 5). After blotting, place the tape piece (with mites attached) on a drop of water at the centre of a glass slide and observe the mites under a compound microscope. At 10× or 40× objective lens, russet mites can be viewed as wedge-shaped with a tapered end and broader head, and they have two

**Fig. 12.12.** A healthy shoot (upper) and a shoot (lower) severely infested by hemp russet mites.

pairs of legs at the anterior end and long hairs at the posterior end (Fig. 12.14).

Although the general wedge-shaped bodies and legs are useful to classify the pest as an eriophyid mite, accurate identification of an encountered mite into a species using morphological characteristics may be challenging, especially when the morphological characteristics are not distinct enough to separate biological species. Sometimes, taxonomic confusions exist for some organisms at the low levels of taxonomic ranks. The DNA barcoding-based identification system can be used for quickly determining the species and phylogenetic relationships with other species (see

Fig. 12.13. A patch of russet mites feeding on the lower side of a hemp leaf between and along the veins.

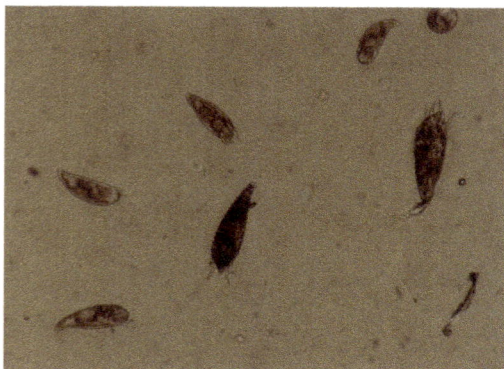

Fig. 12.14. Six russet mites blotted from a hemp leaf and observed under a compound microscope (10× objective lens). Note that adults and different development stages (larva and nymph) are present.

Chapter 5). Briefly, pick a single mite (adult preferred) or a few up to 50 mg from an infested leaf and place them into a 1.5 ml microcentrifuge tube or a sample homogenization tube such as lysing matrix tube. Homogenize the mite sample using a mortar and pestle, a bead beater, a quick sample homogenizer such as Qbiogene FastPrep, or any other available equipment. The sample can be proceeded to DNA extraction using any commercial kit, following the manufacturer's instructions. The protocol described after this paragraph is based on Qiagen's DNeasy® Blood & Tissue Kit and using the Qbiogene FastPrep homogenizer. Other similar kits designed for insect DNA extraction can be used, according to the user's preference. The next step is to perform PCR using universal primers (see Table 5.6 in Chapter 5) or other primers used for mite identification (Khaing et al., 2014). The PCR product can be directly sequenced after purification or after cloning into a T-vector. Obtained DNA sequences can be pasted into NCBI BLASTn search box to find the most closely related species based on sequence similarity and other parameters.

The protocol to extract total genomic DNA from insect or mite specimens is as follows

1. Weigh out approximately 50 mg of mites or insects in a lysing matrix A tube (weigh one tube with the sample and one without sample to obtain the net sample weight).
2. Add 180 µl of PBS and homogenize samples using the Qbiogene FastPrep 24 at speed 6.0 for 40 s. Steps 1 and 2 finish sample homogenization.
3. Add 20 µl of proteinase K and 200 µl of buffer AL, mix and incubate at 56°C for 10 min. This step finishes cell lysis.
4. Add 200 µl of 100% ethanol and mix thoroughly.
5. Pipette the mixture into the DNeasy mini spin column placed in a 2 ml collection tube. Centrifuge at 8000 rpm for 1 min. Discard the flow-through and collection tube. Steps 4 and 5 finish DNA precipitation and capture on the membrane.
6. Place the column in a new 2 ml collection tube and add 500 µl of buffer AW1 and centrifuge at 8000 rpm for 1 min. Discard the flow-through and collection tube.
7. Place the column in a new 2 ml collection tube and add 500 µl of buffer AW2 and centrifuge at 14,000 rpm for 3 min. Discard the flow-through and collection tube. Steps 6 and 7 finish DNA washing.

**8.** Place the column in a new 1.5 ml microcentrifuge tube, pipette 200 µl of AE buffer into the centre of column membrane, wait about 1 min, then centrifuge at 8000 rpm for 1 min. Solution in the tube is final DNA product. This step finishes DNA elution.

**9.** Measure the DNA concentration and store the DNA at –20°C.

Detecting microscopic mites mostly relies on the direct observation of mites on hemp or cannabis plants under a stereo microscope, as this method can examine a large of number of leaves and plants in a short period of time. It can screen a mass of material for russet mite with little chance of missing any infested plants. *Cannabis* cultivators can use a microscope to detect mites as soon as they appear and then use either morphological or molecular methods to determine the species, if necessary. Certain detection methods such as qPCR or NGS can be highly sensitive or specific to an organism, but they may not have sufficient detection coverage. That is because the plant tissue used in testing is on a scale of milligrams to grams. Such a small sample size, even with many repetitive samples, may not catch a low-level infestation in a big hemp field. That said, traditional pest scouting (field observations and microscopic examinations) and entomological diagnosis should always be used first and should not be replaced by any single molecular testing method.

## Spider Mites

Spider mites are a group of organisms in the family Tetranychidae that contains over 1000 species. They are tiny, usually less than 1 mm, and not visible to the naked eye. However, spider mites moult their skins, lay small round eggs and produce silk webs on plant surfaces, all of which make an infested plant organ look fuzzy. The web, a unique structure for protection from potential predators, is visible to the human eye and its presence suggests a spider mite infestation. Spider mites can feed on hundreds of plant species, including *C. sativa*. They usually feed on the undersides of plant leaves and cause foliar damage. The two-spotted spider mite (*Tetranychus urticae*) is a highly polyphagous, cosmopolitan and also well-studied species. It is probably the most damaging spider mite on *Cannabis* and other plants. The mites spread from plant to plant by contact or wind. Under favourable conditions, the mites can reach a population level that significantly damages plants and reduces crop yields.

## Symptoms

Two-spotted spider mites use their needle-like piercing and sucking mouthparts to penetrate plant tissue and suck the fluid from plant cells. As plant fluid is an essential component to support cell survival and structures, the excessive removal of fluid from plant tissue will cause structural and functional damage of cells and ultimately cell collapse. Initial damage may be limited to a few or a group of cells at feeding sites, which is not noticeable (Fig. 12.15), especially when the mite population is low. As the mites continue to feed and reproduce, collective damage manifests as stippling, bleaching, chlorosis, or a mixed colour of yellow, grey and bronze (Fig. 12.16). Early infested tissue may turn brown or completely necrotic, especially tender tissues (Fig. 12.17). In the late infesting stage, fine and grey to white strands of webbing are noticeable on plant surfaces. Infested leaves or flowers have a fuzzy, webbing, or dirty appearance.

## Diagnosis

In general, the direct observation of active spider mites feeding on leaves or flowers under a stereo microscope indicates spider mite infestation, which may easily conclude the diagnosis. Even without

**Fig. 12.15.** A nymph of two-spotted spider mite feeding on a cannabis leaf without noticeable damage.

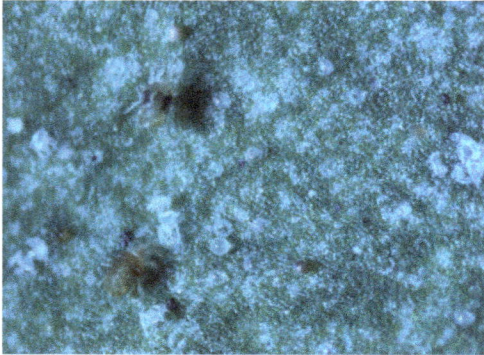

**Fig. 12.16.** Typical leaf symptoms caused by spider mites (observed under a stereo microscope). Note the ball-shaped eggs.

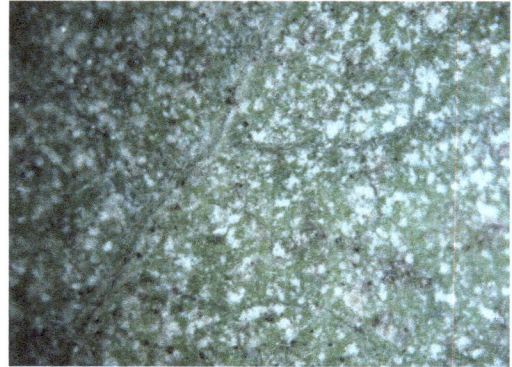

**Fig. 12.18.** A non-*Cannabis* leaf exhibiting whitish spotting on the surface due to two-spotted spider mite infestation.

**Fig. 12.17.** Brown and necrotic younger tissue inside a cannabis flower bud due to spider mite infestation.

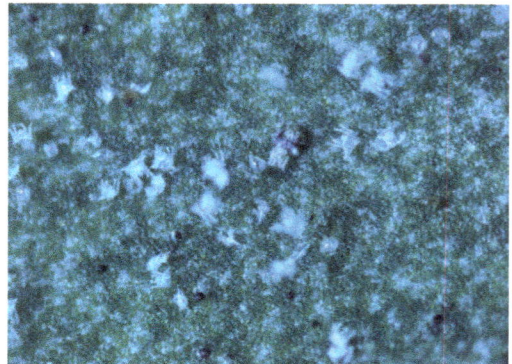

**Fig. 12.19.** Moulted skins of two-spotted spider mites on a non-*Cannabis* leaf surface.

active mites observed during the time of examination, the dirty-looking and sporadic bleaching chlorosis suggest a spider mite infestation (Fig. 12.18). In most cases, moulted skins or exoskeletons and eggs are present on the surface of infested leaves (Fig. 12.19). Spider mites can complete a life cycle in a few weeks; therefore, it is common to see mixed generations, which is helpful in observing all stages of their life cycle (Fig. 12.4). Two-spotted spider mites, both males and females, have a large dark spot on both sides of the central portion of the body (Fig. 12.16). The dark colour is due to the accumulation of body wastes that are visible through the transparent body wall. Two-spotted spider mites are generally oval in shape and about 0.4 mm in size, and their colour may be variable

from green to greenish-yellow, brown, or orange red. The egg is round and hatches to larva usually a couple of days after being laid. The larva initially has six legs and moults once to become a nymph with eight legs. After two additional moults, the nymph becomes an adult. Reproduction requires mating between males and females and a mated female can lay several hundred eggs. To summarize, a diagnosis of spider mite damage can be mostly based on the unique morphological characteristics of adult mites, the presence of all development stages and the unique symptoms on specific plant organs. If a new or unfamiliar spider mite species is observed, and not all life stages are present to warrant a morphological identification, any available mite stage can be collected and proceeded to DNA extraction followed by PCR using a universal primer pair to determine the species (see Chapter 5).

## Leaf Miners

Insects behaving as leaf miners include moths, sawflies and flies. They belong to different orders of Insecta, primarily Lepidoptera, Hymenoptera and Diptera. Some beetles in the order Coleoptera may also behave like leaf miners. In the order Diptera, common leaf miners include the pea leafminer (*Liriomyza huidobrensis*), vegetable leafminer (*L. sativae*), serpentine leafminer (*L. brassicae*) and spinach leafminer (*Pegomya hyoscyami*). These insects affect vegetable crops and some of them may affect hemp and cannabis plants. Although leaf miners are common and often cause minor damage to crops, in some cases leaf miner infestation can be problematic. For example, a hemp crop growing inside a greenhouse was mostly infested by a leaf miner insect, causing substantial foliar damage (Fig. 12.20). Plants infested by leaf miners are also more prone to fungal diseases such as *Alternaria* leaf spot and blight. There is little information on what type of leaf miner insects are associated with *Cannabis* plants, as detailed identification is lacking.

## Symptoms

The larvae of leaf miner insects live between the upper and lower leaf surfaces and eat the mesophyll tissue that is critical for photosynthesis. After this part of tissue is eaten (loss of green colour) and filled with the frass (faeces) produced by larvae, the trace of a larva moving inside the leaf is visible as a pattern of serpentine tunnelling. Early stages of infestation may only show tiny blisters or blotches without extended tunnelling. Inside a tunnel, there are dark or brownish particle-like faeces, which are visible (through the thin and opaque leaf membrane) as a dark and multiple-dash line at the centre of the tunnel trace (Fig. 12.21). Leaf miner invasions are more observable on older leaves of hemp plants (Fig. 12.22). Severely infested hemp plants may exhibit reduced growth vigour, as photosynthesis is compromised.

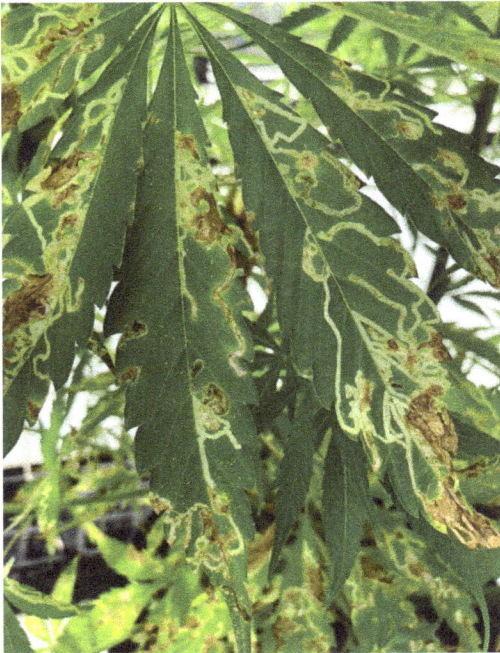

**Fig. 12.20.** A hemp plant severely damaged by an unidentified leaf miner insect.

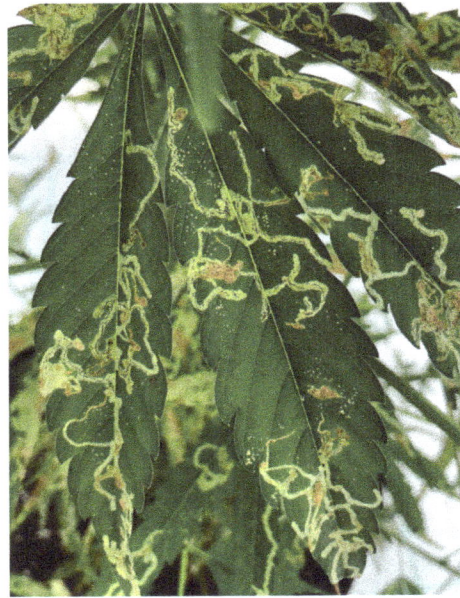

**Fig. 12.21.** A closer look at infested leaves showing the dark line at the centre of the tunnel trace.

## Diagnosis

Serpentine tunnelling in leaves is a diagnostic symptom for leaf miner insects. However, the specific insect species involved in the leaf mining needs to be determined. Under a stereo microscope, use a needle to remove the superficial membrane to reveal the tunnel and find the larvae eating the tissue. Most leaf miner larvae are cylindrical in shape with a tapered head region and are white or yellow in colour. Larvae move forward from the entry point during feeding; therefore, they are more likely to be found at the end of the tunnel. Collect several larvae into a 1.5 ml microcentrifuge tube and proceed to DNA extraction and DNA-based identification. Adults may not be seen after they lay eggs on the leaves. If present, however, collect them for either morphology-based or DNA-based identification.

## Cannabis Aphid (*Phorodon cannabis*)

Aphids are a group of small, sap-sucking insects that belong to the superfamily Aphidoidea of the order Hemiptera. There are many species, some of which are common but serious pests on many agricultural crops and horticultural plants. Many aphid species are effective carriers of plant viruses and transmit viruses from plant to plant. Hemp crops, both field- and indoor-grown, are prone to aphid infestation. The cannabis aphid (*Phorodon cannabis*) is the most common species found on *Cannabis* plants.

## Symptoms

Using their piercing-sucking mouthparts, cannabis aphids suck the plant fluid from hemp phloem tissue (Fig. 12.23). Once the phloem vessel is penetrated, the sap, rich in sugars and other nutrients, passively enters the aphid's food canal due to liquid pressure. The depletion of sap from phloem tissue causes a shortage of nutrients needed for leaf cell growth and development. The damage symptoms may be subtle initially, especially when the population is low, but become noticeable when a large number of aphids feed on plants. Infested leaves become yellowish, bronze to light reddish, or chlorotic (Fig. 12.24). Another significant symptom is the presence of aphid remains such as numerous cast skins on both the upper leaf surface (Fig. 12.25) and lower surface (Fig. 12.26) and the presence of mixed stages of aphids (Fig. 12.27). In some cases, the sticky liquid droplets produced by the aphids are present. The droplets are honeydew secreted by aphids, rich in sugar and nutrients, and often stimulate

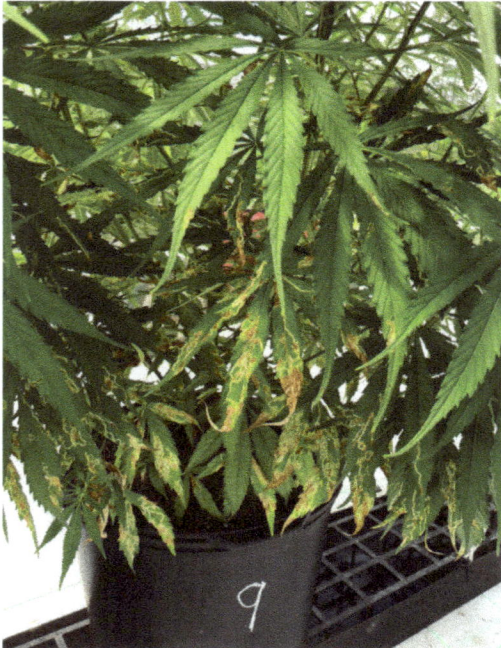

**Fig. 12.22.** A plant with lower portions of foliage infested with a leaf miner insect.

**Fig. 12.23.** A cannabis aphid feeds on the phloem vessel of midvein on the lower side of a hemp leaf. Note that two yellowish eggs were laid.

**Fig. 12.24.** Abnormal colour change of a hemp leaflet due to heavy infestation of cannabis aphid.

**Fig. 12.25.** Numerous cast skins (after moulting) stuck to the upper leaf surface as well as the extensive growth of a sooty mould on hemp leaves.

**Fig. 12.26.** Cannabis aphid nymphs and cast skins present on the lower surface of hemp leaves.

growth of some fungi, causing sooty moulds (see Fig. 7.8). Sooty moulds are a secondary condition and affect normal plant photosynthesis as well as bud quality (Fig. 12.28). The type and severity of symptoms caused by aphids may depend on the aphid and host plant species, population levels, presence or absence of natural enemies and environmental conditions. In general, they can cause leaf chlorosis, green islands, reduced growth of affected plant organs, leaf rolling inwards (usually with a high number of aphids inside the rolled leaf) and leaf distortion. Severe infestations can diminish foliage size and functions.

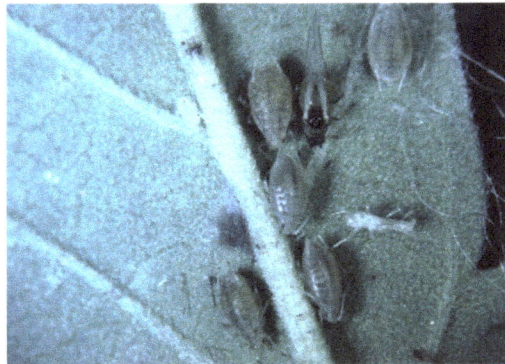

**Fig. 12.27.** A group of winged and wingless forms of cannabis aphids feeding along the veins at the lower surface of a hemp leaf.

**Fig. 12.28.** Damage of a hemp bud caused by the cannabis aphid and sooty mould. Note the brownish necrotic lesions at the tips and edges of new leaves.

## Diagnosis

Diagnosing aphid damage is straightforward. The visual observation of leaves for cast skins and active aphids feeding on plants can quickly conclude a diagnosis. Under a stereo microscope, the eggs, honeydew, small nymphs and adults can be observed in more detail (Fig. 12.29). With some insect taxonomy expertise, an encountered aphid population with most or all life stages can be identified to the species level. However, different aphid species many look similar and it may not be an easy task for those who are not arthropod taxonomists. As described in Chapter 5, DNA barcoding-based identification offers a shortcut for those who need to identify an encountered insect to a species when they lack resources or expertise in morphological identification. The identification process is detailed in Chapter 5. In brief, collect several nymphs into a 1.5 ml microcentrifuge tube and proceed to genomic DNA extraction as described in the hemp russet mite section above. Use the universal primers LCO1490 and HCO2198 (see Table 5.6) to amplify an approximately 709 bp DNA fragment from the mitochondrial cytochrome oxidase subunit I (COI) gene (see Fig. 5.23). Purify the PCR product using an available PCR purification kit and submit the purified PCR product to a service centre for Sanger sequencing. Some service labs offer Sanger sequencing for as low as US$2.50 per sample and usually deliver the result within 24 h. The sequence data from both strands can be assembled manually, if without software, to obtain the full amplicon sequence as illustrated in Fig. 5.27. The cannabis aphid generates a 709 bp PCR amplicon from the part of COI coding sequence as shown in Fig. 12.30. Using this barcode region can quickly determine a specimen to a species within 2 days at a low cost.

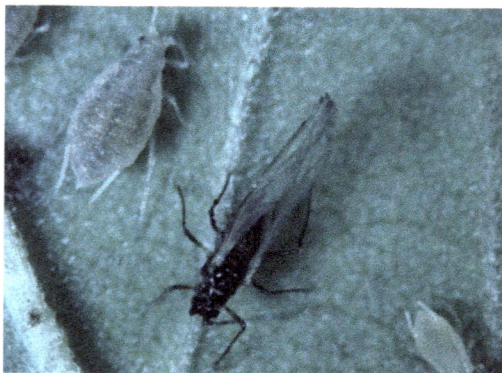

**Fig. 12.29.** Nymphs and a winged adult of the cannabis aphid observed under a stereo microscope.

## Whiteflies

Whiteflies are small, moth-like insects belonging to the family Aleyrodidae. They are coated with an opaque, white waxy powder and can fly, hence the name whitefly. There are many species of whiteflies, some of which cause significant damage to crop plants, especially those growing inside greenhouses. Three species, sweet potato whitefly (*Bemisia tabaci*), greenhouse whitefly (*Trialeurodes vaporariorum*) and banded-winged whitefly (*Trialeurodes abutiloneus*), are commonly found in greenhouses. Whiteflies go through a life cycle starting from the egg to different stages of nymphs (instar I, II and III), to a non-feeding pseudo-pupa stage (instar IV, fourth-stage nymph), and finally to a winged adult. The first-stage nymphs (instar I) hatch from the eggs and become tiny crawlers that move around until locating a feeding site. The second- and third-stage nymphs are flat and scale-like and stay at the same location until they are fully developed into pupa-like fourth-stage nymphs, from which white-winged adults emerge. This significant change from scale-like nymphs to moth-like adults is accomplished through a process called metamorphosis. In greenhouse conditions, the life cycle of whiteflies takes about 3 weeks and this short cycle contributes to the rapid build-up of a high population. Whiteflies cannot survive freezing temperatures, so they are more prevalent in greenhouse or indoor environments. Hemp crops grown in greenhouse are susceptible to whitefly infestation, as well as those grown in warm zones. Whiteflies impact plant health through sap feeding and transmitting many plant viruses.

## Symptoms

Similar to aphids, whiteflies suck fluids from phloem tissue and excrete honeydew. Their feeding activity drains nutrients from plant cells and causes reduced plant growth vigour in general, but specific symptoms appear first on foliage as yellowing or necrosis (Fig. 12.31). The onset of these symptoms is due to the excessive loss of sap and cell damage caused by the prolonged feeding by a high population of whiteflies. Another symptom, though not expressed by plants, is the physical presence of

Primer **LCO1490** sequence: 5′-GGTCAACAAATCATAAAGATATTGG-3′

Primer **HCO2198** sequence: 5′-TAAACTTCAGGGTGACCAAAAAATCA-3′

Reverse complement of **HCO2198**: 5′- TGATTTTTTGGTCACCCTGAAGTTTA-3′

Assembled full PCR product sequence amplified by LCO1490/HCO2198 primers:

GGTCAACAAATCATAAAGATATTGGAACTTTATATTTTTTATTTGGTATTTGATCAGGTATAATTGGATC
ATCTCTTAGAATTTTAATTCGTCTTGAATTAAGTCAAATTAATTCAATTATTAACAATAATCAATTATAC
AATGTTATCGTTACAATTCATGCTTTTATTATAATTTTTTTTATAACAATACCAATTGTAATTGGTGGGT
TTGGAAATTGATTAATCCCTATAATAATAGGTTGTCCTGATATATCTTTCCCACGATTAAATAATATTAG
TTTTTGATTATTACCACCATCATTAATAATAATAATTTGTAGTTTTCTAATTAATAATGGAACAGGAACA
GGATGAACTATTTATCCACCTTTATCAAACAATATTGCACATAATAATATTTCAGTTGATTTAACTATTT
TTTCACTTCATTTAGCAGGAATTTCATCTATTTTAGGAGCAATCAATTTTATTTGTACAATTTTAAATAT
AATACCAAATAATATAAAATTAAATCAAATCCCTCTTTTTCCATGATCAATTTTAATTACAGCTATTTTA
TTAATTTTATCACTACCAGTTTTAGCAGGAGCTATTACAATACTATTAACTGATCGTAATTTAAATACAT
CATTTTTTGATCCAGCAGGAGGAGGAGATCCAATTTTATATCAACATTTATTTTGATTTTTTGGTCACCC
TGAAGTTTA

**Fig. 12.30.** A partial DNA sequence of mitochondrial cytochrome oxidase subunit I (COI) amplified from *Phorodon cannabis* infecting a hemp crop in Nevada. This sequence has a 99.85% similarity to the same COI region of the *P. cannabis* voucher specimen 'INHS:830663' (Accession No.: MH119082.1). The primers used are LCO1490 and HCO2198.

**Fig. 12.31.** Hemp leaves infested with greenhouse whitefly. Upper: whitefly pseudo-pupae on the undersides of leaves. Bottom: chlorosis and yellow spots on the upper sides of leaves.

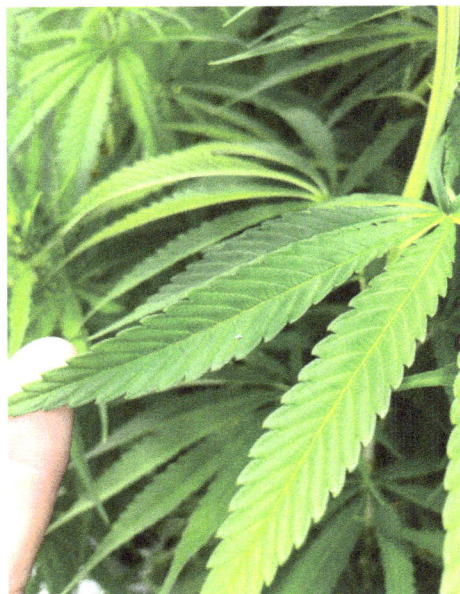

**Fig. 12.32.** A tiny whitefly (at centre of picture) on the upper surface of a hemp leaf (taken with an iPhone).

whiteflies at different development stages. Whiteflies are generally small (less than 2.54 mm), but they are still visible when they stay and feed on hemp leaves (Fig. 12.32). When the population is

low, they may not be noticed, or there may be no observable symptoms on plants. When the population builds up during the middle of the season, a large amount of tiny white flies can be seen on or around the plant. The undersides of leaves often have white or black dots or scale-like spotting. These are whitefly colonies in the form of adults, nymphs, or pseudo-pupae. Meanwhile, the upper sides of leaves appear chlorotic and have yellowish spotting with fewer nymphs (Fig. 12.31). Leaf tips may turn to yellow or brown and become necrotic. Heavily infested plants eventually lose growth vigour and decline. Whitefly infestation may cause black sooty moulds on leaf surfaces, due to the excretion of sticky honeydew.

### Diagnosis

The initial step to diagnose whitefly infestation is to examine hemp leaves visually for tiny white flies on both sides of leaves. Chlorosis and yellowish spotting along with the presence of an active whitefly population often lead to a preliminary diagnosis. Examine several symptomatic leaves under a stereo microscope to confirm that the insects observed are whiteflies. The unique scale-like nymphs and white adults (if present) are diagnostic for whiteflies (Fig. 12.33). Further morphological identification can be done through observations of adults and the last stage of nymphs. The adults of banded-winged whiteflies have a dark zigzag pattern of bands across the wings, which can be easily distinguished from sweet potato and greenhouse whiteflies, both of which have solid white wings (White, 2020). The latter two species can be further separated by the way they hold their wings. The greenhouse whitefly holds its two wings out almost flat, giving the adult a triangular shape, whereas

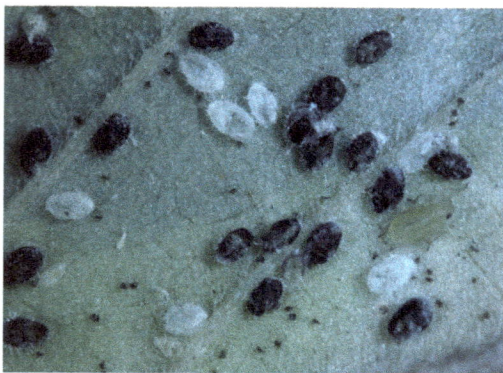

**Fig. 12.33.** Nymphs of greenhouse whitefly on the underside of a hemp leaf. Note that an aphid is present on the right.

sweet potato whitefly holds its wings down at a sharp angle, giving the whitefly the appearance of a tiny grain. If these characteristics are not helpful, the instar IV nymphs can be examined. Greenhouse whitefly nymphs are more three-dimensional with flat perpendicular sides and have very obvious, long waxy filaments on the top. In contrast, nymphs of sweet potato whiteflies are flatter, without the perpendicular sides, and have only a few filaments, which are often hard to see. Nymphs of banded-winged whiteflies look like greenhouse whitefly nymphs by shape, but they have short and curly filaments on the top. These morphological characteristics may be sufficient to distinguish these three species under a microscope, but a DNA-based identification is often needed to identify all encountered whiteflies, especially those that are not so common and are difficult to distinguish morphologically.

### Thrips

Thrips belong to the family Thripidae in the order Thysanoptera (see Fig. 12.1). They are small (usually 1–2 mm long) and may not be visible to the naked eye. Thrips look slender and their adults have pointed wings; and they have piercing and sucking mouthparts to puncture and suck liquid contents from plant cells and tissue. Their life stages include the egg, instar I and II nymphs, two or three non-feeding resting nymph stages and adult. The first-stage nymphs are smallest and wingless, but they start to feed on leaves. They moult to become second-stage nymphs and continue to feed on the plant. The first two nymph stages are actively feeding stages and cause leaf or flower damage, whereas the later instars are considered as non-feeding pupa-like stages. That said, nymphs are often observed on damaged leaves (Fig. 12.34). There are many thrips species, some of which are important pests of many agricultural crops. Some are natural vectors of plant viruses in the genera *Tospovirus*, *Ilarvirus*, *Carmovirus*, *Sobemovirus* and *Machlomovirus* (Jones, 2005). In hemp, onion thrips (*Thrips tabaci*) and western flower thrips (*Frankliniella occidentalis*) are frequently observed with noticeable damage (Cranshaw *et al.*, 2019). In Colorado, onion thrips are observed to be associated more with hemp foliage, whereas western flower thrips appear to be associated more with flowers. Thrips have been observed in hemp crops in other states, but their identities are not confirmed.

## Symptoms

The most characteristic symptoms caused by thrips are the sandblasting or silver appearance of leaves (Fig. 12.35), though the leaf is not rubbed by sand. Leaf chlorosis and distortion are also associated with thrips feeding (Fig. 12.36). Mixed stages of thrips are often found on both sides of leaves and move slowly during feeding. Visually or under a stereo microscope, minute and speck-like black droppings are noticeable on leaves (Fig. 12.37). Because thrips suck fluid from leaf tissue, a high population of thrips infestation can cause browning and necrosis along leaf edges and tips and chlorosis at the centres of leaves (Fig. 12.38). Necrosis likely results from reduced intracellular and intercellular liquid levels as well as sap inside leaf veins. A leaf vein consists of xylem vessels that carry water and

Fig. 12.36. Tomato leaf silvering and distortion caused by onion thrips.

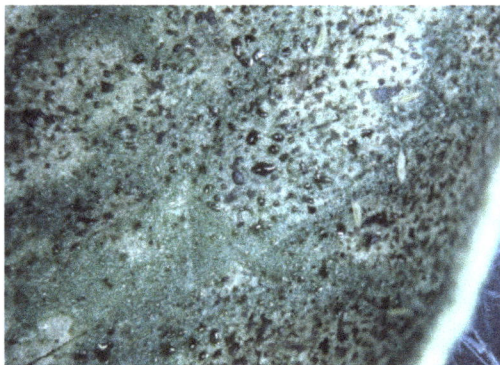

Fig. 12.37. Black droppings and active feeding nymphs on a leaf surface.

Fig. 12.34. Thrips (nymph) feeding on an alfalfa leaf.

Fig. 12.35. Onion leaves damaged by onion thrips.

Fig. 12.38. A hemp leaf exhibiting necrosis and yellow patches caused by thrips infestation. Note that there was minor leaf miner damage in some areas.

phloem vessels that carry nutrients, and both types of vessel function as water and nutrient distribution channels inside the leaf. When water and nutrients are taken away by a large number of thrips through a long period of feeding, distant tissues such as the leaf tip and edges often undergo cell death first, due to nutrient and water deficits. Chlorosis is more likely at the areas fed on by thrips, due to the physical damage to the epidermis and mesophyll tissues by their piercing and sucking feeding behaviour, and mechanically damaged cells can quickly undergo a cell death process. A large number of cell deaths in leaf tissue reduces photosynthesis and causes the leaves to turn a faded green colour. Severe infestation usually hinders hemp plant growth and may trigger secondary fungal infections.

**Fig. 12.39.** A light-coloured thrips nymph feeding on the surface of a hemp leaf.

### Diagnosis

Thrips damage is easily recognizable by the presence of feeding nymphs, black droppings and a characteristic silvering appearance on affected leaves or flowers. These characteristics, though visible to the naked eye, are often observed in more detail under a stereo microscope. Therefore, any fuzzing or dirty-looking, black-spotted, or chlorotic leaves should be first placed under a stereo microscope with a reflective light on to look for adults, nymphs, cast skins, or droplets. The presence of any, if not all, life stages suggests a thrip infestation. The next step is to examine both sides of leaves or flowers (if present and affected) for characteristic tissue damage that is usually exhibited as superficial discoloration or chlorotic spots without definite margins. In symptomatic areas, active thrips are often found, but not always. However, even without active thrips present at the time of examination, the unique damage symptoms sometimes warrant a diagnosis of a pre-existing infestation. Thrips found in infested leaves are often feeding-nymph stages that look slender and are white to light yellow in colour (Fig. 12.39). The adults are dark-coloured and usually have two pairs of feather-like wings. Although adults have more morphological characters used to separate species, both nymphs and adults are useful in morphological identification. The key distinguishable characteristics include colour, the presence or absence of setae, the number of antennal segments and the colour of ocelli, and they all require a 40× magnification to observe. Alternatively, a DNA-based identification can be performed.

### Stem Borers

Borer insects, at their larval or adult stages, are generally not minute or microscopic in size. However, they often attack the stem or stalk and create invisible damage such as tunnelling inside the plant. As the damage is mostly hidden inside the stem, the symptoms that show up on plants can be tricky to diagnose correctly. There are two borer insects that damage hemp crops: European corn borer (*Ostrinia nubilalis*) and Eurasian hemp borer (*Grapholita delineana*) (Cranshaw *et al.*, 2019). Corn earworm (*Helicoverpa zea*) is another pest that causes a similar injury pattern to that of Eurasian hemp borer and it has been diagnosed for damaging hemp in several states. Stem borers injure stems and buds and they can systemically affect plant health. For example, a borer insect, preliminarily diagnosed as European corn borer during a short trip to a hemp field outside the continental USA, attacked the lower stem close to the ground and created tunnelling inside the stem. The injury ultimately caused yellowish foliage, branch flagging and stem breakage.

### Symptoms

The caterpillars (larvae) of European corn borers attack the stem at any location and create holes to get inside the stem. Larvae feed on internal stem tissue and move to fresh portions of the stem, causing destructive tunnels (Fig. 12.40). The feeding process destroys xylem and phloem vessels, which disrupts normal water and nutrient transportation within the plant. Foliage may turn yellow, with

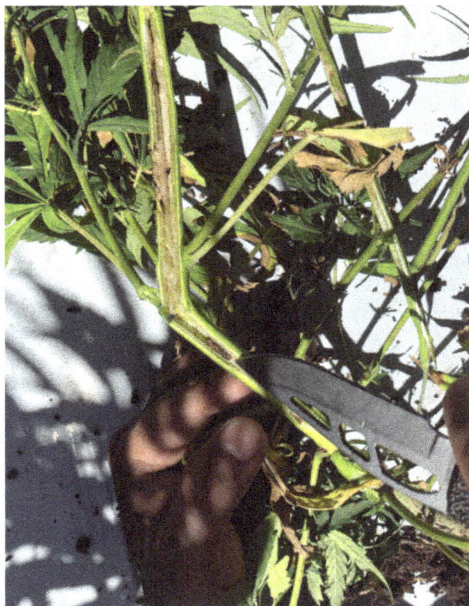

**Fig. 12.40.** A hemp stem dissected to reveal the internal tunnel with a European corn borer caterpillar residing at the junction of damaged and healthy stem tissue.

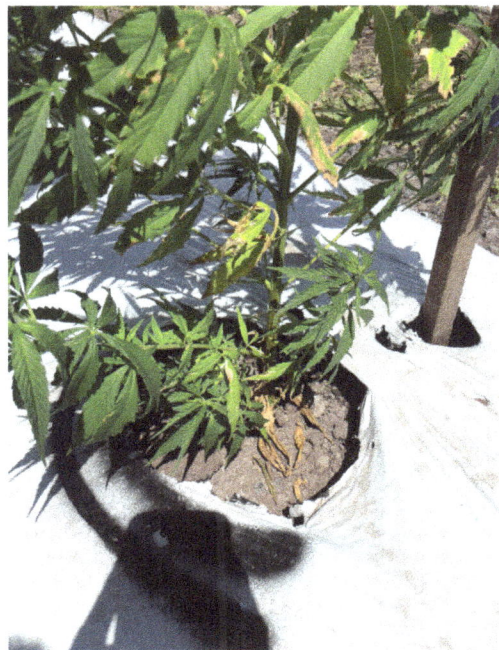

**Fig. 12.41.** Leaf scorch on lower leaves due to the stem damage caused by the European corn borer.

some leaves becoming scorched (Fig. 12.41). Irregular-shaped holes surrounded by brownish and corky tissue on the main stem are often present (Fig. 12.42). Plants with internal tunnelling may break during windy days. Branch dieback occurs when the majority of stem tissue is tunnelled. The larvae of Eurasian hemp borer can enter the upper portion of the stem, especially the area near the base of buds, causing girdling and tunnelling. They can enter the buds and cause direct injury to bud tissue. After buds are harvested, larvae may be dislodged from the bud tissue during drying or other handling processes. Corn earworms also invade and tunnel into the buds and cause serious damage to certain varieties. Developing seeds may be consumed or damaged by the larvae of Eurasian hemp borer or corn earworm.

### Diagnosis

Stem or stalk borer insects attack the hemp stem in their larval stage and create tunnelling inside the pith and vascular tissue. As the insects hide inside, it is difficult to pinpoint foliar symptoms as shown in

**Fig. 12.42.** Multiple holes on a hemp stem created by European corn borer larvae.

Fig. 12.41 as insect damage. In most cases, if not carefully examined, the problem may be misdiagnosed as drought stress or other abiotic factors. Initial diagnosis should be started by performing a complete examination of a symptomatic hemp plant from the leaf and bud to the stem and root. Look closely at scorched leaves to determine if any pathogen signs are present. In this case, there should be no fungal growth or other bacterial signs such as wet-soaked lesions and sticky fluids, unless the plant is also infected by a foliar disease. Check buds and stem sections near the bud base for any signs of holes or mechanical damage. If injuries are present, Eurasian hemp borer or corn earworm may be suspected and the bud and stem should be dissected to reveal larvae. Check the main and branched stems for any abnormalities, including holes, injuries, cankers or lesions, discoloration, or mouldy growth. In the case of a borer insect infestation, visible holes along the stem are compelling evidence (Fig. 12.42). Cut the stem longitudinally to reveal internal symptoms and most likely a larva will be found inside. Examine the crown and roots for discoloration or decay. In the case of a European corn borer infestation, holes and minor tissue decay around the holes may be observed (Fig. 12.43). Note that fungal or bacterial pathogens do not generally create holes when they cause crown rot; rather, they cause tissue decay around the entire crown area with visible mould or slimy bacterial growth. The presence of holes at the crown and upper portion of the stem, even with minor tissue necrosis around the holes, suggests a borer infestation. Finally, examine the root to see if there is any root rot. Generally, the root system is not impacted.

While field diagnosis can determine if observed plant symptoms are caused by a borer pest, a sample of symptomatic plant should be submitted to a lab for further diagnosis, especially when it is not known whether or not a borer insect is associated. In the lab, further dissection may be performed to locate the insect, followed by morphological identification. The larva of the European corn borer is relatively larger than and morphologically different from that of the Eurasian hemp borer. The former is about 1 inch (2.5 cm) long when fully grown and has a creamy to greyish colour with some round and brown spots. The latter is quite small, only 6–8 mm long maximum, and changes colour from cream (with a black head) in the early stage to reddish or orange when fully grown. Both borer insects damage hemp plants by their early or more developed larva stages and they can be found inside the stem or bud tissue. The corn earworm caterpillar is about 2.5 cm long, similar to European corn borer larva, but it is green with a mottled black colour. These morphological appearances might be helpful in differentiating the three species, but these characteristics may overlap among species due to intraspecific morphological variations. In addition, other look-alike species may be encountered. In this case, DNA-based identification can quickly identify the species of an insect at any stage. In brief, collect one larva or adult, whichever is available, into a 1.5 ml microcentrifuge tube and proceed to DNA extraction followed by a PCR using universal primers for arthropods (see Table 5.6).

**Fig. 12.43.** A hole at the crown area of a hemp plant infested by the European corn borer.

Purify the PCR product and submit it to a genomic service centre for DNA sequencing. Use the obtained sequence data to search the NCBI nucleotide databases through the basic local alignment search tool (BLAST) (Altschul *et al.*, 1990) and find closely related species (see Chapter 5).

## Macroscopic Insects

There are many insects that are large enough to be visible to the naked eye. Unlike microscopic insects and mites, these insects are on a scale of centimetres. When they actively chew leaves, suck fluids from plant tissue, crawl on plant surfaces, or simply fly around, they provide evidence of their presence (Fig. 12.44). The damage they cause is usually visible and closely associated with their activities. Diagnosing an injury caused by a macroscopic insect is straightforward and, most of time, the diagnostic process is only a matter of insect identification. Many insect pests found on hemp are common pests of other agricultural crops and they are well known to local producers. In many cases, growers can quickly identify the pest by its common name. As hemp is still a relatively new crop in many regions, it is not known how a given insect chooses hemp plants as its food source during its development and whether or not there is a strict specificity between the hemp plant and insect species. However, it is likely that many pre-existing local or regional pests may infest hemp crops. Fortunately, many of these pests are well documented by the regional programme of integrated pest management (IPM) and the information is mostly available through the land-grant university cooperative extension websites or printed factsheets. Nevertheless, when a crop is severely injured by either a known pest or an unusual pest, representative pest specimens should be collected and submitted to a professional entomologist for identification. These unusual pests may be invasive or new to the region after in-depth identification and they may be particularly harmful to local crops.

## Insect Specimen Collection and Submission

The submission of insect specimens should follow local diagnostic lab guidelines. In general, adequate specimens of different stages, when available, should be collected. Information such as locations, collection date, host plant species and injury symptoms observed should be provided in detail to support laboratory diagnosis and specimen identification. Soft-bodied insects or their stages (such as larvae) should be placed in small screw-top vials or bottles containing at least 70% isopropyl alcohol. Hard-bodied insects or their stages (such as pupae) may be collected either in 70% isopropyl alcohol or in a secured container to protect their structural integrity. If possible, hard-bodied specimens can be placed between layers of soft tissue paper and secured in place before being placed in a sturdy box. All specimens should be packed carefully and securely inside a sturdy box to avoid movement during shipping and handling. A well-preserved or protected specimen is essential for morphology-based identification. However, when a specimen is collected for a DNA-based identification, it can be submitted as it is without being preserved in 70% isopropyl alcohol.

## Arthropod Pest Management

In non-*Cannabis* crops, arthropod pests can be controlled using the registered pesticides. However, many pesticides are not approved for use in hemp

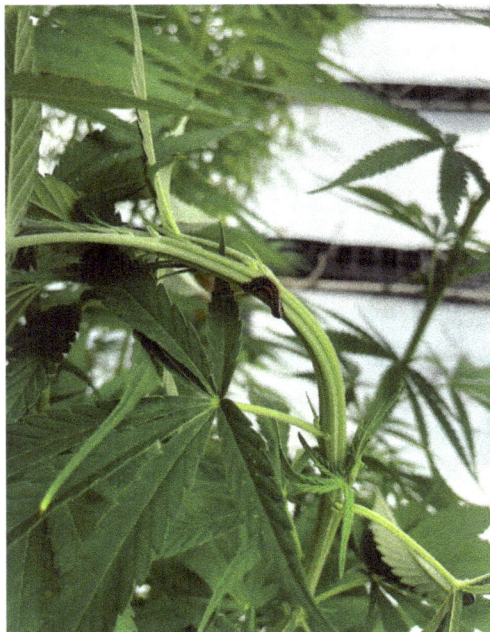

**Fig. 12.44.** A brown caterpillar moves along the stem of a hemp plant.

and cannabis crops. Limitation of pesticide usage can make pest management more challenging. Therefore, the strategy to mitigate damage largely relies on prevention, monitoring, eradication and sometimes more expensive but less effective methods, such as biological controls or alternative remedies. These methods are doable for indoor cultivation but may not be practical for large-scale outdoor production. The specific management strategy depends on the type of pest commonly encountered in local crops, the production model (indoor/outdoor), environments (controlled or uncontrolled), the level of pest infestations and the products approved for use in the crop by state or federal authorities.

### Preventing pest entry to indoor facilities

Indoor facilities, including greenhouses, can be constructed to prevent the entry of small insects by installing screens, covering the ventilation and having a double-door entry. Personnel working with plants outside the facility should change or clean clothes and shoes before entering the *Cannabis* production facility, as small mites can be carried via apparel. Plant stocks or cuttings should be examined for pests before being moved into the cultivation facility. It is fairly achievable to prevent macroscopic pest entry into an indoor facility but preventing microscopic pests can be challenging. In this case, microscopic examination of representative plant materials for mites or tiny insects is important, to ensure that move-in materials are free from harmful pests. Prevention, while possible, is always rewarded with zero or less infestation, which saves time and money used for treatment.

### Monitoring pests throughout production

Once a crop is planted, monitor plants periodically for any signs of infestation, including feeding stages of insects, injury symptoms, or certain materials produced by pests. A number of plants should be selected and examined thoroughly from flower buds to leaves, branches and the stem. Examine both sides of leaves, as some insects and mites predominantly feed on the lower side of leaves. If a leaf becomes curled, roll back the leaf to see if aphids or other insects reside inside. Use a magnifier or stereo microscope to check the leaf surface for mites. The next step is to perform a symptom check and record any insect or mite-specific injuries such as chewed leaves, silver-coloured leaf surfaces, clear or dark droplets, a

colony of exoskeletons, or growth of sooty moulds. Note that an exhibition of visible symptoms generally indicates damage already made by a significant number of pests. The purpose of this monitoring is to detect a pest at its early stage so that action can be taken to eradicate the pest before damage is made. If injury symptoms do occur, increase the monitoring frequency and improve examination methods. For example, use a microscope to screen sampled plant tissue for microscopic pests and increase surveillance frequency from monthly to weekly. As illustrated in Chapter 4, a cannabis cultivation facility can build its own basic plant diagnostic lab for a small investment and this lab can perform microscopic examinations and basic culture of pathogens (see Table 4.1), which greatly enhances plant health monitoring without relying on external labs. Building your own lab not only saves time and money when it comes to diagnosis, but also greatly improves detection coverage and sensitivity, as the size and number of samples can be adjusted as needed to catch a potential infestation as early as possible.

### Eliminating alternative host plants for pests

Insect pests feed on a broad range of host plants, including weeds and other non-crop plants. Some pests complete their life cycle on different plant materials. The elimination of weeds in and around the field helps reduce pest reproduction and population levels. For example, beet leafhopper, the natural carrier of *Beet curly top virus*, feeds on many weed species (see Chapter 10) and causes both pest and disease pressure on outdoor hemp crops. Weeds around or near greenhouses should be cleaned, as they may promote insect and mite reproduction. If possible, a hemp field can be isolated from other crop fields to minimize both disease and pest pressure.

### Planting barrier crops

A barrier crop can be planted around a hemp field to prevent the spread of pests and diseases from other crops. Barrier crops can function in two different ways. One is based on the virus-sink hypothesis, in which barrier crops can remove viruses from insect mouthparts during their feeding before they have a chance to feed on a protected crop, such as a hemp crop. This was demonstrated to be effective for some viruses transmitted in a non-persistent manner by aphids (Fereres, 2000).

When used appropriately, a hemp crop can be protected from some aphid-borne viruses even if the population lands on the crop. Another hypothesis is that a barrier crop can serve as a physical barrier that impedes pest colonization in a protected field. To serve this function, taller crops such as sunflower, maize, or guinea corn (*Sorghum bicolor*) are often considered. Although barrier crops are used as a general cultural practice to avoid pests and diseases, the selection of a specific barrier crop should be based on its effectiveness against local pests.

### Trapping flying insects

Insect traps use light, food, chemical attractants, or sticky substances to capture individual insects and/or kill them. Coloured sticky traps are the most commonly used in greenhouses to catch a large number of flying pests. For example, blue sticky traps are used to catch thrips and yellow sticky traps can catch aphids, whiteflies and leaf miner adults, all of which are important pests in hemp. While many traps are designed to monitor insect occurrence, diversity, or population levels, some can be used directly to suppress population levels of a target pest. For example, yellow sticky traps used in greenhouses significantly suppressed the population increase of adult and immature sweetpotato whitefly (*Bemisia tabaci*) (Lu *et al.*, 2012). In contrast, traps used in the field did not reduce the whitefly population. This study suggested that the use of sticky traps is not an effective method to control a pest for an outdoor crop.

### Using natural enemies

Introducing natural enemies such as parasites, pathogens and predators into a cultivation can lower the population of targeted pests. This is called biological control, which is an important part of integrated pest management. Common natural enemies include lacewings, lady beetles, parasitic wasps and flies, predatory mites and entomopathogenic nematodes, bacteria, fungi and viruses. Some of these are available commercially as biological or microbial pesticides, such as *Bacillus thuringiensis* (Bt), entomopathogenic nematodes and granulosis viruses. These pathogens directly kill individual pests when applied, so they function as biopesticides. On the other hand, predatory or parasitic insects may not be as effective as microbial pesticides when they are used in the field. Some commercial

microbial products are authorized for use in *Cannabis* crops by certain states and cultivators should consult the state agency for the list of approved products.

### Applying alternative products

Some chemical- or plant-based products are claimed to be effective in pest and disease control. For example, neem oil is used to control insect pests. Other plant-based oil products are also listed as approved products in certain states. However, growers may need to read the label carefully and verify the effectiveness of the active ingredients listed before deciding to use a product on a *Cannabis* crop.

## References

Altschul, S.F., Gish, W., Miller, W., Myers, E.W. and Lipman, D.J. (1990) Basic local alignment search tool. *Journal of Molecular Biology* 215, 403–410. doi: 10.1016/S0022-2836(05)80360-2

Anonymous (2020) Hemp Insect Fact Sheets. Available at https://hempinsects.agsci.colostate.edu/hemp-insects-text/ (accessed 19 November 2020).

Cranshaw, W., Schreiner, M., Britt, K., Kuhar, T.P., McPartland, J. *et al.* (2019) Developing insect pest management systems for hemp in the United States: a work in progress. *Journal of Integrated Pest Management* 10, 1–10. doi: 10.1093/jipm/pmz023

Fereres, A. (2000) Barrier crops as a cultural control measure of non-persistently transmitted aphid-borne viruses. *Virus Research* 71, 221–231. doi: 10.1016/s0168-1702(00)00200-8

Jones, D.R. (2005) Plant viruses transmitted by thrips. *European Journal of Plant Pathology* 113, 119–157. doi: 10.1007/s10658-005-2334-1

Khaing, T.M., Shim, J.-K. and Lee, K.-Y. (2014) Molecular identification and phylogenetic analysis of economically important acaroid mites (Acari: Astigmata: Acaroidea) in Korea. *Entomological Research* 44, 331–337. doi: 10.1111/1748-5967.12085

Lu, Y., Bei, Y. and Zhang, J. (2012) Are yellow sticky traps an effective method for control of sweetpotato whitefly, *Bemisia tabaci*, in the greenhouse or field? *Journal of Insect Science* 12, 113. doi: 10.1673/031.012.11301

Ortega-Hernández, J. (2016) Making sense of 'lower' and 'upper' stem-group Euarthropoda, with comments on the strict use of the name Arthropoda von Siebold, 1848. *Biological Reviews of the Cambridge Philosophical Society* 91, 255–273. doi: 10.1111/brv.12168

White, J. (2020) Whiteflies in the Greenhouse. Available at: https://entomology.ca.uky.edu/files/efpdf2/ef456.pdf (accessed 13 November 2020).

# 13 Holistic Diagnosis and *Cannabis* Microbiome

The term 'holistic' is often used in medicine but rarely in plant diagnostics. It defines an approach that treats a thing as a whole instead of considering it as a sum of different unrelated parts. The concept of holistic diagnosis used in this chapter has three layers of meaning. First, a plant is a system and all parts of the system should be viewed as a whole. Diagnosis should be based on an entire plant instead of on a single piece of plant material. Secondly, the approach of 'all things considered' should be practised when finding pests, pathogens, or factors that contribute to an observed plant problem. This approach considers the diagnosis process to be an investigation instead of a specific test for a single pathogen. Thirdly, all factors found in a disease are related and each of them may have a unique role in disease initiation and progression. The relationships among various pathogens, pests and the plant system form a unique network that may define the type, complexity, or severity of a crop disease (Fig. 13.1). A holistic approach looks into this network and often finds broader or more in-depth causes and such findings help growers to develop more effective pest management strategies.

## Pest and Disease Complexes

A hemp crop grown in farmland is exposed to an array of adverse biological and environmental factors, such as pathogenic microorganisms, invertebrate pests and erratic weather conditions. When plants become diseased, there may be more than one pathogen or factor involved. This is why many plant samples submitted to a plant clinic are often found to be infected by multiple pathogens or pests. This demonstrates that, in nature, the phenomenon of 'one pathogen/factor – one disease' is not very common. Rather, a disease complex, in which multiple organisms co-infect the same host

plants to cause more severe damage (May *et al.*, 2009), may be the norm. A disease complex can be caused by several related species (Schoener *et al.*, 2017) or by different types of organisms, including those across different kingdoms (Wheeler *et al.*, 2019). For example, many root diseases are caused by synergistic interactions between nematodes, an organism in the Animalia Kingdom, and soilborne pathogens in the Fungi Kingdom (Back *et al.*, 2002). More often, viral diseases of hemp are linked to the infestation of certain insect vectors, for example beet leafhopper infestation resulting in *Beet curly top virus* disease. Similarly, hemp sooty mould is sometimes associated with an aphid infestation. A holistic diagnosis can dissect any given complex to find those organisms responsible for the disease.

## Multi-pest infestations

Hemp plants can be infested by a number of insects and mites during a certain time period. In a hemp field, both thrips and leaf miner can attack leaves, causing severe tissue damage (see Fig. 12.38) with some plants also being infested with spider mites. Under greenhouse conditions, both aphids and whiteflies can co-infest hemp plants (Fig. 13.2). In some cases, whiteflies and leaf miner are major pests throughout the production period. Multi-pest infestations are also encountered in cannabis cultivation facilities and the most common pests are hemp russet mites and two-spotted spider mites (see Figs 12.9 and 12.17). Unlike the disease complex that may involve specific synergistic mechanisms established by participating pathogens, a multi-pest infestation may occur randomly without obvious interactions among the pests. Therefore, their impact on plant health is additive rather than synergistic. However, a study using tomato plants infested

DOI: 10.1079/9781789246070.0013

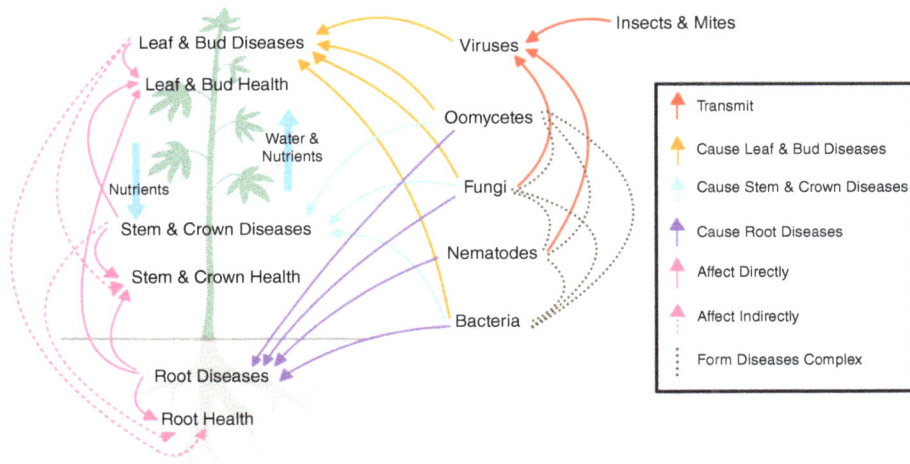

**Fig. 13.1.** Overall plant health impacted by pathogen types, organ diseases, pests and their interactions. As the plant vascular system transports water and nutrients from the root to the rest of the plant and distributes nutrients made in the leaves to other parts, any diseased plant part will impact other parts. Different types of pathogens may interact to cause disease complexes. Insects and mites, as well as some pathogens, transmit viruses to plants in addition to the damage they cause directly to the plants.

**Fig. 13.2.** The underside of a hemp leaf heavily infested by aphids and whiteflies.

with both spider mites (*Tetranychus urticae*) and aphids (*Myzus persicae*) suggested that multi-pest infestation can affect plant biochemistry very differently from single-pest infestation (Errard *et al.*, 2015). To diagnose a multi-pest infestation, plants or their samples should be examined visually and under a microscope to reveal both large and microscopic pests.

## Multi-pathogen infections

Multi-pathogen infections can involve several species in the same genus or different species across different groups of pathogens. The number of pathogens involved depends on the hemp variety and susceptibility, as well as environmental stress levels.

### Multi-species in the same genus

It is common to find multiple closely related pathogens infecting a hemp crop at the same time. For example, a hemp crop can be infected by up to five different *Fusarium* species during the growing season (see Chapter 9 and Fig. 9.9). In this case, each species plays a particular role sequentially during the disease progression. In the middle stage of disease development, 82% of plants were infected by *Fusarium oxysporum*; 1 month later, the number dropped to 51%. Meanwhile, the percentage of plants infected by *F. tricinctum* increased from 18% to 55% (Fig. 13.3). This species shift is explained by the fact that *F. oxysporum* is the primary pathogen that initially infects and kills the root tissue, while *F. tricinctum* is a weak pathogen that commonly invades injured tissue. As more plants

## Hemp Root Rot Caused by Five *Fusarium* Species

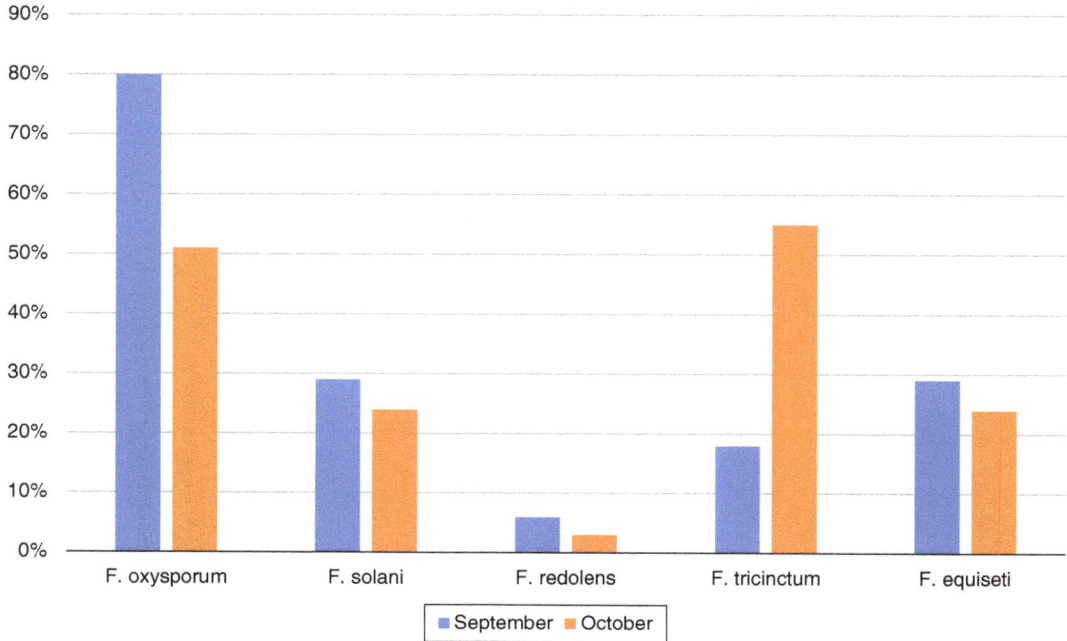

**Fig. 13.3.** Frequencies of recovery of each *Fusarium* species from hemp roots during two sampling periods. In September, *F. oxysporum* was isolated from most plants. In October, when the majority of plants were killed, *F. tricinctum* was recovered at a higher rate. Recovery rates for *F. solani*, *F. redolens* and *F. equiseti* remained stable during the two sampling periods.

are killed by the primary pathogens (*F. oxysporum* and *F. solani*), the weak pathogens (*F. redolens*, *F. tricinctum* and *F. equiseti*) take over and colonize on weakened or dead tissue.

### Multi-type pathogens infecting a plant

A hemp plant can be co-infected by different types of pathogens, each of which causes its own symptoms. There is no evidence that these pathogens interact with each other or synergistically impact plant health, but a plant co-infected by them exhibits more symptoms. For example, a hemp plant infected by both *Beet curly top virus* (BCTV) and clover proliferation phytoplasma ('*Candidatus* Phytoplasma trifolii') exhibited both leaf curl and witches' broom (Fig. 13.4). A hemp plant can be co-infected by both a fungal pathogen and phytoplasmas. For example, *Alternaria* and phytoplasmas can attack the same hemp plant, causing both leaf spot and witches' broom (Fig. 13.5). *Alternaria* is a common fungal pathogen (sometimes opportunistic)

while phytoplasmas are an obligate bacterium mainly living inside phloem vessels. Biologically and pathologically, these two organisms are distantly related. However, a hemp plant stressed by phytoplasma infection can be more prone to fungal infections. In addition to fungus–phytoplasma co-infections, an unidentified virus may have contributed to the mosaic symptoms (Fig. 13.6). It is not uncommon to see virus–phytoplasma dual infections, as both types of pathogen are transmitted by insects, particularly leafhoppers. In some cases, virus–phytoplasma dually infected hemp plants can be additionally infected by *Pythium* species and symptoms caused by each pathogen are still distinguishable (Fig. 13.7).

Diagnosing multiple types of pathogens involved in a particular crop or plant requires systematic inspection of a number of plants in the field. Based on the symptom types and their frequencies, plant tissue samples representing each symptom can be taken and submitted for laboratory analysis. As shown in Fig. 10.13, growers should perform a

**Fig. 13.4.** Typical leaf curl caused by *Beet curly top virus* and witches' brooms caused by '*Candidatus* Phytoplasma trifolii' on a hemp plant. Note that this plant was severely stunted and underdeveloped, due to dual infections.

**Fig. 13.5.** A hemp plant exhibiting both *Alternaria* leaf spot (right) and witches' broom diseases (left).

**Fig. 13.6.** Leaf mosaic, *Alternaria* leaf spot (left), and witches' broom (right) on the same hemp plant.

symptom polling and inform a diagnostic lab of the potential pathogens to be tested for. In the lab, samples from the same case may be directed to multiple diagnostic procedures to test for presumptive pathogens. For example, crown rot tissue may be plated on a PARP medium to isolate *Pythium* species, curly leaf may be proceeded to DNA extraction followed by PCR to detect BCTV, leaf vein tissue from witches' broom may be collected to detect phytoplasma DNA and mosaic leaf tissue may undergo an ELISA test for a specific virus. As such, the symptoms of a plant are diagnostically classified, tissue samples are analysed separately and suspected pathogens are tested specifically. The results of this holistic approach often lead to multi-pathogen detections, which provide a better understanding of the disease aetiology.

### A list of pathogens infecting a crop

While dual- or tri-infections are often found in a hemp crop or even on a single plant, there are cases where a list of plant pathogens is detected from a crop at a single time point. This section uses a real case to illustrate the process of holistic diagnosis that led to the discovery of an array of pathogens associated with a hemp crop.

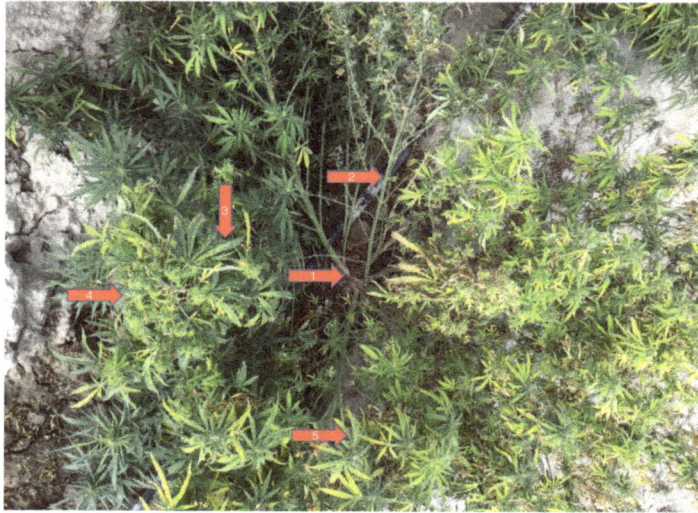

**Fig. 13.7.** A hemp plant infected by *Pythium aphanidermatum* causing crown rot (arrow 1) and branch wilt (arrow 2), by an unidentified virus causing leaf mosaic and chlorosis (arrow 3), by '*Candidatus* Phytoplasma trifolii' causing witches' broom (arrow 4) and by *Beet curly top virus* causing leaf narrowing and curling (arrow 5).

FIELD DIAGNOSIS A hemp crop planted in 2019 appeared to be unproductive and unhealthy. The crop mostly failed, with many plants exhibiting systemic wilt, severe stunting and rapid death prior to harvest. To perform a preliminary field diagnosis, several sections of the field were inspected, with representative healthy and diseased plants examined from the top to the roots. Some plants were severely stunted with typical witches' brooms (Fig. 13.8), which suggested phytoplasma infection. Some had mild stem discoloration, but on cutting into the stem the internal decay was revealed (Fig. 13.9) and most diseased plants had extended decay from the crown to the middle portion of the stem (Fig. 13.10). Many plants had stem canker with defined lesions and sometimes patches of white mould (Fig. 13.11). A large percentage of diseased plants had root rot, with white mycelia visible on their root balls (Fig. 13.12); however, some had apparently healthy roots even though the crown and stem were rotten. Plants with root, crown and stem rot lost their root system completely (Fig. 13.13). Some plants with minor stem canker also had distinct vascular discoloration (Fig. 13.14). These symptomatic observations revealed the complexity and severity of the disease that was not able to be explained through the 'one pathogen – one disease' concept. Rather, this case

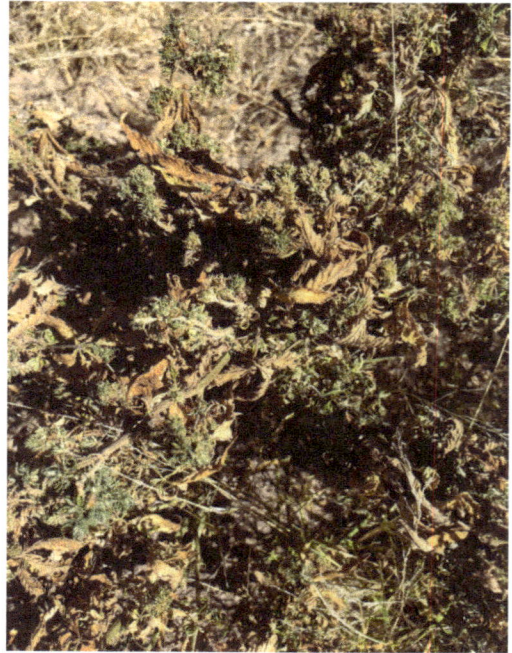

**Fig. 13.8.** A hemp plant severely dwarfed due to witches' broom disease caused by '*Candidatus* Phytoplasma trifolii'.

**Fig. 13.9.** Internal rot revealed inside a hemp stem that appeared normal on the outside.

**Fig. 13.10.** A hemp crown rot revealed through cutting. Note that stem and crown rots were connected.

represented collective damage caused by multiple pathogens sequentially. There is no doubt, however, that abiotic stresses might have contributed to the disease development, as the crop was planted in desert soil. To find all associated pathogens, 24 whole plants, including the root system and rhizosphere soil, were collected for further lab diagnosis.

**LAB DIAGNOSIS** Sample plants were examined again visually and under a microscope to check for insects and mites and disease symptoms on leaves, stem, and roots. Pieces of root, crown and stem tissue were plated on PDA (with or without amendment with streptomycin) and PARP plates to isolate fungi and oomycetes. Although a single species was

often observed in some plates (Fig. 13.15), it was very common to see multiple organisms growing out from the same tissue pieces (Fig. 13.16). Isolates obtained from both PDA and PARP plates were recorded, sorted by colony morphology and further sub-cultured. Species of isolates were identified using the ITS regions of rDNA following protocols described in Chapter 5. Plants with witches' broom were processed with DNA extraction from the midveins and petioles, followed by universal qPCR detection for phytoplasmas. The phytoplasma species was identified by its 16S rRNA gene sequence (see Chapter 10).

From 24 sample plants, seven *Fusarium*, six *Pythium*, one *Rhizoctonia* and one phytoplasma

**Fig. 13.11.** A hemp stem canker with visible white moulds (pointed by gloved finger). Note that the mould can be a primary or secondary/weak pathogen and it is a useful sign for diagnosis.

**Fig. 13.12.** Hemp root rot with growth of white mould.

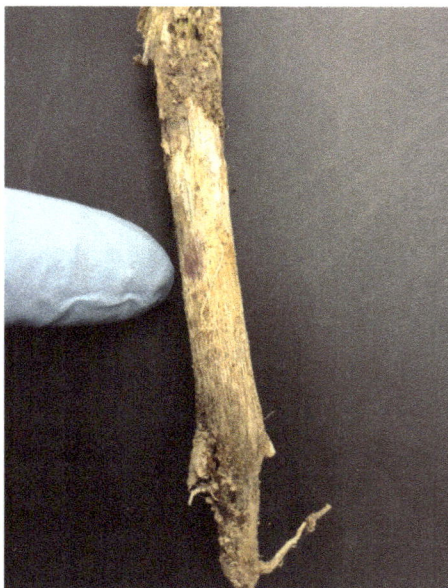

**Fig. 13.13.** Hemp plants killed by root, crown and stem rots. Note that the root systems of both plants were destroyed with white mould (left) and pink stain (right, pointed by gloved finger) on decayed stems.

**Fig. 13.14.** Vascular discoloration (dark brown) of a hemp stem.

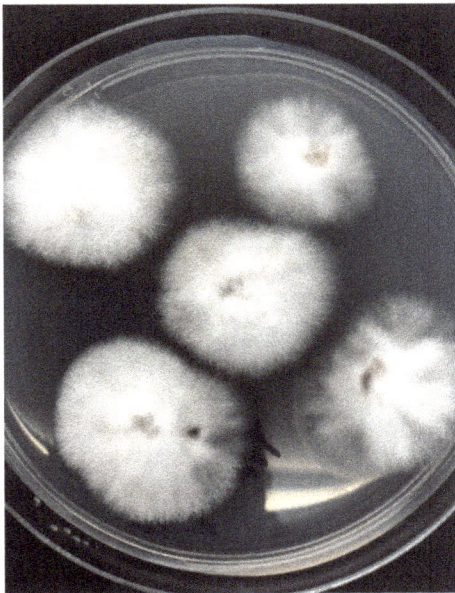

**Fig. 13.16.** Both *Fusarium* and *Pythium* growing out from hemp stem tissue on a PDA plate. Note that *Fusarium* grew slower with dense white colonies while *Pythium* grew faster with loose overgrown colonies.

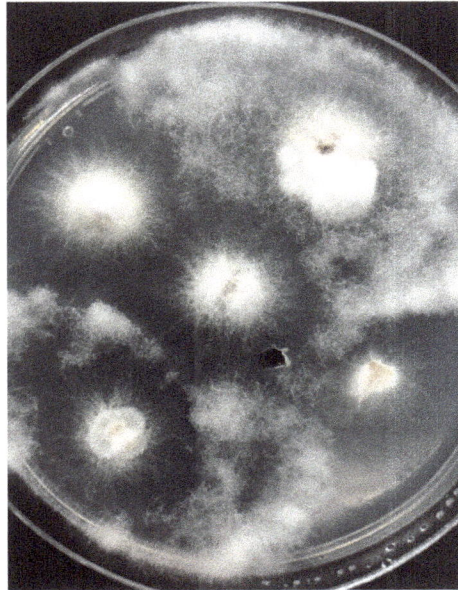

**Fig. 13.15.** *Fusarium equiseti* growing out from hemp stem tissue on a PDA plate.

species were identified from diseased plant tissue (Table 13.1), among which *F. equiseti* and *P. ultimum* were the most dominant at the time of isolation. Because these fungi and oomycetes were isolated from the diseased rather than healthy plants, their presence in diseased tissue suggests that each of them had a certain degree of aggressiveness and played a role in the development of

root, crown and stem rots as well as vascular wilt. Interestingly, at least 50% of plants were infected by two organisms, particularly *Fusarium* and *Pythium* species (Table 13.2). The results indicate that hemp plants of certain varieties are susceptible to a wide range of *Fusarium* and *Pythium* species. In fact, many *Pythium* species are found in agricultural fields and have significant pathogenicity to crop plants such as soybean and maize (Radmer *et al.*, 2017). Similarly, *Fusarium* species, especially *F. oxysporum*, *F. equiseti* and *F. solani*, survive in cultivated soil up to 20 cm depth even under multiple rotation schemes (Silvestro *et al.*, 2013). Therefore, a field with pre-existing inocula of any of these pathogens in the soil may not be feasible for hemp production, as there is a chance for a hemp crop to develop a disease complex caused by *Fusarium* and *Pythium* species. The holistic approach used in this case not only reveals the full spectrum of organisms involved in the disease but also provides better understanding of the causes. In return, a more efficient disease management programme can be developed based on the biology of these organisms.

As plant-parasitic nematodes play a significant role in soilborne disease development (Back *et al.*, 2002), it was speculated that nematode-induced wounds could have created opportunities for a range of *Pythium* and *Fusarium* species to invade.

**Table 13.1.** Plant pathogens detected from diseased hemp plants collected from a crop grown in Nevada.

| Organism identified | Number of sample plants infected | Percentage of sample plants infected |
|---|---|---|
| *Fusarium equiseti* | 14 | 58.3% |
| *Pythium ultimum* | 11 | 45.8% |
| *Fusarium oxysporum* | 7 | 29.2% |
| *Rhizoctonia solani* | 3 | 12.5% |
| 'Candidatus Phytoplasma trifolii' | 3 | 12.5% |
| *Pythium conidiophorum* | 3 | 12.5% |
| *Fusarium solani* | 3 | 12.5% |
| *Pythium salpingophorum* | 2 | 8.3% |
| *Fusarium tricinctum* | 2 | 8.3% |
| *Fusarium incarnatum* | 1 | 4.2% |
| *Fusarium redolens* | 1 | 4.2% |
| *Pythium heterothallicum* | 1 | 4.2% |
| *Pythium perplexum* | 1 | 4.2% |
| *Pythium arrhenomanes* | 1 | 4.2% |
| *Fusarium proliferatum* | 1 | 4.2% |

**Table 13.2.** Frequencies of single-, dual- and multi-pathogen infections in hemp plants.

| Infection by organisms | Number of sample plants | Percentage of sample plants |
|---|---|---|
| Single-organism infection | 6 | 25.0% |
| Dual-organism infection | 12 | 50.0% |
| Tri-organism infection | 4 | 16.7% |
| Quad-organism infection | 4 | 16.7% |
| Single *Fusarium* species infection | 5 | 20.8% |
| Dual *Fusarium* species infection | 1 | 4.2% |
| Single *Pythium* species infection | 1 | 4.2% |
| Dual *Pythium* species infection | 1 | 4.2% |
| Dual infection by *Fusarium* and *Pythium* | 13 | 54.2% |
| Tri-infection by *Fusarium*, *Pythium* and *Rhizoctonia* | 3 | 12.5% |

To verify if any nematode species were present, soil taken from the rhizosphere of 24 sampled plants was analysed for total nematode counts. Surprisingly, no plant-parasitic nematodes were found from any samples; instead, a very high population of a mycophagous or fungus-feeding nematode, *Aphelenchus avenae,* was found (Fig. 13.17). These soil samples had an average of 6622 nematodes per 100 ml of soil, with some containing up to 18,202 nematodes. The presence of an unusually high population of this nematode explained its mycophagous activity, because there were plenty of fungi and oomycetes around the decaying roots serving as its food. This phenomenon is similar to the presence of massive number of bacteria-feeding nematodes in bacterium-infected tissue (see Fig. 8.32). However, in this case, hemp roots were not infected by a bacterium and soil samples had a normal density of bacteria-feeding nematodes.

## Disease complexes involving nematodes and soilborne pathogens

There are many disease complexes reported from a diverse range of crops that involve both nematodes and soilborne pathogens. These diseases are induced through four different synergistic mechanisms between the nematodes and pathogens

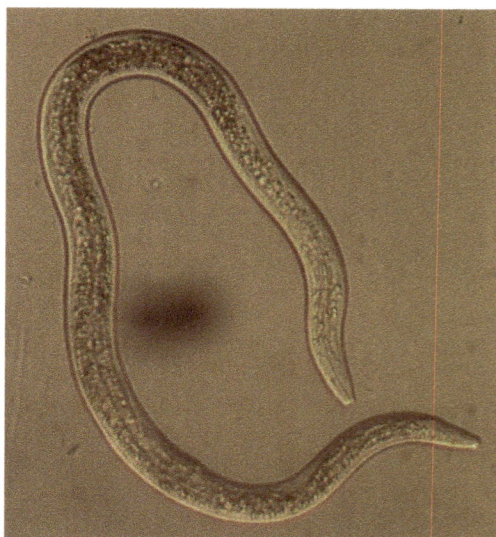

**Fig. 13.17.** *Aphelenchus avenae*, a fungus-feeding nematode found around roots of diseased hemp plants.

(Back *et al.*, 2002). First, soilborne pathogens use wounds created by nematodes to invade plants. Many ectoparasitic nematodes use their stylets to penetrate root epidermal cells, causing minor to moderate wounds. Endoparasitic nematodes can migrate inside the root to cause much more severe disruption to internal root tissue. Both types of injuries can lead to infection by certain pathogens. Secondly, nematode parasitism can change the host plant's physiology to be more prone to fungal infections and colonization. For example, giant cells induced by root-knot nematodes and syncytia induced by cyst nematodes are rich in nutrients and serve as an attractive food source for fungal colonization. Thirdly, nematode infections may increase root exudates or change the profiles of exudates, which makes the rhizosphere environment more attractive to fungal or oomycete pathogens. Finally, nematode infections can weaken or break down the host plant's resistance to certain soilborne pathogens. The most common nematode genera involved in disease complexes are *Meloidogyne*, *Heterodera*, *Globodera*, *Pratylenchus* and *Rotylenchulus*, while the most commonly involved soilborne pathogens are *Fusarium*, *Rhizoctonia*, *Verticillium*, *Pythium* and *Phytophthora*. Both types of organisams are destructive to plant health and, when working synergistically, they cause much more severe damage to plants. For instance, a disease complex caused by the root-knot nematode (*Meloidogyne* sp.) and *Pythium aphanidermatum* can completely destroy the root system of cucumber plants (see Fig. 9.30). There has been no reported case of a nematode–soilborne pathogen disease complex from hemp crops as of yet; however, the prevalence of nematodes and soilborne pathogens in hemp fields as well as hemp's susceptibility may lead to the development of root disease complexes. Diagnosing these complex diseases requires a holistic approach to analyse both nematodes and pathogens and reveal all aetiological components.

### Pest–pathogen complexes

The typical pest–pathogen complexes include pests (insects and mites) transmitting viruses into hemp plants, causing viral diseases (see Chapter 10) and pests producing honeydew or nutritional droplets that trigger sooty moulds (see Chapter 7). In both scenarios, the impact of the complex on plant health is a sum of each organism's individual damage (1 + 1 = 2) instead of synergistic (1 + 1 > 2),

even though there is a demonstrated relationship between the vector and the virus, and that insect infestation increases sooty mould severity. Diagnosing a pest–pathogen complex involves both pest identification and pathogen analyses, followed by further investigation into whether the disease is developed under the condition of a specific pest infestation.

## Microbiota and Microbiome

This book has described *Cannabis* crop diseases and their diagnoses. Many of them are caused by well-known pathogens that also cause diseases in other crops. These pathogens have been demonstrated to cause symptoms on plants by satisfying Koch's postulates, a criterion under a 'one pathogen – one disease' paradigm. Then there are some diseases that were found to be associated with two, three, or a list of microorganisms, implying that a disease can be a result of the activities of many organisms under given environmental conditions. In fact, a plant does not live alone, nor does any microbe associated with a host plant. A plant interacts with vast communities of known and unknown microorganisms at its root-zone (the rhizosphere), organ surface (phyllosphere) and internal tissue (endosphere). Some of them are epiphytic and some are endophytic, both of which may have neutral, beneficial, or sometimes harmful impacts on plant health. The assemblage of these microorganisms in a plant or a specific environment is called microbiota, but the concept is further expanded to another term, called microbiome. Unlike microbiota, microbiome refers to not only the microorganism communities (bacteria, archaea, lower and higher eukaryotes and viruses) but also their genomes and the surrounding environments (Marchesi and Ravel, 2015). The microbiome is a popular research topic in both agriculture and medicine and their advances have led to the 'One Health Concept' that emphasizes the interconnection of all areas of life through their respective microbiota (Berg *et al.*, 2020). As a result, the concept of disease aetiology is also evolving from a single and unsocial organism to a holistic view of a disease being a 'pathobiome' state defined by imbalance and low diversity of a host's microbiota.

### Epiphytes

There are some microorganisms that live and grow on the surface of plants without causing visible harm. For example, the leaves of coffee plants (*Coffea arabica*) host an abundance of epiphytic and endophytic fungi (Santamaría and Bayman, 2005). Epiphytes add to the diversity of plant-associated microbiota and the most common microbes found on plant surfaces are fungi and bacteria. Epiphytes may initially land on the plant surface through air, water, or insects. Once the moisture and nutrients on the plant surface are sufficient to support their survival, they may colonize and grow. Epiphytic microbes are often overlooked, as they do not cause any symptoms or grow as vigorously as pathogenic microbes. They are affiliated with plant surfaces with limited nutrient supplies and restricted microenvironments. The mechanisms of the relationship between epiphytic microbes and host plants are largely unknown, but it is believed that epiphytes may offer some intangible benefits to the hosts. Some epiphytes may initially survive on plant surfaces as epiphytes, but they may infect plant tissue as opportunistic pathogens under certain conditions, such as when the host plant is stressed.

There are epiphytic microbes on *Cannabis* plant surfaces. For example, in one of our preliminary studies, *Cladosporium* spp., *Aspergillus* spp., *Alternaria tenuissima*, *A. alternata*, *Ascochyta* sp. and the bacterium *Serratia* sp. were found to be prevalent organisms associated with healthy hemp buds. These organisms are detected through a simple swabbing procedure (Fig. 13.18) and none of them caused any visible symptoms at the time of testing. Surface swabbing is an effective method to transfer epiphytic microorganisms to a culture plate for isolation and further identification. Briefly, use a sterile moistened cotton swab gently to wipe the plant surface (e.g. a bud) and then immediately wipe the surface of an all-purpose culture plate. Within a few days, bacteria and fungi grow into colonies. Individual colonies can be purified and identified by their morphological characteristics and DNA sequences.

### Endophytes

Unlike epiphytes, endophytes are a group of microorganisms that live within healthy plant tissue. *Cannabis sativa* plants harbour a diversity of microbial communities inside their tissues and organs and these microbes are believed to contribute to plant health, i.e. facilitating mineral nutrient

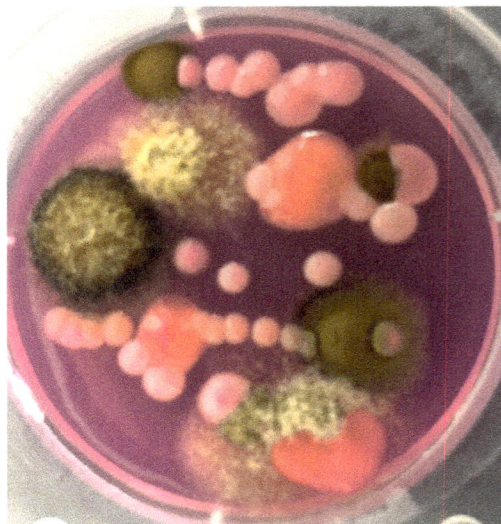

**Fig. 13.18.** A swab test on a culture plate showing diverse fungi and bacteria associated with a healthy hemp bud.

uptake, inducing resistance against pathogens and modulating plant secondary metabolite production (Taghinasab and Jabaji, 2020). A recent study explored the abundance and diversity of culturable endophytes residing in the petioles, leaves and seeds of several industrial hemp varieties and found 134 bacterial and 53 fungal strains in 18 bacterial and 13 fungal taxa, respectively (Scott *et al.*, 2018). The study found that *Pseudomonas*, *Pantoea* and *Bacillus* were the most abundant bacterial genera while *Aureobasidium*, *Alternaria* and *Cochliobolus* were the most common fungal genera. Although some of these endophytes are speculated to be useful in promoting plant growth and controlling diseases, the mechanisms of endophytic associations and their effectiveness in real-world applications remain to be proved.

From a disease diagnostic perspective, endophytes are often encountered during routine culture-based diagnosis. When an organism is frequently isolated from both symptomatic and asymptomatic plant tissue, one may wonder whether the organism is a pathogen or endophyte. In general, a microorganism isolated from internal healthy tissue is considered an endophyte, even though it may potentially act as a pathogen in other cases. For example, *Fusarium equiseti* acts as an endophyte that can promote pea growth and alter the interaction between pea and other pathogens to reduce root rot disease severity (Šišić *et al.*, 2017). Similarly, *F. equiseti* colonizes barley roots endophytically, effectively competes with fungal pathogens and confers beneficial effects to the host plant (Macia-Vicente *et al.*, 2009). On the other hand, *F. equiseti* was reported to cause seedling damping-off on Aleppo pine (Lazreg *et al.*, 2014) and crown rot on cucumber (Aldakil *et al.*, 2019). This double-edged organism is also found frequently in hemp plants. As shown in Fig. 13.3 and Table 13.1, *F. equiseti* was a predominant organism associated with diseased hemp roots, crowns and stems. However, *F. equiseti* was also frequently isolated from healthy hemp crown and stem tissue (Fig. 13.19). In this case, hemp plants exhibited slight enlargement at the crown area (Fig. 13.20) but the internal tissue was perfectly healthy (Fig. 13.21). Obviously, *F. equiseti* acted as an endophyte instead of a pathogen, but it is not known whether it actually promoted the growth of crown tissue.

Some endophytic fungi serve as natural biocontrol agents. Notable examples are the fungi in the genus *Trichoderma*. Many species in this genus are commonly found in soil and root systems. They are avirulent plant symbionts and can colonize root surfaces or become established inside root tissue. *Trichoderma* provides proven benefits to host plants: producing compounds that induce plant's resistance responses, parasitizing on pathogenic fungi to suppress diseases and promoting root growth and crop productivity (Harman *et al.*, 2004). For example, *Trichoderma* isolates induced strong and divergent plant defence reactions and delayed the development of disease caused by

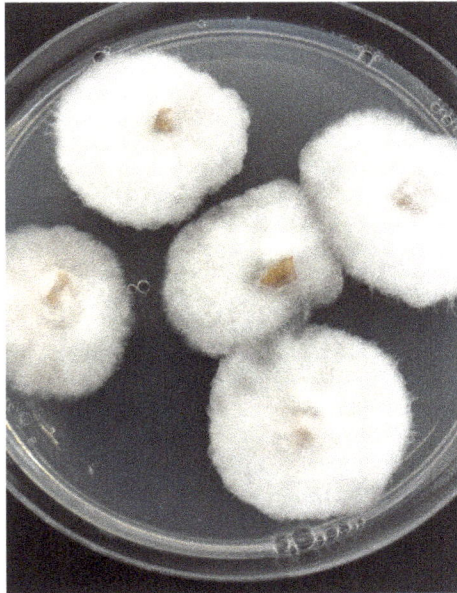

**Fig. 13.20.** A hemp plant having healthy crown and roots from which endophytic *Fusarium equiseti* was predominantly isolated. Note the slight enlargement between the basal stem and the root.

**Fig. 13.19.** Colonies of *Fusarium equiseti* growing from healthy hemp crown tissue.

**Fig. 13.21.** Healthy tissue inside the crown and root of the plant shown in Fig. 13. 20.

*P. capsici* in hot pepper (*Capsicum annuum*) (Bae *et al.*, 2011). In *C. sativa* plants, *Trichoderma* was frequently found as an endophyte (Punja *et al.*, 2019). Some *Trichoderma* species appear to be associated naturally with health hemp roots and crown tissue. For example, *Trichoderma harzianum* was predominantly isolated from a cultivar of hemp plants that appeared healthy except for slight colouring at the crown area (Fig. 13.22). From the crown tissue, *T. harzianum* was the only organism isolated (Fig. 13.23). This suggests that *T. harzianum* colonizes inside the crown tissue as a beneficial endophyte. In diseased hemp roots, *Trichoderma* species may also be isolated along with pathogens. For example, *Trichoderma virens* was frequently

**Fig. 13.22.** An asymptomatic hemp plant from which *Trichoderma harzianum* was isolated from crown tissue.

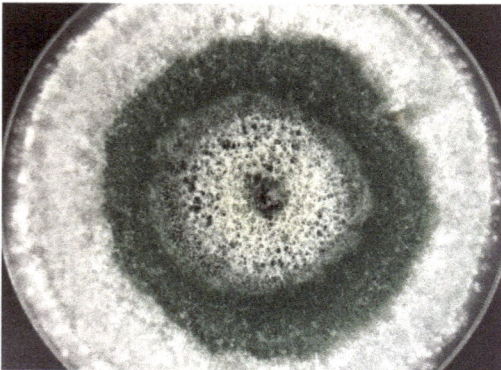

**Fig. 13.23.** A fully-grown colony of *Trichoderma harzianum* isolated from healthy crown tissue shown in Fig. 13.22. Note that the colony of *T. harzianum* on PDA is light grey to off-white in the early stages (first 72 h) but gradually turns to green later (Jahan *et al.*, 2013).

isolated from hemp roots infected by multiple *Fusarium* species (see Chapter 9). The presence of *T. virens* in diseased roots signals that it may actively suppress the pathogens through parasitism and inhibition (see Fig. 9.40). Other endophytic fungi such as *Chaetomium*, *Trametes*, *Penicillium* and *Fusarium* were also found in cannabis crown, stem and petiole tissues (Punja *et al.*, 2019), but their roles in cannabis plant health are unknown.

## The benefits and drawbacks of *Cannabis* microbiota

When it comes to plant–microbe interactions in a natural ecosystem, it is not a question of whether or not hemp, or any other crop plant, harbours a collection of microbial endophytes and epiphytes. Rather, the more meaningful questions are what species are associated with host plants, when they begin to colonize, how they evolve and interact with plant growth, and what potential benefits or impacts the host plants receive from their colonization. Researches have demonstrated the beneficial roles of plant-associated microbes (Berg *et al.*, 2020; Taghinasab and Jabaji, 2020) and these plant-beneficial microbes can be directly delivered to agricultural crops through seed-coating technology to protect crops from diseases and abiotic stress, as well as to promote crop growth and yield (Rocha *et al.*, 2019). Some microbes, particularly plant growth-promoting rhizobacteria (PGPR), have been used in agricultural practice for many years. For example, PGPR has been used in controlling plant diseases and promoting crop productivity since the early 1980s in China (Liu *et al.*, 2016). Recently, PGPR was proposed to be used in *Cannabis* plants to promote yield, product quality and disease resistance (Lyu *et al.*, 2019).

However, the presence of plant-associated microbiota may complicate disease diagnosis. Many fungal or bacterial genera contain not only species or strains that are well known to be pathogenic but also others that may be neutral or beneficial to their host plants. For example, *Fusarium oxysporum* is one of the major pathogens causing diseases in various crops, but some strains behave as endophytes and potentially benefit host plants by controlling diseases through endophyte-mediated resistance (de Lamo and Takken, 2020). That said, microbes isolated from diseased plant tissue may include those that are neutral or beneficial. As such, inoculation tests are often needed to determine the relationship between the host plant and a given microorganism.

Another drawback of *Cannabis* microbiota is its implication on the consumption of *Cannabis* products. Thompson *et al.* (2017) assessed a microbiome of medical marijuana and found that some bacterial and fungal pathogens associated with medical marijuana may potentially pose a risk to those who consume it, particularly immunosuppressed patients. The occurrence of opportunistic human pathogens in *Cannabis* products, either through incidental contaminations or through natural microbiota assemblage, leads to stricter tests for product quality and safety.

# References

Aldakil, H., Jaradat, Z.W., Tadros, M. and Alboom M.H. (2019) First report of *Fusarium equiseti* causing crown rot on cucumber in Jordan Valley. *Plant Disease* 103, 2692. doi: 10.1094/PDIS-11-18-2073-PDN

Back, M.A., Haydock, P.P.J. and Jenkinson, P. (2002) Disease complexes involving plant parasitic nematodes and soilborne pathogens. *Plant Pathology* 51, 683–697. doi: 10.1046/j.1365-3059.2002.00785.x

Bae, H., Roberts, D.P., Lim, H.-S., Strem, M.D., Park, S.-C. *et al.* (2011) Endophytic *Trichoderma* isolates from tropical environments delay disease onset and induce resistance against *Phytophthora capsici* in hot pepper using multiple mechanisms. *Molecular Plant-Microbe Interactions* 24, 336–351. doi: 10.1094/MPMI-09-10-0221

Berg, G., Rybakova, D., Fischer, D., Cernava, T., Vergès, M.C. *et al.* (2020) Microbiome definition re-visited: old concepts and new challenges. *Microbiome* 8, 103. doi: 10.1186/s40168-020-00875-0

de Lamo, F.J. and Takken, F.L.W. (2020) Biocontrol by *Fusarium oxysporum* using endophyte-mediated resistance. *Frontiers in Plant Science* 11, 37. doi: 10.3389/fpls.2020.00037

Errard, A., Ulrichs, C., Kühne, S., Mewis, I., Drungowski, M. *et al.* (2015) Single- versus multiple-pest infestation affects differently the biochemistry of tomato (*Solanum lycopersicum* 'Ailsa Craig'). *Journal of Agricultural and Food Chemistry* 25, 10103–10111. doi: 10.1021/acs.jafc.5b03884

Harman, G.E., Howell, C.R., Viterbo, A., Chet, I. and Lorito M.T. (2004) *Trichoderma* species – opportunistic, avirulent plant symbionts. *Nature Reviews. Microbiology* 2, 43–56. doi: 10.1038/nrmicro797

Jahan, N., Sultana, S., Adhikary, S.K., Rahman, S. and Yasmin, S. (2013) Evaluation of the growth performance of *Trichoderma harzianum* (rifai.) on different culture media. *IOSR Journal of Agriculture and Veterinary Science* 3, 44–50.

Lazreg, F., Belabid, L., Sanchez, J., Gallego, E., Garrido-Cardenas, J.A. *et al.* (2014) First report of *Fusarium equiseti* causing damping-off disease on Aleppo pine in Algeria. *Plant Disease* 98, 1268. doi: 10.1094/PDIS-02-13-0194-PDN

Liu, D.-D., Li, M. and Liu, R.-J. (2016) Recent advances in the study of plant growth-promoting rhizobacteria in China. *Chinese Journal of Ecology* 35, 815–825. doi: 10.13292/j.1000-4890.201603.033

Lyu, D., Backer, R., Robinson, W.G. and Smith, D.L. (2019) Plant growth-promoting rhizobacteria for cannabis production: yield, cannabinoid profile and disease resistance. *Frontiers in Microbiology* 10, 1761. doi: 10.3389/fmicb.2019.01761

Macia-Vicente, J.G., Rosso, L.C., Ciancio, A., Jansson, H.-B. and Lopez-Llorca, L.V. (2009). Colonisation of barley roots by endophytic *Fusarium equiseti* and *Pochonia chlamydosporia*: effects on plant growth and disease. *Annals of Applied Biology* 155, 391–401. doi: 10.1111/j.1744-7348.2009.00352.x

Marchesi, J.R. and Ravel, J. (2015) The vocabulary of microbiome research: a proposal. *Microbiome* 3, 31. doi: 10.1186/s40168-015-0094-5

May, C.E., Potage, G., Andrivon, D., Tivoli, B. and Outreman, Y. (2009) Plant disease complex: antagonism and synergism between pathogens of the Ascochyta blight complex on pea. *Journal of Phytopathology* 157, 715–721. doi: 10.1111/j.1439-0434.2009.01546.x

Punja, Z.K., Collyer, D., Scott, C., Lung, S., Holmes, J. *et al.* (2019) Pathogens and molds affecting production and quality of *Cannabis sativa* L. *Frontiers in Plant Science* 10, 1120. doi: 10.3389/fpls.2019.01120

Radmer, L., Anderson, G., Malvick, D.M., Kurle, J.E., Rendahl, A. *et al.* (2017) *Pythium*, *Phytophthora*, and *Phytopythium* spp. associated with soybean in Minnesota, their relative aggressiveness on soybean and corn, and their sensitivity to seed treatment fungicides. *Plant Disease* 101, 62–72. doi: 10.1094/PDIS-02-16-0196-RE

Rocha, I., Ma, Y., Souza-Alonso, P., Vosátka, M., Freitas, H. *et al.* (2019) Seed coating: a tool for delivering beneficial microbes to agricultural crops. *Frontiers in Plant Science* 10, 1357. doi: 10.3389/fpls.2019.01357

Santamaría, J. and Bayman, P. (2005) Fungal epiphytes and endophytes of coffee leaves (*Coffea arabica*). *Microbial Ecology* 50, 1–8. doi: 10.1007/s00248-004-0002-1

Schoener, J.L., Wilhelm, R., Rawson, R., Schmitz, P. and Wang, S. (2017) Five *Fusarium* species associated with root rot and sudden death of industrial hemp in Nevada. (Abstr.) *Phytopathology* 107, S5.98. doi: 10.1094/PHYTO-107-12-S5.98

Scott, M., Rani, M., Samsatly, J., Charron, J.B. and Jabaji, S. (2018) Endophytes of industrial hemp (*Cannabis sativa* L.) cultivars: identification of cultural bacteria and fungi in leaves, petioles, and seeds.

*Canadian Journal of Microbiology* 64, 1–17. doi: 10.1139/cjm-2018-0108

Silvestro, L., Stenglein, S., Forján, H.J., Dinolfo, M.I., Arambarri, A. *et al.* (2013) Occurrence and distribution of soil *Fusarium* species under wheat crop in zero tillage. *Spanish Journal of Agricultural Research* 11, 72–79. doi: 10.5424/sjar/2013111-3081

Šišić, A., Baćanović, J. and Finckh, M.R. (2017) Endophytic *Fusarium equiseti* stimulates plant growth and reduces root rot disease of pea (*Pisum sativum* L.) caused by *Fusarium avenaceum* and *Peyronellaea pinodella*. *European Journal of Plant Pathology* 148, 271–282. doi: 10.1007/s10658-016-1086-4

Taghinasab, M. and Jabaji, S. (2020) Cannabis microbiome and the role of endophytes in modulating the production of secondary metabolites: an overview. *Microorganisms* 8, 355. doi: 10.3390/microorganisms8030355

Thompson, G.R. 3rd, Tuscano, J.M., Dennis, M., Singapuri, A., Libertini, S. *et al.* (2017) A microbiome assessment of medical marijuana. *Clinical Microbiology and Infection* 23, 269–270. doi: 10.1016/j.cmi.2016.12.001

Wheeler, D.L., Scott, J., Dung, J.K.S. and Johnson, D.A. (2019) Evidence of a trans-kingdom plant disease complex between a fungus and plant-parasitic nematodes. *PLoS ONE* 14, e0211508. doi: 10.1371/journal.pone.0211508

# Index

Pages numbers in *italics* refer to Figures.